Springer Proceedings in Materials

Volume 15

Springer Proceedings in Materials publishes the latest research in Materials Science and Engineering presented at high standard academic conferences and scientific meetings. It provides a platform for researchers, professionals and students to present their scientific findings and stay up-to-date with the development in Materials Science and Engineering. The scope is multidisciplinary and ranges from fundamental to applied research, including, but not limited to:

- Structural Materials
- Metallic Materials
- Magnetic, Optical and Electronic Materials
- Ceramics, Glass, Composites, Natural Materials
- Biomaterials
- Nanotechnology
- Characterization and Evaluation of Materials
- Energy Materials
- Materials Processing

To submit a proposal or request further information, please contact one of our Springer Publishing Editors according to your affiliation:

European countries: **Mayra Castro** (mayra.castro@springer.com)
India, South Asia and Middle East: **Priya Vyas** (priya.vyas@springer.com)
South Korea: **Smith Chae** (smith.chae@springer.com)
Southeast Asia, Australia and New Zealand: **Ramesh Nath Premnath** (ramesh.premnath@springernature.com)
The Americas: **Michael Luby** (michael.luby@springer.com)
China and all the other countries or regions: **Mengchu Huang** (mengchu.huang@springer.com)

More information about this series at https://link.springer.com/bookseries/16157

U. Kamachi Mudali · S. T. Aruna ·
H. P. Nagaswarupa · Dinesh Rangappa
Editors

Recent Trends in Electrochemical Science and Technology

Proceedings of Papers Presented at NSEST-2020 and ECSIRM-2020

Springer

Editors
U. Kamachi Mudali
VIT Bhopal University
Bhopal, Madhya Pradesh, India

H. P. Nagaswarupa
Department of Chemistry
Davangere University
Tholahunase, Karnataka, India

S. T. Aruna
Surface Engineering Division
National Aerospace Laboratories
Council of Scientific and Industrial Research
Bengaluru, Karnataka, India

Dinesh Rangappa
Department of Applied Sciences
Visvesvaraya Technological University
Chickballapur, Karnataka, India

ISSN 2662-3161 ISSN 2662-317X (electronic)
Springer Proceedings in Materials
ISBN 978-981-16-7553-9 ISBN 978-981-16-7554-6 (eBook)
https://doi.org/10.1007/978-981-16-7554-6

This Springer imprint is published by the registered company Springer Nature Singapore Pte Ltd.
The registered company address is: 152 Beach Road, #21-01/04 Gateway East, Singapore 189721, Singapore

Contents

About the Editors

Dr. U. Kamachi Mudali Vice-Chancellor of VIT Bhopal University & Honorary Professor of Practice, IIT Madras is formerly Distinguished Scientist of Department of Atomic Energy (DAE), and Chief Executive & Chairman of Heavy Water Board, a flagship industrial unit of DAE. Dr. Mudali made pioneering contributions to advanced materials & coating technology development; corrosion science, engineering and technology; materials, process and equipment development for reprocessing applications; failure analysis, consultancy and societal contributions. He has 465 journal publications, five patents, and 9740 citations with an h-index of 44 and i-10 index of 246. He is recognized in the World's Top 2% scientists from India in the field of Materials, and is a Fellow of 12 professional associations. He has guided/coordinated 162 UG, PG and PhD students, and has published 21 Books/Proceedings. Dr. Mudali is decorated with several distinguished recognitions from India and abroad including: Distinguished Alumnus Award from IIT Bombay where he did M.Tech. (Corrosion Science and Engineering) in 1984, and PSG College of Technology, Coimbatore where he did M.Sc. (Materials Science) in 1982, GD Birla Gold Medal and Metallurgists of the Year Award from IIM & GoI, MASCOT National Award, Vasvik Award, INS Medal, Outstanding Faculty Award of HBNI, Homi Bhabha Science and Technology Award of Department of Atomic Energy, Fellow & Frank Newman Speller Award of NACE International, USA, the highest recognition and first Indian to get in the field of corrosion.

Dr. S. T. Aruna is currently working as a Senior Principal Scientist at Council of Scientific and Industrial Research—National Aerospace Laboratories (CSIR-NAL), Bangalore, India. She has a Master's Degree in Chemistry from Mysore University (1994) and she secured the first rank and won four gold medals. She obtained her Ph.D. from the Indian Institute of Science (IISc), Bangalore in 1998. She worked as a postdoctoral fellow at Bar-Ilan University, Israel in the area of dye-sensitized solar cells from 1998–1999. She has authored ~106 research papers in international peer-reviewed journals, nine patents, co-authored a book and written ten book chapters and four encyclopedia chapters. Her research papers have been well-cited and her publications have an h-index of 40. Her name featured among 1494 Indian scientists in the list of top 2% of the world scientists released by Stanford University in January 2021. Her current research interests include the development of tapecast ceramic substrates for electronics application, nanomaterials, plasma sprayed thermal and environmental barrier coatings, oxygen sensor and solid oxide fuel cells (SOFCs). She has guided five Ph.D. and 25 Masters' Students.

Dr. H. P. Nagaswarupa is currently an Associate Professor in the Department of Studies in Chemistry, Davangere University, India. He obtained his Master's degree in Chemistry from Bangalore University (2002) and Ph.D. degree from Bharathidasan University, Trichy (2012). He has been teaching Chemistry for the undergraduate and postgraduate level since 18 years. He has guided three Ph.D. students and currently guiding six Ph.D. students. He has published over 160 research papers in national/international journals and holds three patents. His papers are well cited with an h-index of 31. He has organized more than 25 workshops/conferences at the national/international level. He is also a member of various committees for the inspection of new as well as old institutions. His areas of interest include Nano Science & Technology, Materials Science and Corrosion Science.

Prof. Dinesh Rangappa received his M.Sc. and Ph.D. in Environmental Science from the University of Mysore, India. He obtained his second doctoral degree Dr. Engg., (Materials Science and Engineering, 2004) from Tokyo Institute of Technology, Japan. He has received JSPS Research Fellow award at AIST, Japan. He was deputed to Visvesvaraya Technological University (VTU), as a Special officer and PG Coordinator for VTU Bangalore Region. Since 2013, he is serving as Professor and Chairman, Department of Nanotechnology, VTU. Professor Rangappa has significantly contributed to developing a novel process for bulk production of nanomaterials, engineering nanomaterials for energy storage, sensors and photovoltaic, photocatalysis, photoluminiscent, inorganic Nano-pigment, wastewater treatment and biomedical applications. He has published more than 120 papers in national and international journals (h-index-23), eight book chapters and two patents granted. He has been awarded Prof. Satish Dhawan Karnataka State Award for Young Engineers—2014, Pavan Nagapal Memorial Award 2015 and Gandhian Young Technological Innovation Award 2016, and Japan Society for the Promotion of Science (JSPS) Bridge Fellowship Award 2017.

Benzotriazole Encapsulated Nanocontainer-Based Self-healing Coatings for Corrosion Protection of Mild Steel

Aarti Gautam, K. R. C. Soma Raju, K. V. Gobi, and R. Subasri

1 Introduction

Mild steel is one of the extensively used materials in industrial sectors because of its ease of availability and low cost. Mild steel is regarded as one of the desired materials for construction because of its excellent mechanical properties such as ductility, tensile strength, etc. [1]. The presence of low carbon content makes it more efficient to be used in areas where high-temperature conditions are required. Mild steel is highly malleable and hence finds its use in oil and pipeline industries. However, it is prone to get oxidized and to undergo corrosion as soon as it comes in contact with air, moisture, acids, etc., and all these could lead to changes in physical appearance, reduced mechanical strength, and structural deformation, which could cause severe accidents [2].

Hence, protection is necessary to prevent the surface from being corroded. Hexavalent chromate conversion coatings (CCC) and phosphate coatings are effective for protecting the metal surfaces, as they impart excellent corrosion resistance properties and are cost-effective [2, 3]. Even though they are considered the best protection

A. Gautam (✉) · K. R. C. Soma Raju · R. Subasri
Centre for Sol-Gel Coatings, International Advanced Research Centre for Powder Metallurgy and New Materials (ARCI), Balapur, Hyderabad, Telangana 500005, India

K. R. C. Soma Raju
e-mail: somarajuk@arci.res.in

R. Subasri
e-mail: subasri@arci.res.in

A. Gautam · K. V. Gobi
Department of Chemistry, National Institute of Technology Warangal, Warangal, Telangana 506004, India
e-mail: kvgobichem@nitw.ac.in

U. K. Mudali et al. (eds.), *Recent Trends in Electrochemical Science and Technology*, Springer Proceedings in Materials 15,
https://doi.org/10.1007/978-981-16-7554-6_1

methods, they are restricted for further use hereafter due to their toxicity and carcinogenic nature [4]. The most promising alternate method is the use of hybrid organic and inorganic silane-based sol–gel coatings [5]. Sol–gel coatings are highly stable, eco-friendly, and possess good adhesion and barrier properties, and they require only simple curing at relatively low temperatures [6]. Even though the sol–gel coatings form thick and homogeneous coatings on the metal substrates, these coatings get damaged and lose their barrier properties during long exposure times to aggressively corrosive media [7]. To overcome this drawback, imparting additional functionality to such simple, stable and easy-forming hybrid sol–gel coatings is highly preferred. Incorporation of corrosion inhibitors into sol–gel coatings is a recent strategy to produce self-healing coatings, which possess enhanced barrier properties for a longer duration in addition to anticorrosion properties [5]. In order to fabricate self-healing coatings, the self-healing agent can either be added directly into the matrix sol or nanocontainers entrapped (sealed) with corrosion inhibitors can be added into the sol–gel matrix. The direct addition of corrosion inhibitors can cause an excess of leaching, leading to unwanted reactions with the matrix. Hence, encapsulation is the better alternative. Nanocontainers encapsulated with corrosion inhibitors show controlled release and avoids excess leaching [8].

Different kinds of nanocontainers such as layered double hydroxides, layer-by-layer polyelectrolyte shells, nanoclays such as montmorillonite and halloysite, carbon nanotubes, polymeric nanocontainers, etc. have so far been investigated [9, 10]. Halloysite nanotubes (HNT) are widely used nanocontainers because they are easily available, eco-friendly, and cost-effective. Halloysite nanotubes are naturally occurring aluminosilicates, $Al_2Si_2O_5(OH)_4.2H_2O$, having a hollow lumen for easy entrapment of corrosion inhibitors. Halloysite nanoclay has a layered structure consisting of one alumina sheet and one silica sheet, as shown in Fig. 1a [11].

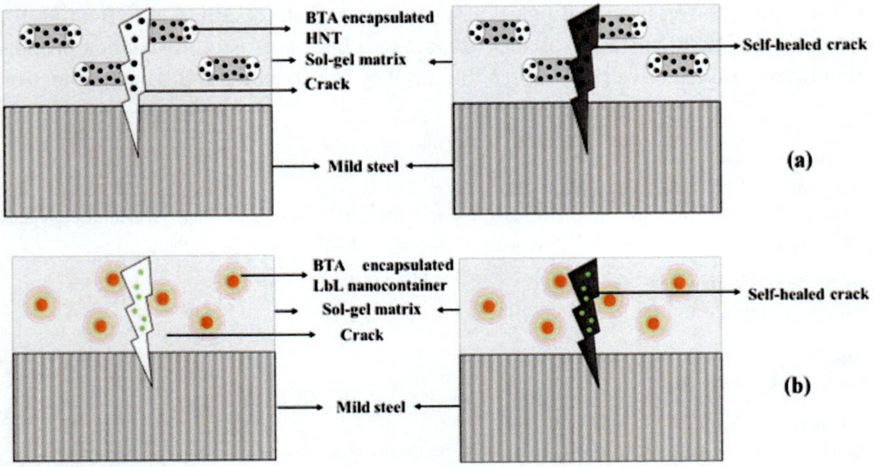

Fig. 1 Schematic of expected autonomic-healing process: **a** BTA encapsulated in HNT, **b** BTA encapsulated in LbL nanocontainers

Layer-by-layer (LbL) polyelectrolyte shell-based nanocontainers are also promising materials for entrapping the corrosion inhibitor within the polymeric shells. LbL approach involves the use of nanoparticles as a support system onto which polyelectrolyte layers of opposite charges are deposited. Corrosion inhibitors are entrapped in between the layers for their controlled release and thus provide more stability to the coating matrix, as depicted in Fig. 1b [12, 13].

The purpose of this research work is to examine the anticorrosion and self-healing behavior of eco-friendly hybrid sol–gel matrix dispersed with benzotriazole (BTA) encapsulated into two different types of nanocontainers, namely HNT and LbL polyelectrolyte shell. The encapsulation of BTA into different types of nanocontainers depends on the charge of the BTA molecule. Being an anionic inhibitor, it can bind easily with the hollow lumen of HNT which is positively charged because of the presence of the alumina layer inside. A similar effect is seen when BTA is loaded into LbL nanocontainers where BTA is entrapped in between the polyelectrolyte layers based on charge difference [14]. The effects of encapsulation of corrosion inhibitors dispersed into the sol–gel matrix on its barrier and mechanical properties are analyzed and discussed.

2 Experimental

2.1 *Materials*

For the synthesis of hybrid silane-based sol–gel matrix (3-Glyidyloxypropyl) trimethoxy silane (GPTMS, Alfa Aesar, Assay-97%) and tetraethyl orthosilicate (TEOS, purchased from Sigma Aldrich, USA, Assay-98%) precursors were used. Halloysite nanoclay (HNT, Sigma Aldrich, USA) was used as an encapsulating agent for corrosion inhibitors. Anionic polyacrylic acid (PAA) and cationic polydiallyl dimethyl ammonium chloride (PDADMAC) (Sigma Aldrich, USA) were used as the polyelectrolyte layered nanocontainers on Fe_2O_3 nanoparticles (supplied by Sigma Aldrich, USA) for encapsulating the self-healing material. Benzotriazole (SRL, Purity-97%) was used as self-healing material in both HNT and LbL nanocontainers. For coating deposition, mild steel coupons of dimensions 2.5 cm × 2.5 cm × 0.3 cm were used. The chemical composition of mild steel was found to be C: 0.18 wt%, Mn: 0.58 wt%, Si: 0.25 wt%, P: 0.018 wt%, S: 0.037 wt% and Fe: balance.

2.2 *Preparation of Assembled Hematite-based LbL Nanocontainers*

Standard polymer solutions of PAA and PDADMAC were prepared by dissolving polymer solution in deionized water of concentration 10 mg/mL. For layer-by-layer assembly, PAA and PDADMAC were adsorbed on hematite particles at a concentration of 0.1 mg/mL. The first negatively charged layer of PAA was deposited on hematite particles by centrifuging at the speed of 6000 rpm for 15 min. The obtained powder was then re-dispersed and stirred in 0.1 M NaCl solution for 10 min and sonicated for 1 min in order to remove loosely bound particles. A similar procedure was followed for deposition of BTA (2 g/100 mL) which was followed by the deposition of the second layer of PAA. The fabrication of nanocontainers was terminated using a cationic polyelectrolyte layer of PDADMAC. The pH of all the solutions during the fabrication was maintained at 5.5 [14].

2.3 Loading of BTA in HNTs

For the synthesis of inhibitor-loaded HNT, 357 mg of BTA was added to 0.1 L of isopropyl alcohol (IPA) and subsequently, 5 g of HNT was added. The mixture was stirred for 20 min and evacuated at room temperature for 3 h. The obtained suspension was collected by centrifugation followed by oven drying at 50 °C for 20 min [15].

2.4 Sol Preparation and Coating Procedure

Matrix sol was synthesized by hydrolysis of GPTMS with TEOS in a mole ratio of 3.5:1 using hydrogen chloride (HCl) as catalyst followed by the addition of an optimized amount of self-healing agent. The sols were named as matrix (only for matrix sol), BLLS (loading the sol with BTA entrapped LbL nanocontainers), BS (loading the sol with BTA), BHS (loading the sol with BTA encapsulated HNT). For polishing the mild steel (MS) coupons 1000 grade sandpaper was used and was degreased using acetone in an ultrasonicator for 15 min. A dip-coater (EPG GmbH, Germany) was used to generate the coatings at a withdrawal speed of 0.1 cm/s and was cured at 130 °C for 60 min in a drying oven.

2.5 Characterization and Testing

Surface morphology and elemental composition of as-received hematite nanoparticles, HNT, and inhibitor-loaded nanocontainers were investigated using ZEISS Gemini 500 scanning electron microscope equipped with energy-dispersive X-ray spectroscopy (EDS) attachment. X-ray diffraction (XRD) was performed for as procured hematite nanoparticles using X-ray diffractometer (Bruker AXS D8 Advance, USA) in the 2θ range of 6°–100° with a step size of 0.1°. Zeta potential measurements for each deposited polyelectrolyte layer for LbL nanocontainers were carried out using Zetasizer (MALVERN AT-0050007) instrument. Coating thickness was measured using gauge PosiTector® 6000 (De Felsko Corporation, USA). Multi-hatch gauge (BEVS 2203, China) according to ISO standard 2409 was used to perform the adhesion test and the substrates were then observed under an optical microscope (Olympus BX51M).

Electrochemical impedance and polarization studies for bare as well as coated mild steel coupons were carried out using the electrochemical analyzer from CH Instruments (Model CHI 604E, USA). The electrochemical cell consists of a three-electrode system with a saturated calomel electrode (reference electrode), platinum electrode (counter electrode), and the metal surface (working electrode). AC signal of 0.01 V amplitude was applied with frequency ranging from 100 kHz to 0.01 Hz to record the electrochemical impedance data. The bare and coated MS coupons were immersed in corrosive media of concentration 0.6 mol/L for 1 h before performing the polarization studies. A potential of $\pm$300 mV was applied in accordance with open circuit potential at a scanning rate of 1 mV/s for conducting polarization measurements. The electrochemical impedance and polarization data were examined with the help of ZSimpWin and CH604E software. The salt spray test was performed by exposing the scribed coated and uncoated MS coupons to a highly corrosive environment for 24 h based on ASTM B117.

3 Results and Discussion

3.1 *Scanning Electron Microscopy Analysis*

SEM analysis for as-received α-Fe_2O_3 nanoparticles, corrosion inhibitor-loaded LbL nanocontainers, as procured halloysite nanotube and inhibitor-loaded halloysite nanotube was carried out. Elemental composition was determined using EDS. The FESEM image shown in Fig. 2 for as-received α-Fe_2O_3 nanoparticles reveals that the particles are in clusters having spherical structures with an average diameter of 30 nm. Fe and O were found as constituent elements from the EDS data as depicted in Fig. 2. The polyelectrolyte layer deposited α-Fe_2O_3 nanoparticles show large agglomerates of approximately 200 nm diameter, and EDS analysis indicates the presence of Fe, O, C, Na, Cl, and N confirming the formation of BTA entrapped polyelectrolyte layers on α-Fe_2O_3 nanoparticles. FESEM analysis of as-received HNT shows that the nanoclay has a tubular structure with transparent hollow lumen, and the elemental analysis reveals the presence of Al and Si elements in larger proportions. The inhibitor-loaded HNT shows denser and darker tubular structures, confirming the loading of BTA in the hollow lumen of HNT. Further, the elemental composition determined by EDS analysis shows the presence of Al, Si, C, and N, confirmed the entrapment of BTA in HNT.

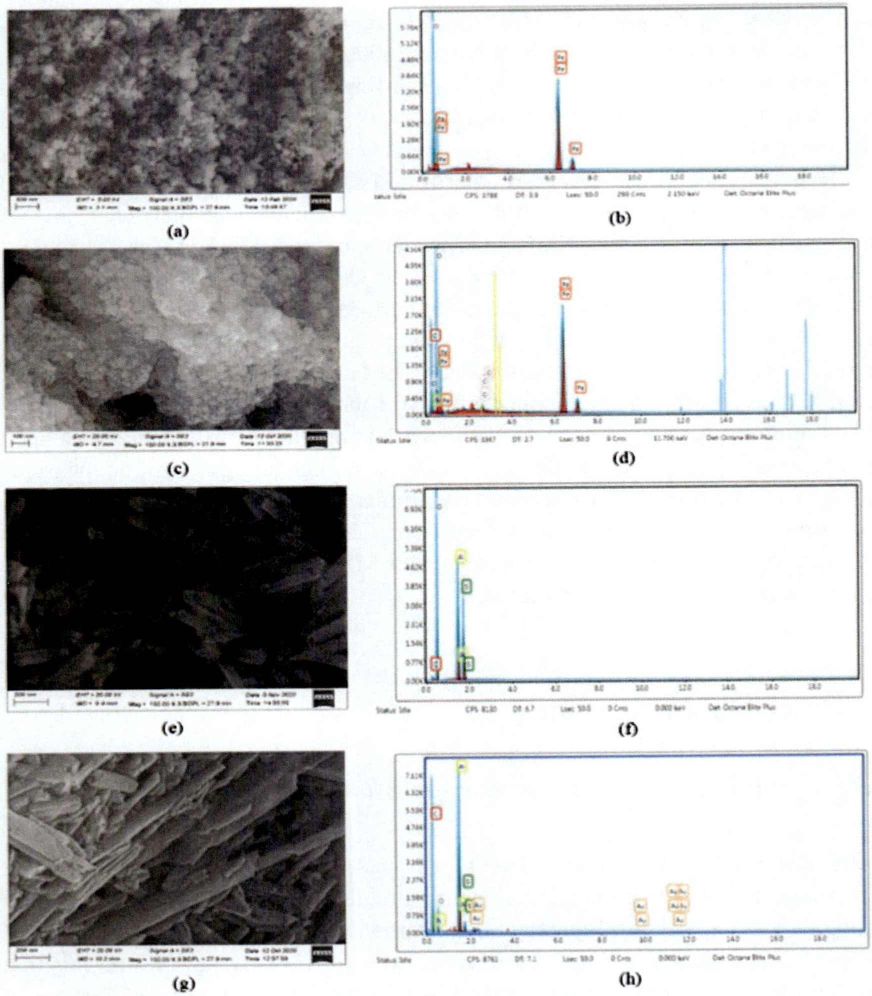

Fig. 2 FESEM images and EDS profiles: **a**, **b** Fe_2O_3 nanoparticles; **c**, **d** BTA entrapped polyelectrolyte layers deposited α-Fe_2O_3 nanoparticles; **e**, **f** HNT clay; **g**, **h** BTA encapsulated HNT nanoclay

3.2 X-Ray Diffraction

The XRD pattern for as-received α-Fe_2O_3 nanoparticles is shown in Fig. 3. The XRD analysis shows the intense peak at the 2θ value of 35°, indicating that the crystal structure of α-Fe_2O_3 is tetragonal, according to the International Centre for Diffraction Data (ICDD File no. 04-014-7214).

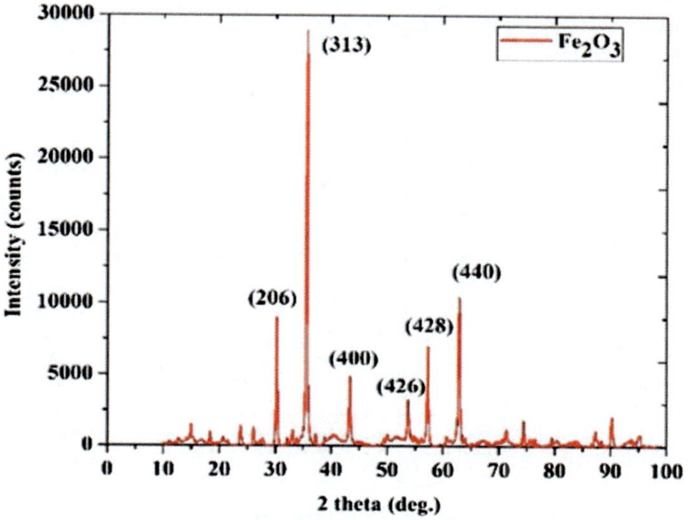

Fig. 3 XRD pattern for as-received Fe_2O_3 nanoparticles

3.3 *Zeta Potential and Electrophoretic Mobility Measurements*

Zeta potential and electrophoretic mobility were carried out for α-Fe_2O_3 nanoparticles along the several stages of stepwise depositions of polyelectrolyte layers using the Zetasizer instrument. Zeta potential values of α-Fe_2O_3 nanoparticles at all stages were found to be in the range of ± 15 to ± 30 mV, thereby confirming that the particles are highly stable without the formation of any agglomerates. The electrophoretic mobility confirmed that the α-Fe_2O_3 nanoparticles were found to be positively charged (Fig. 4). The negative electrophoretic mobility for PAA layer deposited nanoparticles confirmed that the negative surface charge surrounds the nanoparticles, confirming the firm deposition of the PAA layer on α-Fe_2O_3 nanoparticles. Entrapment of BTA on the PAA layer decreased the negative charge mobility values. The loosely bound particles of BTA were removed by depositing a second negatively charged layer of PAA. With the further deposition of a second PAA layer, the electrophoretic mobility value became a little more negative. The nanocontainers were terminated using the cationic polyelectrolyte layer of PDADMAC, and hence the electrophoretic mobility for this layer was found to be positive, as depicted in Fig. 4. The zeta potential measurements confirmed that the surface charge of each deposited layer was in accordance with the charge of each layer and the particles are stable in nature. The observed results of zeta potential and electrophoretic mobility studies supported the stepwise deposition of polyelectrolyte layers and the fabrication of LbL nanocontainers, and the results are in good agreement with the literature report [14].

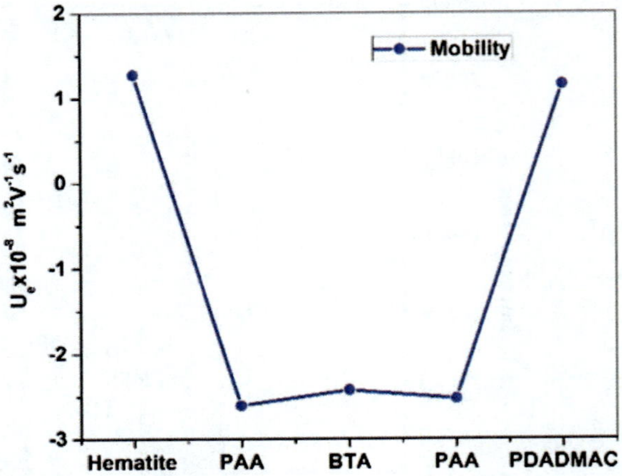

Fig. 4 Comparison of electrophoretic mobilities of each deposited layer

3.4 *Thickness and Adhesion Measurements*

The thicknesses of matrix, BS, BLLS, and BHS sol–gel coatings were found to be in the range of 6–8 microns. Adhesion test was carried out using a multi-hatch gauge where crosshatches were marked manually. The images obtained using an optical microscope before and after the adhesion test are shown in Fig. 5. For all the coatings, the adhesion test showed that no coating was removed from the corners of each intersection point of the crosshatch. The corners and edges of the squares were smooth. Hence, the observations confirm that all the coatings possess good adhesion properties with the substrate in accordance with the ISO standard 2409, where such types of coatings are classified into 0 class (means 0% removal).

3.5 *Electrochemical Impedance Spectroscopy (EIS) and Potentiodynamic Polarization Studies (PPS)*

EIS along with PPS was conducted for bare and coated MS substrates exposed to 0.6 mol/L NaCl solution for 1 h. The data obtained after the electrochemical measurements were studied using two different equivalent electric circuits, shown in Fig. 6. The equivalent circuit impedance parameters are R_s (resistance due to solution), CPE_{coat} (capacitance due to coating) R_{coat}, (resistance due to coating), C_{edl} (capacitance due to electrical double layer), and R_{ct} (resistance due to charge transfer). Figure 6a conveys the equivalent electric circuit that fitted best for bare substrate, where R_s is connected in series with CPE_{edl} and R_{ct} which are in parallel with each other. In order to obtain the EIS parameters for coated substrates with and without encapsulating BTA into the nanocontainers, the equivalent electric circuit shown in Fig. 6b fitted best. Here, the R_s is in series with CPE_{coat} parallel to R_{coat},

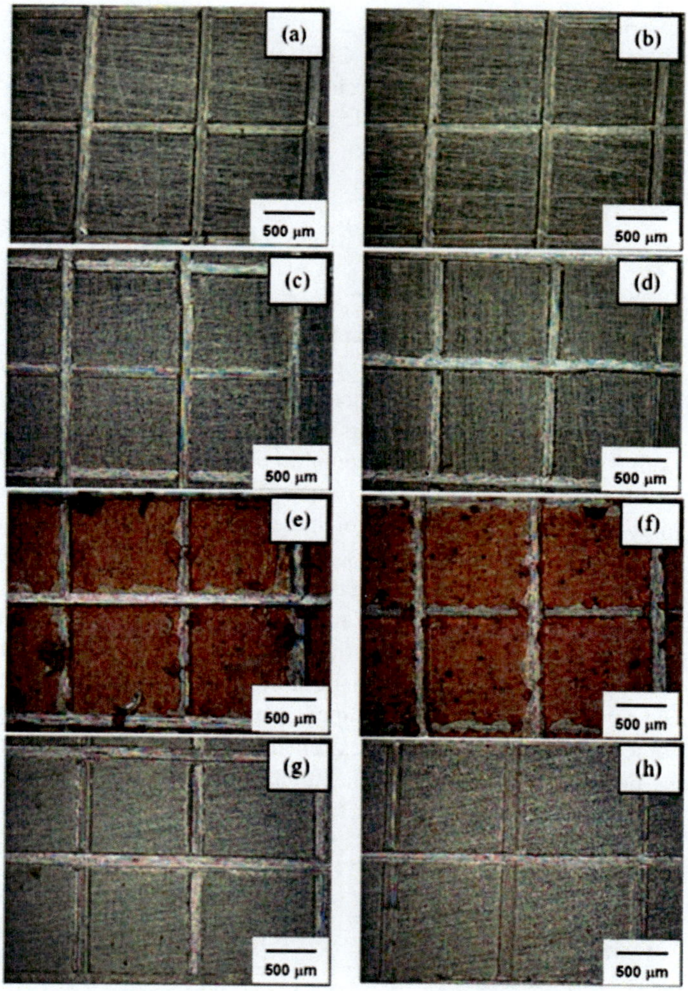

Fig. 5 Tape adhesion test images of different coatings **a**, **b** matrix sol, **c**, **d** BS, **e**, **f** BLLS, and **g**, **h** BHS coupons

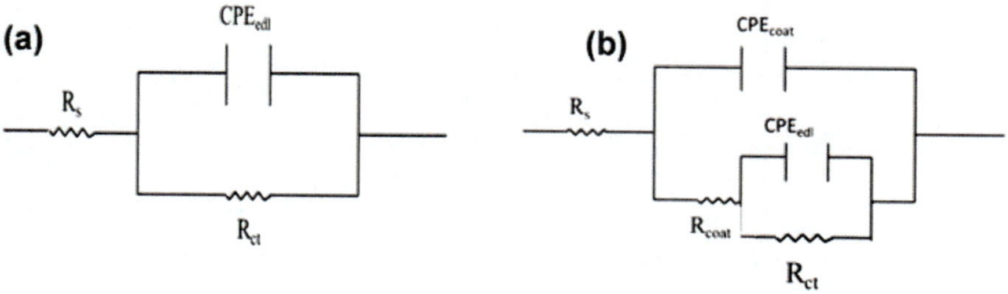

Fig. 6 Equivalent electric circuits for **a** bare and **b** coated coupons

which are in series with C_{edl} parallel to R_{ct}. Since the Nyquist plot obtained deviates from the standard semicircle shape, the CPE (constant phase element) was used instead of a pure capacitor for all the circuit fittings. Presence of intrinsic surface non-uniformity of mild steel is responsible for the non-ideal nature of the electrical double layer. The pseudocapacitance value can be obtained using the formula:

$$C = (Q_o * R)^{1/n}\left({}^{1}/_{R}\right)$$

where C is pseudocapacitance in F/cm^2, Q_o is constant phase element in S-sec^n/cm^2, n is frequency factor and R is resistance in Ω.

The Nyquist plots obtained after 1 h exposure to NaCl solution of concentration 0.6 mol/L are displayed in Fig. 7. The plots obtained for bare and coated coupons are nearly semicircular in shape, showing good corrosion resistant properties. The impedance value for bare substrate was found to be very low, from which it can be inferred that it has very low corrosion resistance. Hybrid sol–gel coating and added corrosion inhibitor can be responsible for increased impedance values of matrix and BTA loaded matrix coated substrates, attributing to the larger diameter of the semicircle. The sol–gel coating consisting of BTA encapsulated LbL nanocontainers (BLLS) showed higher impedance values and larger semicircle compared to the only matrix coated and BS coated substrates, thereby exhibiting higher corrosion resistance and lower corrosion rate. The BTA-HNT-sol coated substrates (BHS) exhibited enhanced corrosion resistance with impedance values several orders higher than those of all other coatings, indicating the extremely superior corrosion resistance and barrier properties of BHS coatings. The electrochemical impedance data along with

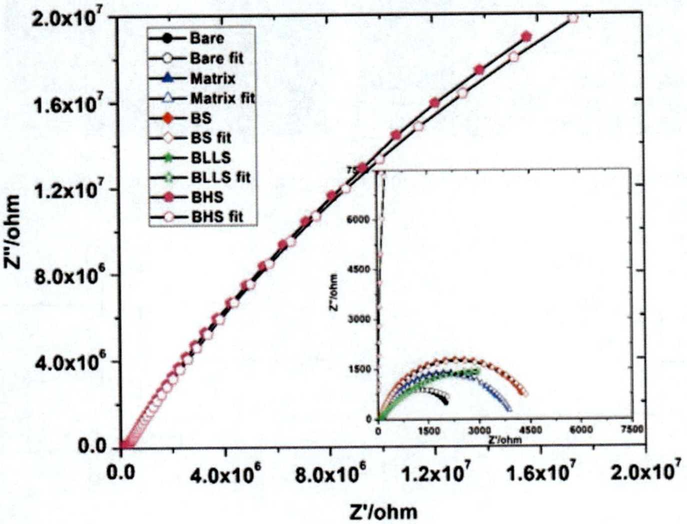

Fig. 7 Nyquist plots for bare and coated coupons

charge transfer resistance values are shown in Table 1. The data obtained show that the capacitance value for BLLS coated substrates has increased when compared to the only matrix coated substrates which can be because of the presence of polyelectrolyte layers that swell upon absorption of water and acts as a barrier by blocking the entry of foreign species. With time, these polyelectrolyte layers may burst with an increase in water uptake initiating the release of encapsulated corrosion inhibitor which self-heals the damage on the metal surface thereby confirmed from the coating resistance value. The BS coated substrates have shown an increase in coating capacitance because of the uptake of water along with an increase in coating resistance which can be attributed because of the presence of freely loaded BTA into the matrix sol which adsorbs on the mild steel surface at the sites where Cl^- ions of the electrolyte solution was attached. This results in the formation of a protective film that restricts the entry of water molecules to some extent. The BHS coated substrates have shown a decrease in capacitance value suggesting that the coating is more compact which blocks further penetration of water molecules. This barrier film formation protects the metal surface from corrosion. The coating and charge transfer resistance values have increased drastically for BHS coated substrates when compared to other coatings. The corrosion inhibitor molecules get protonated in presence of electrolyte solution which forms a bond with the negatively charged surface of the metal. This interaction helps in corrosion mitigation by self-healing the defects on the metal surface [16].

Potentiodynamic polarization profiles of bare and coated substrates after 1 h exposure to 0.6 mol/L NaCl solution were recorded and are shown in Fig.

8. The values of corrosion potential and corrosion current are shown in Table 2. The data reveal that the corrosion potential values changed a meager by 60 mV from bare electrodes and became more positive for sol-gel coatings (Matrix and BS) and more negative for nanocontainer encapsulated substrates (BLLS and BHS). However, the value of corrosion current was high for bare, matrix, and BS substrates, but the nanocontainer encapsulated coatings (BLLS and BHS) exhibited low corrosion currents. This extreme decrease in corrosion current established the enhanced self-healing ability provided by BTA entrapped HNT sol coating. The encapsulated BTA-HNT sol (BHS) coated substrates exhibited extremely low corrosion current density and hence very low corrosion rate, indicating that the controlled release of BTA from HNT provided better self-healing properties and good corrosion resistance when compared to uncoated and other coated substrates.

Table 1 EIS fit data for bare and coated substrates after 1 h exposure to 0.6 mol/l NaCl solution

Substrate	R_{coat} (Ω.cm^2)	R_{ct} (Ω.cm^2)	C_{coat} (F/cm^2)	C_{edl} (F/cm^2)	χ^2
Bare	–	2558	–	4.280×10^{-4}	5.369×10^{-3}
Matrix	158.1	3813	1.599×10^{-5}	5.614×10^{-5}	3.579×10^{-3}
BS	726.8	3955	1.012×10^{-4}	1.11×10^{-5}	3.076×10^{-3}
BLLS	438.4	5954	7.343×10^{-5}	1.189×10^{-5}	4.801×10^{-3}
BHS	2.287×10^5	$1.03\,3 \times 10^8$	2.178×10^{-10}	7.478×10^{-7}	6.490×10^{-3}

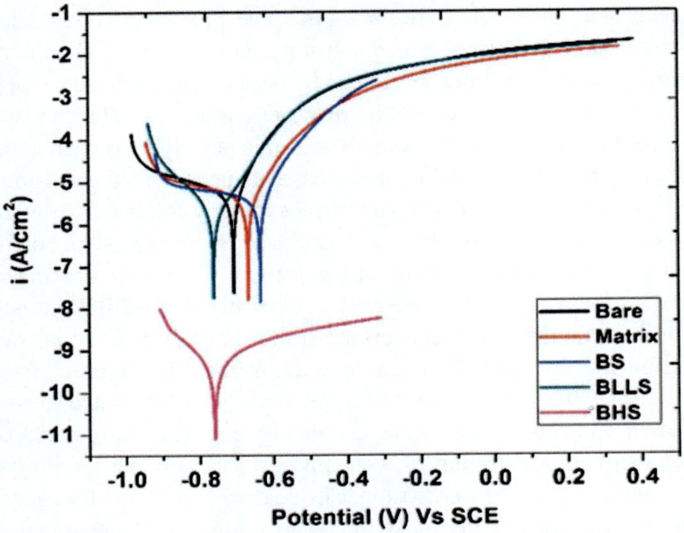

Fig. 8 Tafel plots for bare and coated coupons

Table 2 Electrochemical polarization data for bare and coated coupons

Substrate	E_{corr} (V vs. SCE)	I_{corr} (A/cm^2)
Bare	−0.707	5.54×10^{-6}
Matrix	−0.668	5.37×10^{-6}
BS	−0.634	4.86×10^{-6}
BLLS	−0.761	2.25×10^{-6}
BHS	−0.760	5.35×10^{-10}

3.6 *Salt Spray Test*

Photographs of uncoated and coated substrates after salt spray test are shown in Fig. 9. Artificial scribes were created on the coated substrates in order to visualize the self-repairing nature of the coatings. The substrates were kept in a salt spray compartment and exposed to 5 wt% NaCl solution for different time durations namely, 1, 2, 4, 6 and 24 h. The images obtained after 1 h exposure show that the corrosion products are formed on the bare substrate, whereas, for the coated substrates, no significant change was observed. After 2 h exposure, the bare substrate was 50% corroded but the matrix coated substrate showed the rust formation at the edges of the defect and nearby area. The BTA loaded coatings still possessed good corrosion inhibition properties. Images obtained after 4 and 6 h exposure clearly showed that the bare substrate got corroded completely, whereas matrix, BS, and BHS coated substrates were nearly 50% corroded, which indicates that the barrier feature of the matrix-coated substrates prevents the penetration of foreign electrolyte species enabling the protection of

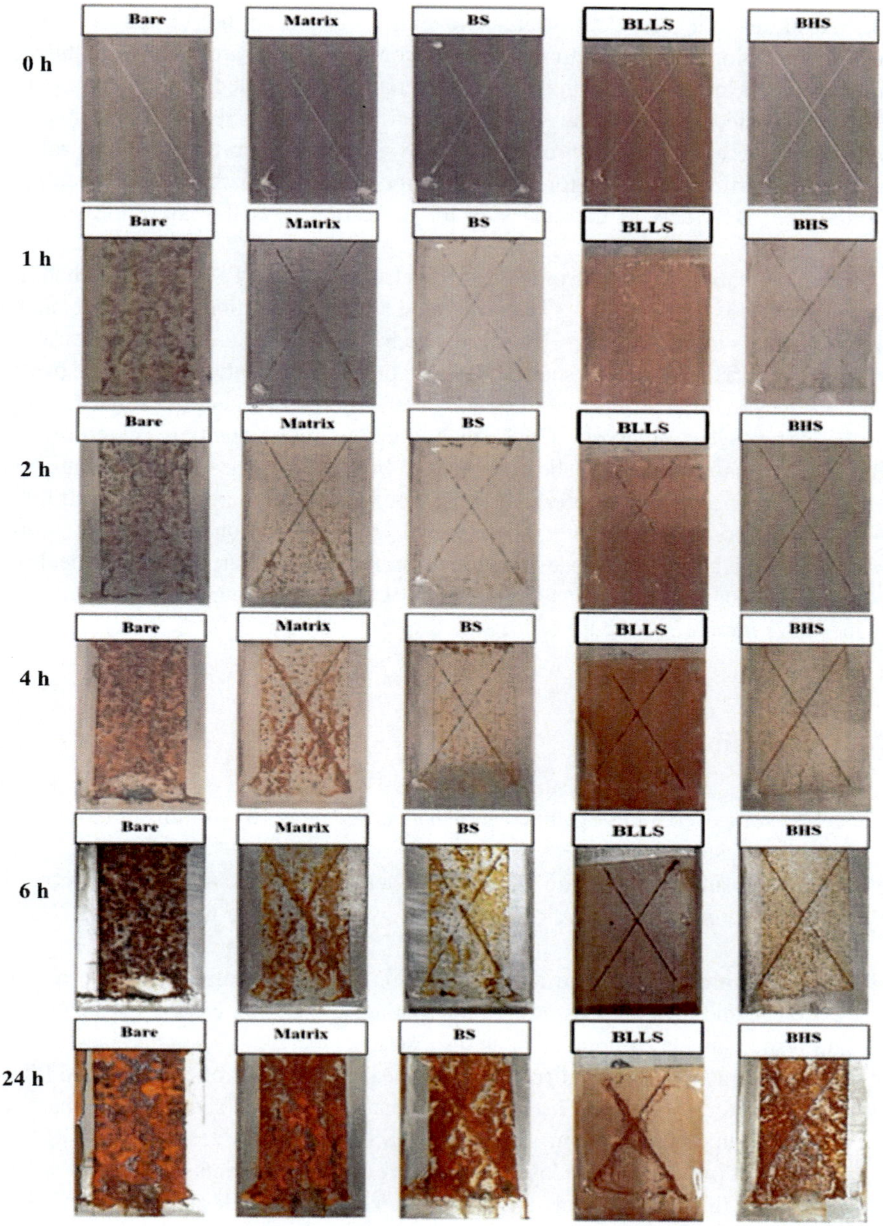

Fig. 9 Images of bare and coated MS substrates when exposed to salt spray

MS substrates to some extent whereas, loading of BTA into the sol delivered self-healing behavior as the inhibitor is released from the scribed area. Images obtained after 24 h exposure illustrate that all the substrates got corroded completely but the BLLS-sol coated substrate showed enhanced barrier properties because of the layered structure, where the polyelectrolyte layers swell up on absorbing the electrolyte solution imparting protection for longer durations when compared to other coatings. In addition, BTA acts as the next step for protection by self-healing the surface defects.

The information obtained from SST differs from EIS and PPS. The reason behind this difference is that the SST is an accelerated test where artificial defects are made on the coatings to observe the self-healing mechanism of bare and coated substrates, whereas, in EIS technique, no such damage occurs and information is retrieved based on the adhesion of coatings onto the metal surface. The PPS measures the stability of the system when polarized with respect to open circuit potential. Therefore, the BLLS coated substrates have shown enhanced barrier properties when compared to other coatings because of the presence of a layered structure where the polyelectrolyte layers swell upon uptake of water. These polyelectrolyte layers on increase in pressure may burst which in turn releases the corrosion inhibitor imparting the self-healing behavior for longer durations. Similar effect was observed where the capacitance values have increased because of the water uptake.

4 Conclusions

- Hybrid sol–gel coatings exhibited good barrier properties by forming homogenous coatings on substrates.
- BTA encapsulated halloysite nanotubes when mixed in sol–gel matrix showed enhanced corrosion protection properties when exposed to 0.6 mol/L NaCl solution.
- Electrochemical impedance and polarization studies revealed higher corrosion resistance and lower corrosion rate for coatings containing BTA encapsulated in halloysite nanoclay among all coated substrates.
- The salt spray analysis confirmed the self-healing behavior of encapsulated BTA into HNT for up to 6 h. The BLLS-sol coated substrates showed good barrier properties up to 24 h when subjected to SST (salt spray test) because of the presence of polyelectrolyte layers which swell up and restricts the entry of water molecules. In addition to this, the encapsulated BTA provides protection by self-healing the damage.

Acknowledgements The authors are thankful to Director, ARCI for all support throughout the investigation carried out. The authors would like to acknowledge Dr. K. Suresh and Balaji Padya for XRD and Zeta potential measurements.

References

1. Basik M, Mobin M (2020) Chondroitin sulfate as potent green corrosion inhibitor for mild steel in 1 M HCl. J Mol Struct 1214:128231. https://doi.org/10.1016/j.molstruc.2020.128231
2. Hamidon TS, Hussin MH (2020) Susceptibility of hybrid sol-gel (TEOS-APTES) doped with caffeine as potent corrosion protective coatings for mild steel in 3.5 wt.% NaCl. Prog Org Coat 140:105478. https://doi.org/10.1016/j.porgcoat.2019.105478
3. Zhao J, Xia L, Sehgal A, Lu D, McCreery RL, Frankel GS (2001) Effects of chromate and chromate conversion coatings on corrosion of aluminum alloy 2024–T3. Surf Coat Technol 140:51–57. https://doi.org/10.1016/S0257-8972(01)01003-9
4. Zheludkevich ML, Tedim J, Ferreira MGS (2012) "Smart" coatings for active corrosion protection based on multi-functional micro and nanocontainers. Electrochim Acta 82:314–323. https://doi.org/10.1016/j.electacta.2012.04.095
5. Figueira RB, Fontinha R, Silva CJR, Pereira EV (2016) Hybrid sol-gel coatings: smart and green materials for corrosion mitigation. Coatings 6:12. https://doi.org/10.3390/coatings6010012
6. Figueira RB, Silva CJR, Pereira EV (2014) Organic–inorganic hybrid sol–gel coatings for metal corrosion protection: a review of recent progress. J Coat Technol Res 12:1–35. https://doi.org/10.1007/s11998-014-9595-6
7. Li Y, Wu C, Xue M, Cai J, Huang Y, Yang H (2019) Preparation of sol-gel derived anticorrosive coating on Q235 carbon steel substrate with long-term corrosion prevention durability. Materials (Basel) 12:1960. https://doi.org/10.3390/ma12121960
8. Habib S, Khan A, Nawaz M, Sliem MH, Shakoor RA, Kahraman R, Abdullah AM, Zekri A (2019) Self-healing performance of multifunctional polymeric smart coatings. Polymers (Basel) 11:1519. https://doi.org/10.3390/polym11091519
9. Nazeer AA, Madkour M (2018) Potential use of smart coatings for corrosion protection of metals and alloys: A review. J Mol Liq 253:11–22. https://doi.org/10.1016/j.molliq.2018.01.027
10. Ghazi A, Ghasemi E, Mahdavian M, Ramezanzadeh B, Rostami M (2015) The application of benzimidazole and zinc cations intercalated sodium montmorillonite as smart ion exchange inhibiting pigments in the epoxy ester coating. Corros Sci 94:207–217. https://doi.org/10.1016/j.corsci.2015.02.007
11. Izadi M, Mohammadi I, Shahrabi T, Ramezanzadeh B, Fateh A (2019) Corrosion inhibition performance of novel eco-friendly nanoreservoirs as bi-component active system on mild steel in aqueous chloride solution. J Taiwan Inst Chem Eng 95:555–568. https://doi.org/10.1016/j.jtice.2018.09.012
12. Rydzek G, Ji Q, Li M, Schaaf P, Hill JP, Boulmedais F, Ariga K (2015) Electrochemical nanoarchitectonics and layer-by-layer assembly: From basics to future. Nano Today 10:138–167. https://doi.org/10.1016/j.nantod.2015.02.008
13. Sonawane SH, Bhanvase BA, Jamali AA, Dubey SK, Kale SS, Pinjari DV, Kulkarni RD, Gogate PR, Pandit AB (2012) Improved active anticorrosion coatings using layer-by-layer assembled ZnO nanocontainers with benzotriazole. Chem Eng J 189–190:464–472. https://doi.org/10.1016/j.cej.2012.02.076
14. Kamburova K, Boshkova N, Boshkov N, Radeva T (2016) Design of polymeric core-shell nanocontainers impregnated with benzotriazole for active corrosion protection of galvanized steel. Colloids Surf A Physicochem Eng Asp 499:24–30. https://doi.org/10.1016/j.colsurfa.2016.03.067
15. Abdullayev E, Lvov Y (2011) Halloysite clay nanotubes for controlled release of protective agents. J Nanosci Nanotechnol 11:10007–10026. https://doi.org/10.1166/jnn.2011.5724
16. Njoku DI, Cui M, Xiao H, Shang B, Li Y (2017) Understanding the anticorrosive protective mechanisms of modified epoxy coatings with improved barrier, active and self-healing functionalities: EIS and spectroscopic techniques. Sci Rep 7:1–15. https://doi.org/10.1038/s41598-017-15845-0

Effect of Metallic Ion Species on Nitric Acid Corrosion of Type 304L Stainless Steel

Gopinath Shit, A. Pooonguzhali, and S. Ningshen

1 Introduction

Austenitic stainless steel (SS) is extensively used as a structural material in chemical, nuclear reactor, reprocessing, and waste management plants. In the Plutonium Uranium Reduction Extraction (PUREX) reprocessing process, nitric acid (3–14 M HNO_3) is the primary medium for the dissolution of spent fuel and extracting useful fissile material, which also generates a vast amount of higher valence oxidative fission and corrosion products. Austenitic stainless steels (SS) are structural materials with desirable mechanical properties, formability, weldability, toughness, and resistance to corrosion [1–5], and hence widely used in several chemical industries, power, transportation, and nuclear industrial sectors [1–6]. Further, AISI type 304L SS is explicitly used in nuclear reprocessing and waste management plants, especially where HNO_3 is the primary process medium because of its excellent corrosion-resistant property. In the nuclear fuel reprocessing plant PUREX process, nitric acid is used as a medium from room temperature (RT) to boiling conditions [1–5]. Process medium contains fission products generated during nuclear fission and oxidizing and corrosive ions. These result in intergranular corrosion (IGC) attacks despite using non-sensitized steel [1, 2]. In stainless steel weldment, the heat-affected zone (HAZ) is prone to IGC attack because of the additional sensitization effect, where IGC

G. Shit (✉) · S. Ningshen
Homi Bhaba National Institute, Mumbai 400094, India
e-mail: gopims@igcar.gov.in

S. Ningshen
e-mail: ning@igcar.gov.in

G. Shit · A. Pooonguzhali · S. Ningshen
Corrosion Science and Technology Division, Indira Gandhi Centre for Atomic Research, Kalpakkam 603102, India
e-mail: apoongs@igcar.gov.in

U. K. Mudali et al. (eds.), *Recent Trends in Electrochemical Science and Technology*, Springer Proceedings in Materials 15,
https://doi.org/10.1007/978-981-16-7554-6_2

attack occurs in and around the depletion of chromium near grain boundaries. From the electrochemical aspect, the chromium depleted zone causes the stainless steel anodic behavior from the global anodic current in stainless steel [2]. According to Fauvet et al. [1] three possible reasons behind the increase in corrosion potential in the PUREX reprocessing environment are: (1) in pure nitric acid water medium cathodic reaction is the reduction reaction of HNO_3; (2) the addition of metallic oxidizing species, where reduction reaction involves the reduction of oxidizing ion; and (3) nitric acid media containing metallic ions are nobler than steel which results in galvanic coupling. The electrochemical response of the SS in the nitric acid medium as a function of potential has been discussed [1]. The anodic process generally involves the oxidation of chemical elements such as iron, chromium, and nickel. The process can be separated into different zones: (1) the active zone where stainless steel's uniform dissolution occurs at suitably reducing conditions, (2) in oxidizing conditions, stainless steel is in a passive state, and the dissolution rate is considerably low, and (3) in highly oxidizing medium, passive film (insoluble Cr_2O_3) dissolution occurs due to the oxidation of Cr (III) to Cr (VI) (CrO_3), resulting in trans-passive corrosion [1]. Only the transition domain appears in less severe conditions, and in highly oxidative medium severe IGC attack occurs. The shift of corrosion potential from the passive region beyond the trans-passive region needs to be avoided. Potential can be shifted to the trans-passive region when the nitric acid concentration is very high and in the solution containing oxidizing ions [1–6]. Primarily, IGC is characterized by the formation of triangular grooves at the grain boundaries. The enhancement of IGC mainly depends on the specific reactivity of the grain boundaries and segregation of grain boundaries due to P, B, C, and Si, which impacts the kinetics of the IGC [1–3]. The grain boundaries specific reactivity depends on the grain boundaries metallurgical conditions such as relative misorientation of the grain boundaries and the IGC attack kinetics [7]. IGC attack results in accelerated mass loss, extensive grain dropping, or increased surface contact with media [8]. The extent of IGC can be precisely measured by the morphology of the groove width and depth [7].

In general, the weight loss measurement using ASTM A-262 practice-C (Huey) test determines IGC susceptibility in 65 wt% boiling nitric acid medium. The weight loss measure in each 48 h interval for five cycles (total 240 h) defines the corrosion rate and the average corrosion rate should be around <24 mpy [4]. But for a harsh nitric acid condition such as in nuclear reprocessing conditions, it should be ≤ 18 mpy, and for critical components, the corrosion rate of ≤ 5 mpy is desirable [6, 8]. Thereby, proper control of trace impurities, steel making practices (Electro Slag Remelting, which greatly purifies the alloy), and processing parameters optimization of the trace elements is essential to achieve improved corrosion resistance for SS in the nitric acid application. For long-term operation reprocessing plants, stainless steel with high resistance to IGC in nitric acid is required [7]. Nitric acid grade (NAG) stainless steel type 304L shows more resistance towards sensitization and IGC. This type of stainless steel has modified chemical compositions such as ultra-low C, high Cr, and optimum Ni [6, 7] and restricted minor elements such as Si and P [6, 7]. However, sulfide inclusion can also lead to the end grain corrosion attack which is a decisive form of damage in service plants [6, 9].

In the corrosion of SS in the nitric acid medium, the factors which influence its corrosion include the acid concentration and the oxidizing metallic ions such as Nd^{3+}, Mo^{6+}, Cs^{1+}, Mn^{7+}, Ce^{4+}, etc. which are present as fission products formed by the dissolution of spent fuel, and Cr^{6+} and Fe^{3+} as corrosion products from structural material [1, 10]. Additionally, the temperature (~120 °C) of boiling nitric acid also plays a crucial role in stainless steel corrosion [1, 10]. The oxidizing ions concentration also influences the corrosion of stainless steel by influencing the cathodic reduction reaction of the electrochemical system [1, 3]. Pelle et al. [7] have shown that Fe (III) plays a significant role in reducing concentrated HNO_3 reaction by generating reduced species such as NO_2. Bhise et al. [10] reported that the Ce^{4+} and V^{5+} affect the corrosion rate of stainless steel and the shifting of the corrosion potential towards the trans-passive region in the nitric acid medium. The effect of oxidizing ions in nitric acid medium on the corrosion rate of stainless steel indicated that corrosion accelerates due to oxidizing metal ions and temperature [8]. Hasegawa et al. [11] studied the effect of corrosion rate of Cr ion in boiling and non–boiling nitric acid solutions and reported that the oxidizing effect of Cr (VI) does not depend on the nitric acid concentration.

In this work, the corrosion resistance behavior of type 304L SS in simulated non-radioactive PUREX environment with different oxidizing species (Nd^{3+}, Mo^{6+}, and Cs^{1+} metallic ions present as fission product (FP), corrosion product of Fe^{3+}, and surrogate element V^{5+}, etc.) in different concentrations (2, 6 and 8 M) of nitric acid has been examined. The three-phase corrosion behavior has also been evaluated in nitric acid in simulated PUREX conditions to evaluate its corrosion resistance.

2 Experimental

2.1 Materials and Methods

The chemical composition (wt %) of AISI type 304L SS is shown in Table 1.

For electrochemical corrosion test as received SS 304L sample of size 10 × 10 × 6 mm was molded using Araldite and hardener. Samples were polished up to 0.25 μm using diamond paste. After this, the samples were ultrasonically cleaned with acetone and washed with double distilled water, and then the molded sides were masked with elastomer. For the three-phase corrosion study, the exact dimension of the sample was polished up to 2000 grit SiC abrasive paper. The experiment was similar to the ASTM A-262 practice-C test, where up to 800 SiC grit polishing is recommended [4]. To obtain the corrosion effect of vapor and condensate phase, the samples were

Table 1 Chemical composition (in wt %) of AISI type SS 304L

Cr	Ni	C	Si	Mo	P	Mn	Fe
18.2	9.2	0.02	0.32	0.05	0.03	0.59	Balance

Table 2 The concentration of metallic ions added for electrochemical corrosion test

Ions added	Concentration (g/l)
Nd^{3+}	1.185
Mo^{6+}	1.115
Cs^{1+}	1.40
Fe^{3+}	0.085
V^{5+}	0.1

further polished up to 2000 SiC grit abrasive paper and cleaned ultrasonically before the experiments. To simulate a non-radioactive PUREX environment on AISI type 304L SS, the ions with different oxidizing species (Nd^{3+}, Mo^{6+} and Cs^{+}, corrosion product Fe^{3+}, and surrogate element V^{5+}, etc.) in different concentrations (2, 6 and 8 M) of nitric acid were chosen. The concentration of the metal ions Cs (I), Mo (VI), and Nd (III) solutions used are listed in Table 2 and they were prepared from their respective nitrate salts.

Electrochemical corrosion (open circuit potential and potentiodynamic polarization) test was performed using an electrochemical workstation (Biologic SP 300 model) (Fig. 1a), and the data analysis was performed with EC-lab software. The electrochemical corrosion test was conducted by using a three-electrode setup, Pt as a counter electrode; saturated Ag/AgCl as a reference electrode; and sample as the working electrode. The three-phase corrosion (liquid, vapor, and condensate) in different concentrations (2, 6, and 8 M) of nitric acid was evaluated and the setup is shown in Fig. 1b.

The three-phase corrosion test, similar to ASTM A-262 practice C (Huey test) [4, 9] was conducted in nitric acid solutions with and without metallic ions. After the corrosion test, the samples were characterized using an inverted optical microscope (OM) (Leica model). The metallic ions concentrations used for the boiling three-phase corrosion test are shown in Table 3.

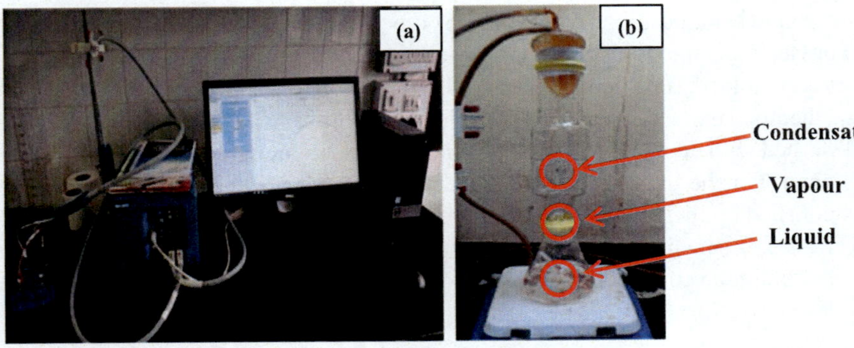

Fig. 1 Photographs of **a** electrochemical workstation and **b** set up used for determination of three-phase corrosion

Table 3 Metallic ions concentrations that were used for the boiling three-phase corrosion test

Metal	Concentration (g/l)	Metal	Concentration (g/l)
Ag	0.097	Rb	0.071
Ba	0.485	Rh	0.350
Ce	0.769	Sm	0.3419
Cs	1.405	Sr	0.146
La	0.417	Te	0.160
Mo	1.115	Y	0.084
Nd	1.186	Zr	0.9
Pd	1.07	Fe	0.085
Pr	0.410		

XPS spectra were recorded using SPECS make (Germany) spectrometer in constant energy analyzer mode, using monochromatized Al Kα radiation (hν = 1,486.74 eV) probe and a hemispherical analyzer for energy analysis with a pass energy of 20 eV and the optimum energy resolution of 0.6 eV was used. Binding energy calibration was carried out using Ag 3d and C 1 s as the reference [9].

3 Results and Discussion

3.1 Electrochemical Corrosion Test

3.1.1 Open Circuit Potential

The open circuit potential (OCP) was performed in 2, 6, and 8 M HNO_3 for 1 h (3600 s). In 2 M nitric acid solution, the initial voltage which was 0.10 V stabilized at 0.17 V. After the addition of Fe^{3+}ion, this value raised to 0.44 V and increased up to 0.51 V. Further, with FP (Nd^{3+}, Mo^{6+}, and Cs^{1+}), the value varied from 0.15 to 0.60 V. In presence of FP and Fe^{3+} ions, the OCP shifted towards the more noble direction and it ranges between 0.50 and 0.54 V. With the addition of V^{5+}, the potential shifts towards the more noble direction. The initial value of 0.55 V is shifted to 0.57 V. In 6 M solutions, a similar trend was observed. As the acid concentration increased, the OCP value increased to 0.47 V. With the addition of Fe^{3+}, the OCP shifted towards the more noble direction (0.75 V) and with the addition of FP and FP with Fe, the OCP shifts towards the nobler direction at around 0.77 V. However, the addition of V^{5+} a marginal decrease in OCP to 0.65 and 0.7 V. This shift of potential towards nobler direction increased with the nitric acid concentration as well as with higher valence metallic ions which is undesirable. With 8 M nitric acid solution, the OCP value shows an initial value of 0.79 V. It reaches up to 0.81 V and after adding Fe^{3+} ions, it shifted to 0.84 V. The addition of FP increased the OCP value to 0.85 V while with the addition of FP and Fe, the OCP was 0.84 V. With the addition of V^{5+}, the

Table 4 The OCP values (V vs. Ag/AgCl) observed for the substrate when exposed to different concentrations of nitric acid and different metal ions

Solution	2 M	6 M	8 M
PURE	0.17	0.48	0.81
Fe	0.51	0.75	0.85
Fission product (FP)	0.60	0.77	0.85
FP + Fe	0.54	0.77	0.84
V	0.57	0.7	0.84

OCP value shifted up to 0.84 V. The measured OCP values are listed in Table 4 and plots are shown in Fig. 2.

Bhise et al. [10] observed similar behavior for SS 304L in presence of oxidizing (V^{5+}) metallic ions in nitric acid medium. The OCP measurement can also indicate the stability of passive films. With the addition of metallic oxidizing ions, the OCP values are shifted towards the noble direction, indicating the risk of shift near or over the trans-passive region's potential, which is undesirable that may lead to IGC [6, 12].

3.1.2 Potentiodynamic Polarization Measurement

The potentiodynamic anodic polarization test was conducted in 2, 6, and 8 M nitric acid solutions, and the curves are shown in Fig. 2, and the obtained parameters are listed in Table 5. In 2 M nitric acid, the obtained E_{corr} is 0.17 V. E_{corr} is shifted to the nobler direction after the addition of Fe^{3+}(0.085 g/l) to 0.49 V. With the addition of FPs (Nd, Mo, and Cs), E_{corr} further shifted to the noble direction to 0.61 V. Further, the addition of Fe^{3+} with FP again shifted the potential towards the active direction, it reached to 0.56 V. The addition of V^{5+}(0.1 g/l) also shifted the potential to 0.56 V. In 6 M nitric acid solution, E_{corr} value is 0.49 V. After the addition of Fe^{3+}, E_{corr} value has shifted towards more noble direction to 0.76 V. With the addition of FP, the potential is shifted towards more nobler direction to 0.84 V, and the E_{corr} value was 0.75 V after the addition of Fe^{3+}. The effect of V^{5+} shows the potential at 0.68 V. In 8 M nitric acid solution, the obtained E_{corr} value is 0.82 V. This value has further shifted towards the noble direction to 0.83 V after the addition of Fe^{3+}, and the value (0.838 V) is almost same as that after the addition of FP. With the addition of Fe^{3+} along with FP, there was no change in the potential, and the obtained value was 0.84 V. But with the addition of V^{5+}, the potential shifted to the more noble direction to 0.85 V in 8 M nitric acid solution. In highly oxidizing and concentrated nitric acid or nitric acid containing oxidizing metallic ion species, the potential shifted towards a trans-passive domain, which resulted in severe IGC in type 304L SS. The higher valance metal ions in general increase the cathodic reaction. The higher valance metal ions thereby enhance the autocatalytic reduction reaction of nitric acid. This leads to the anodic metal dissolution from the stainless steel at the trans-passive region. In this region, the acid is unstable and oxidizes, resulting in oxygen and hydrogen ions [1] and thereby enhancing nitric acid reduction. The addition of Nd (III), Mo

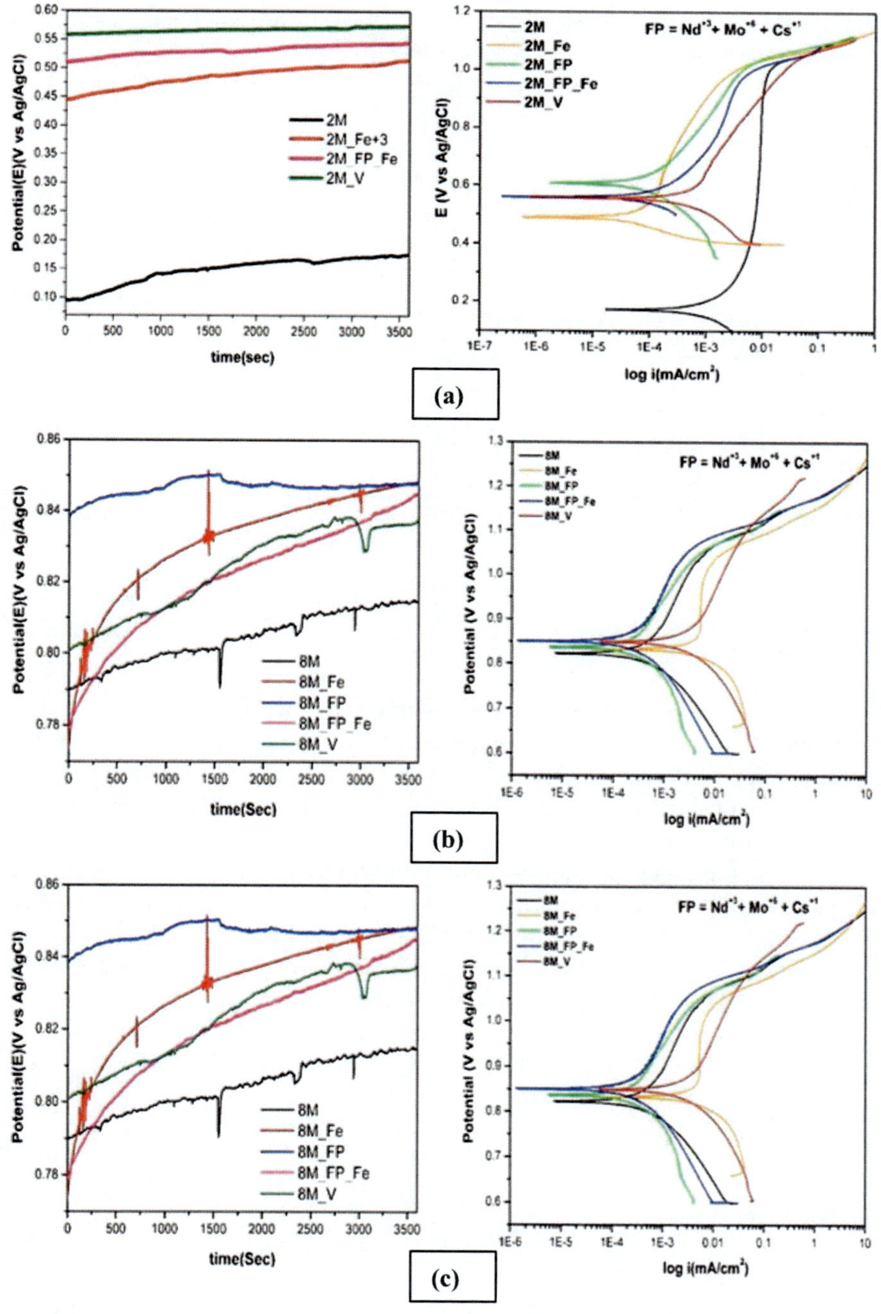

Fig. 2 OCP and potentiodynamic polarization plots of samples in different concentrations of nitric acid **a** 2 M, **b** 6 M, and **c** 8 M in presence of various metals ions

Table 5 Corrosion parameters obtained from potentiodynamic polarization test in 2, 6, and 8 M nitric acid medium and in presence of oxidizing metal ions

Solution	E_{corr} (V vs. Ag/AgCl)	Trans-passive breakdown potential	I_{pass} (μA)
2 M	0.17	1.01	9.1
2 M + Fe	0.49	1.02	0.48
2 M + FP	0.61	1.03	0.89
2 M + FP + Fe	0.56	1.0	1.7
2 M + V	0.56	1.04	3.0
6 M	0.49	1.07	33.3
6 M + Fe	0.76	1.04	0.97
6 M + FP	0.84	1.07	1.34
6 M + FP + Fe	0.75	1.07	1.34
6 M + V	0.68	1.08	5.56
8 M	0.82	1.07	1.7
8 M + Fe	0.83	1.05	5.4
8 M + FP	0.838	1.07	1.3
8 M + FP + Fe	0.84	1.06	0.9
8 M + V	0.85	1.08	0.009

(VI), and Cs (I) increased the oxidation rate of Cr (III) of the steel in nitric acid by the hydration of the cations [11]. According to Hasegawa et al. [11], the addition of nitrate salt of different metal ions in nitric acid solution resulted in the change of corrosion rate of stainless steel, mainly the oxidation rate of Cr (III) not only depend on the NO_3^- ion concentration but also it varies with the added cations.

To support this argument, vapor–liquid equilibrium calculation was carried out in consideration of the hydration of nitrates. Accordingly, the rate constant of Cr (III) increases linearly with the increase of partial pressure of nitric acid. Motooka et al. [13] proposed a mechanism for the increase in the generation of Cr (VI) due to the formation of NO_2 with the thermal decomposition of nitric acid.

$$NO_3^- + 3H^+ + 2e^- = HNO_2 + H_2O;\ E_0 = 0.934\,\text{V vs. SHE} \tag{1}$$

$$NO_2 + H^+ + e^- = HNO_2;\ E_0{=}1.093\,\text{V vs. SHE} \tag{2}$$

$$E = E^0 - 2.303\frac{RT}{F}\text{pH} + \frac{RT}{F}\ln\frac{pNO_2}{[HNO_2]} \tag{3}$$

Thermal decomposition of nitric acid generated NO_x ($x = 1, 2 \ldots$). From Eq. (3) it is evident that the potential E increased with the increase in partial pressure of NO_2. Therefore the oxidation behavior of nitric acid increased with the increase in partial pressure of NO_x component [11].

$$2Cr^{3+} + 3NO_3^- + 5H_2O = 2HCrO_4^- + 5H^+ + 3HNO_2 \quad (4)$$

$$E = E^0 - 2.303 \times 5\frac{RT}{6F}pH + \frac{RT}{6F}\ln\frac{[HCrO_4^-]^2[HNO_2]^3}{[Cr^{3+}]^2[NO_3^-]^3} \quad (5)$$

The generation of Cr (VI) increased with the decrease in the concentration of HNO_2 by the generation of NO*x* gas with HNO_3 gas in boiling nitric acid solution as per the thermodynamic calculation.

Thus, the oxidation of Cr (III) is affected by the vapor–liquid equilibrium in the nitric acid solution which is attributed to the hydration of cations in nitrates. The corrosion rate of stainless steel is accelerated by the co-existence of other cations in nitric acid solution. But this effect is not predominant as it is determined by the increase in nitrate ion concentration instead of cation concentration [11].

Further, the Fe^{3+} ions in nitric acid solution increase the cathodic current, which shifts the corrosion potential (E_{corr}) towards the noble direction. However, the mechanism of nitric acid reduction by Fe^{3+} ions is not well established. According to Pelle et al. [7], two hypotheses were proposed involving the direct reduction of Fe^{3+} or chemical reaction where Fe^{3+} ion is involved in the reduction mechanism of nitric acid as follows [7]:

$$Fe^{3+} + e^- \longleftrightarrow Fe^{2+}(aq) \quad (6)$$

$$Fe^{2+}(aq) + \frac{3}{2}H^+(aq) + \frac{1}{2}NO_3^- \longleftrightarrow Fe^{2+}(aq) + \frac{1}{2}HNO_2(aq) + \frac{1}{2}H_2O(l) \quad (7)$$

$$HNO_2(aq) + H^+(aq) + e^- \longleftrightarrow NO(ads) + H_2O(l) \quad (8)$$

The effect of Fe^{3+} ion on the E_{corr} value in the presence of other ions is not prominent, which could be due to the dissolution of Fe^{3+} ions in the solution. Such dissolution resulted in a lower E_{corr} value in some cases [7, 11]. The standard electrochemical potential of V (V)/V (IV) was considered for corrosion rate evaluation in different concentrations of nitric acid due to its redox potential ($E^0_{25°C} = 1000\,mv/SHE$) in a similar range to the Pu (VI)/Pu (IV) system ($E^0_{25°C} = 1040\,mv/SHE$). In general, the SS corrosion depends on the redox potential of the ions and not on the nature of the ions. Even though the redox potential of the V (V)/V (IV) system is lower than the redox potential of the Pu (VI)/Pu (IV) system, still the corrosion of the latter is higher than the former. In the kinetics of the redox reaction, V (IV) instantaneously oxidized into V (V) in boiling nitric acid medium [1]. The following equation represents the redox reaction of V (V)/V (IV)

$$V(OH)_4^+ + H^+ + e^- \longleftrightarrow VO^{2+} + 3H_2O \quad (9)$$

According to Irisawa et al. [14], in a nitric acid solution containing V (V), corrosion rate remains constant for all immersion durations, but this cannot be explained by the polarization curve and oxidation potential of the metallic cations. The reduction rate influences the oxidizing metallic ions effect. After the addition of the oxidizing metallic ions, the oxidation rate is reduced to a lower oxidation state due to the corrosion reaction called re-oxidization [14]. When stainless steel corrodes, this re-oxidization rate is very high for V (V), even if it has been reduced to V (IV), and the time dependency corrosion rate does not decrease with time and remains constant [7, 11].

After the addition of the metallic oxidizing elements in 2 and 6 M solution, the shift corrosion potential value has a drastic shift. In 8 M nitric acid solution, the corrosion potential value shift is not significant with the addition of metallic ions. This is attributed to the dissolved oxidizing metal ions in the higher 8 M nitric acid solution that shows no prominent effects on the cathodic reaction. In the cathodic reaction for the nitric acid solution of concentration <6 M, the generation of HNO_2 is enhanced, and nitrogen monoxide (NO) is generated in the charge transfer step. The NO accumulates and reacts with HNO_3 to form HNO_2 because NO_2 is not thermodynamically stable in a solution of <8 M nitric acid concentration [1]. In higher concentrated nitric acid solutions such as in 8 M, the generation of HNO_2 is faster, which leads to the generation of thermodynamically stable gaseous NO_2. The NO_2 (N_2O_4) acts as a reducing agent; thus, oxidizing metallic ions can be reduced. The cathodic reactions shift significantly less, despite the solution contains oxidizing metal ions [1, 15].

$$2NO_3^- + 4H^+ + 2e^- \longleftrightarrow N_2O_4 + 2H_2O;\ E_{25^\circ C} = 803\,\text{mV/SHE} \quad (10)$$

3.2 *Three-Phase Corrosion Results*

Figure 3 shows the histograms of three-phase corrosion in different nitric acid concentrations with and without metallic ions. The results of the three-phase corrosion rate have been tabulated in Tables 6 and 7. It is evident that the liquid phase corrosion rate is higher than the vapor and condensate phase. Vapor phase corrosion rate is in the condensate phase in all nitric acid concentrations except in 2 M nitric acid solution. This is attributed to the lower renewable rate in the condensate phase to vapor phase nitric acid. Furthermore, in the vapor phase, the accumulation of dissolved ions is hindered due to replenishment at frequent intervals. Also, the concentration of HNO_3 is lower in vapor and condensate phases.

Thereby, depending, on these factors, the corrosion rates can be lower in the condensate or vapor phase of nitric acid. With the oxidizing metallic fission products, the corrosion rate increased in all phases. This is attributed to the enhancement of corrosion rate catalysis by the higher valence oxidizing metallic ions and reducing products such as nitrous acid and NO_x, which are generated in the liquid medium. According to Balbaud et al. [16], the boiling point of 8 M nitric acid is around

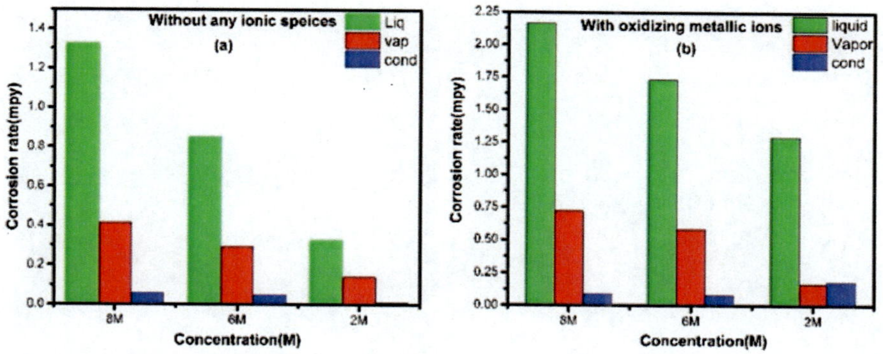

Fig. 3 **a** Variation of corrosion rate (mpy) with different concentrations of nitric acid and **b** variation of corrosion rate (mpy) with different concentrations of nitric acid with metallic oxidizing ions

Table 6 Three-phase corrosion of SS 304L in different concentrations of nitric acid medium without any addition of oxidizing metal ions

Concentration of nitric acid	Corrosion rate (mpy)		
	Liquid	Vapor	Condensate
8 M	1.33	0.41	0.057
6 M	0.85	0.29	0.048
2 M	0.33	0.14	0.00973

Table 7 Three-phase Corrosion of SS 304L in different concentrations of nitric acid medium with oxidizing metal ions

Concentration of nitric acid	Corrosion rate (mpy)		
	Liquid	Vapor	Condensate
8 M	2.16	0.72	0.087
6 M	1.73	0.58	0.078
2 M	1.29	0.163	0.18

111 °C, and the vapor phase temperature is 3 °C below the boiling point, i.e., 108 °C, which can be obtained from the nitric acid water phase diagram. Moreover, in the vapor phase, the oxidizing metallic species effects are also significantly less; thereby, resulting in a lower corrosion rate. Similarly, for the same nitric acid solution condensate phase, in 6 M HNO_3, the temperature is around 108 °C. Subsequently to the frequent renewal of the liquid phase, the metallic oxidizing species is replenished, and the concentration is significantly less, leading to a lower corrosion rate. In 2 M nitric acid with oxidizing metallic ions, condensate phase corrosion is marginally higher than that of the vapor phase due to the non-renewal of nitric acid, slightly lower than the other phases [16]. Corresponding optical microscopy (OM) images of these specimens are shown in Figs. 4, 5, 6, 7, 8 and 9. The OM images show moderate IGC in 8 M boiling liquid nitric acid solution; in the vapor phase, the attack of IGC is comparatively less though the etched grain boundaries are visible. In the condensate phase, the mode of corrosion attacks is not distinguishable, it is

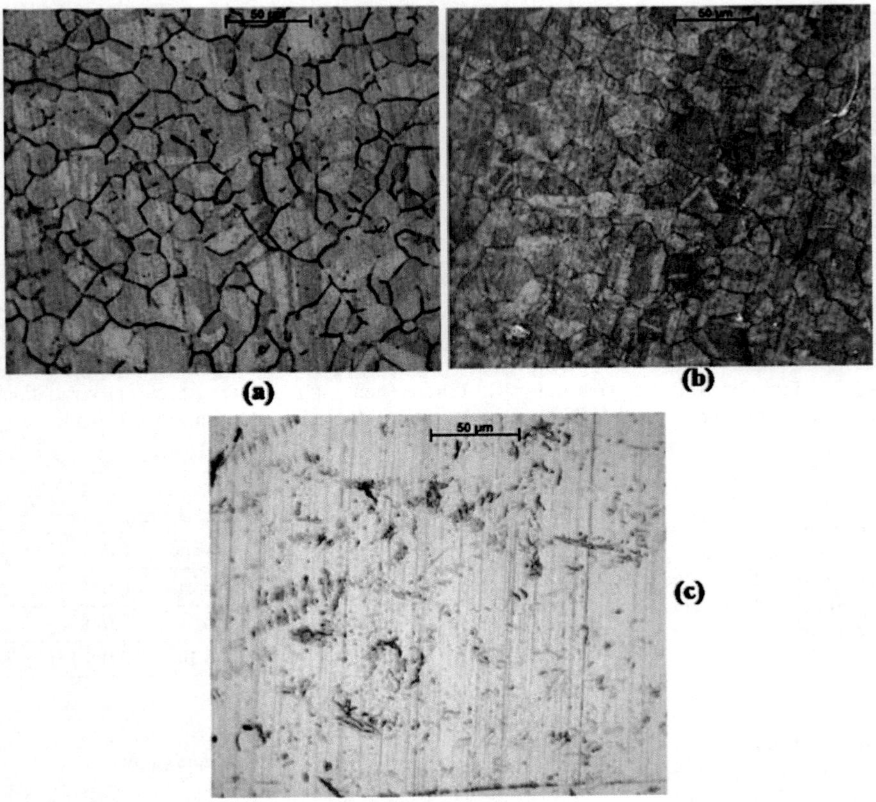

Fig. 4 Optical microscopy images of type 304L stainless steel specimen subjected to three-phase corrosion test in 8 M nitric acid solution without any additional oxidizing metal ions in **a** liquid, **b** vapor, and **c** condensate phases

preferably uniform with selective leaching. Similarly, for 6 M nitric acid solutions in boiling liquid medium, the mode of IGC attack is less severe than in 8 M solutions, but the etched grain boundary morphology is clearly observed. But the vapor and condensate phase corrosions are very negligible, and attacks are very mild with less uniform corrosion. In 2 M nitric acid solution, there was no such IGC attack thereby clearly indicating the influence of nitric acid concentration on the IGC. In the boiling liquid, only small pit-like features are observed. In the vapor and condensate phase, no significant microstructural changes were observed. However, with oxidizing ions, the IGC attack is more prominent than without any ions in 8 M nitric acid solution. In boiling liquid medium, the grain boundaries are observed with comparatively more IGC attacks. But in the vapor and condensate phase, attacks are not that much prominent, but it is more severe compared to the one without any oxidizing ions. In 6 M and 2 M nitric acid solutions, similar kind of microstructural features was observed, but the extent of the attack is greater than without any oxidizing metallic ions.

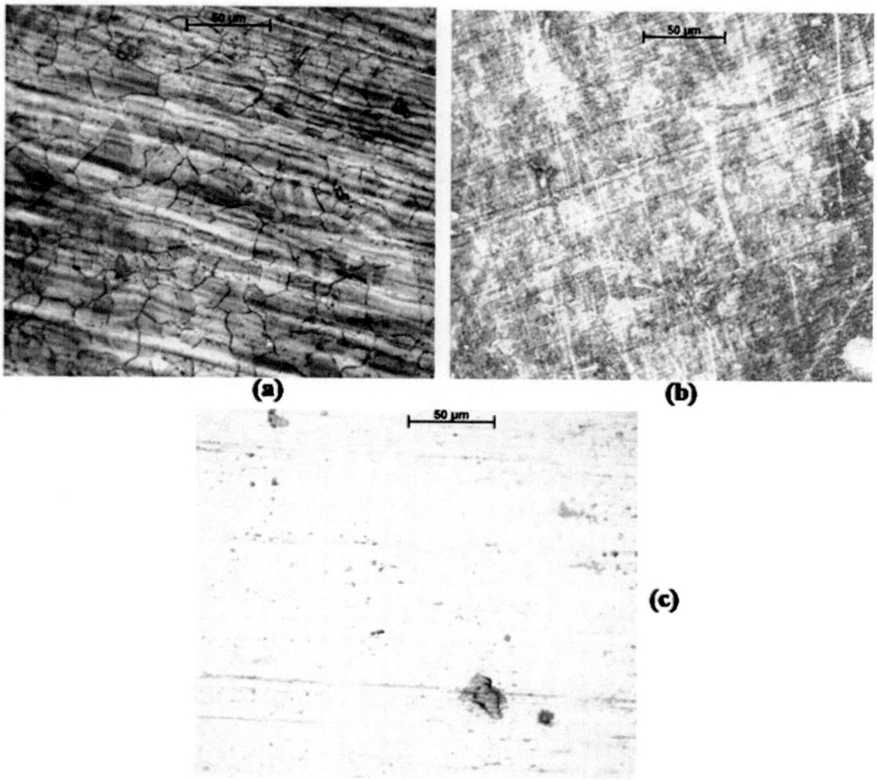

Fig. 5 Optical microscopy images of type 304L stainless steel specimen subjected to three-phase corrosion test in 6 M nitric acid solution without any additional oxidizing metal ions in **a** liquid, **b** vapor, and **c** condensate phases

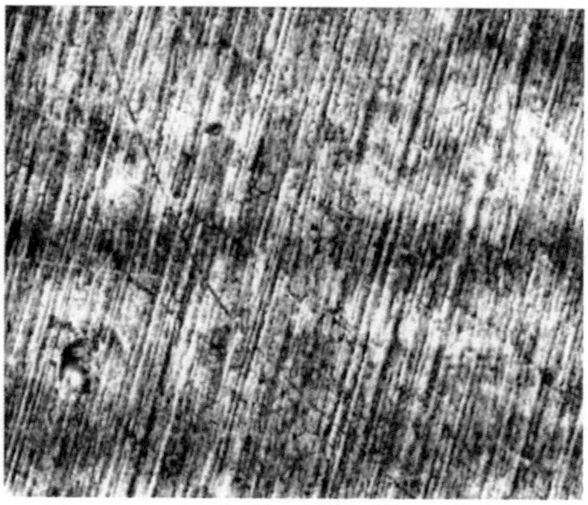

Fig. 6 Optical microscopy image of type 304L stainless steel specimen subjected to three-phase corrosion test in 2 M boiling liquid nitric acid solution without any additional oxidizing metal ions in liquid phase (In vapor and condensate phase there was no significant corrosion attack)

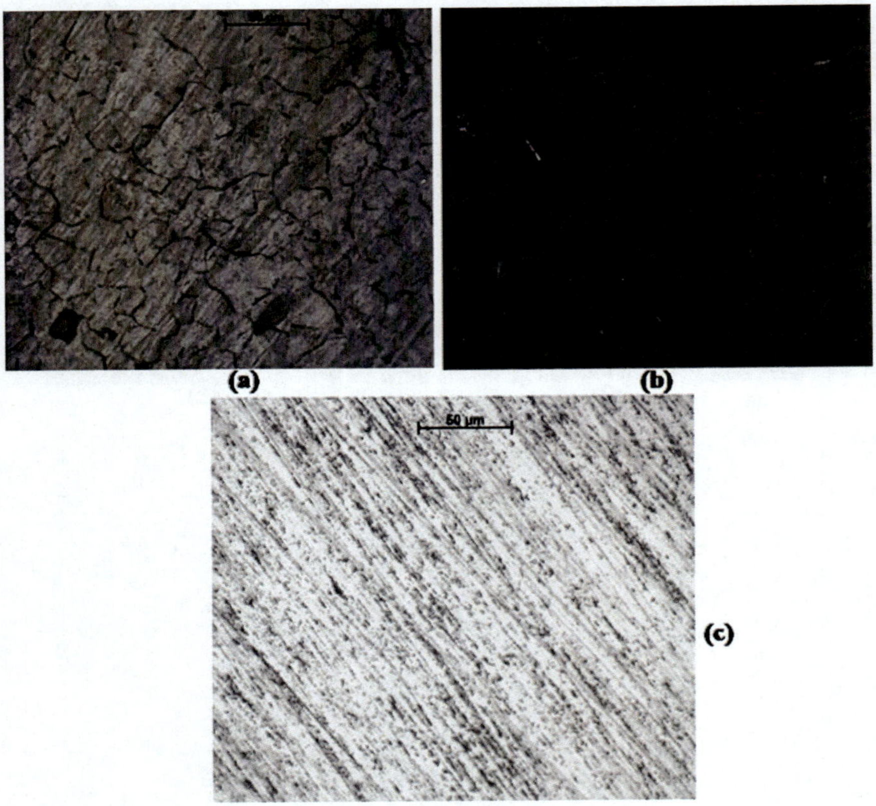

Fig. 7 Optical microscopy images of type 304L stainless steel specimen subjected to three-phase corrosion test in 8 M nitric acid solution in the presence of additional oxidizing metal ions in **a** liquid, **b** vapor, and **c** condensate phases

3.3 Surface Characterization by XPS

The XPS spectra of SS 304L after passivation for 1 h in 8 M, 8 M + FP, and 8 M + V solution is shown in Fig. 10. XPS analysis provides the surface composition of type SS304L after passivation in nitric acid. Survey scan provides elemental composition such as Fe, Cr, and O peaks in the spectra; other elements such as S, Ni, and Mn are not present in the film. The oxidizing metal ions are possibly not absorbed on the surface during passivation. The high affinity towards Cr and Fe with oxygen resulted in the diffusion from the bulk to the surface, resulting in chromium oxide and iron oxide formation. Due to that, only Ni is not present in the film though Ni is present in significant amounts in the bulk of the samples. The high resolution peaks of Fe, Cr, and O show Fe-metal, FeO, and Fe_2O_3; for Cr only Cr_2O_3 is found and oxygen peaks corresponding to O^{2-} and H_2O are observed.

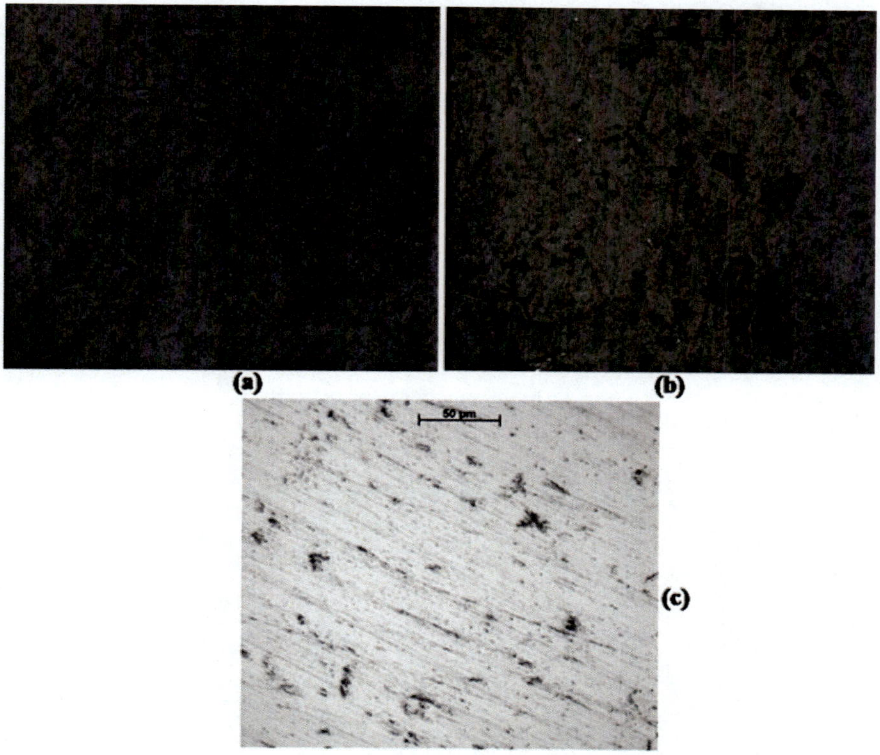

Fig. 8 Optical microscopy images of type 304L stainless steel specimen subjected to three-phase corrosion test in 6 M nitric acid solution in the presence of additional oxidizing metal ions in **a** liquid, **b** vapor, and **c** condensate phases

With the increase in metallic ion concentration from 8 to 8 M + FP and 8 M + V, the intensity of the peak decreases, attributed to the dissolution of the surface elements with the increase in the redox potential and concentration of the redox metal ions. According to Laurent et al. [17], stainless steel undergoes severe dissolution at high potential, especially at the grain boundaries than at the grain faces, which leads to the localized corrosion attack at the trans-passive region.

4 Conclusions

The effect of nitric acid concentrations on the electrochemical corrosion behavior and three-phase corrosion of type 304L SS with and without any oxidizing metallic ion species has been evaluated. The following are the observations:

- With the increase in nitric acid concentration, the corrosion potential shifts towards the noble direction.

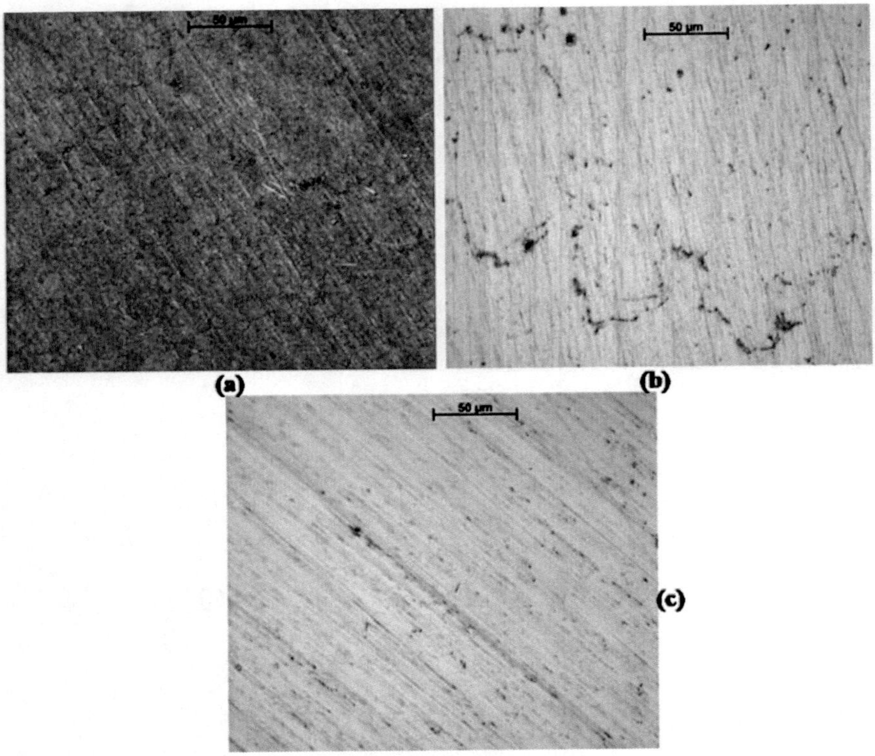

Fig. 9 Optical microscopy images of type 304L stainless steel specimen subjected to three-phase corrosion test in 2 M nitric acid solution in the presence of additional oxidizing metal ions in **a** liquid, **b** vapor, and **c** condensate phases

- The higher valence oxidizing metallic cations catalyzed the cathodic reduction and shifted the corrosion potential towards the nobler direction.
- The influence of Fe^{3+} ion in other metallic ions was not observed because the combined metal ions dissolution in solution is higher than the Fe^{3+} ions.
- V (V)/V (IV) redox potential was higher among all other metallic cations. For the same amount of V (V) ions, the shifting of potential proportionally increases with the increase in nitric acid concentration.
- XPS results revealed the formation of a passive film mainly consisting of Cr_2O_3 and Fe_2O_3. The presence of oxidizing metallic ions does not show a significant influence on the film composition.
- IGC attack on 304L SS was found to be more severe in boiling liquid medium with oxidizing ions at all concentrations of nitric acid compared to vapor and condensate media.

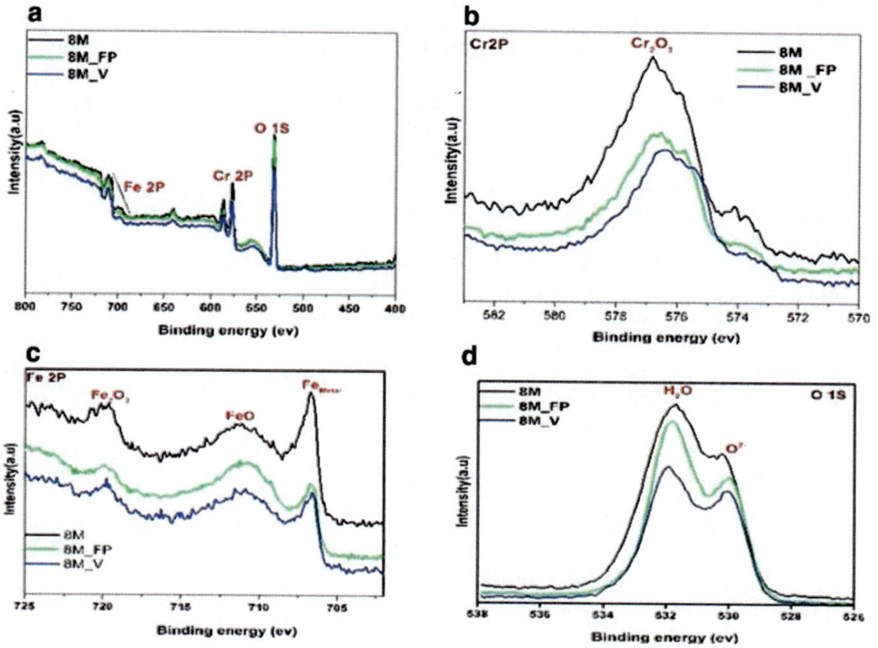

Fig. 10 Surface analysis by XPS **a** survey spectra, **b** high resolution Cr 2P spectra, **c** high resolution Fe 2P spectra, **d** high resolution O1S spectra of SS 304L passivated for 1 h in 8 M nitric acid solution and with various metallic ions

Acknowledgements The authors like to acknowledge IGCAR management for active support towards this research work.

References

1. Fauvet P, Balbaud F, Robin R, Tran QT, Mugnier A, Espinoux D (2008) Corrosion mechanisms of austenitic stainless steels in nitric media used in reprocessing plants. J Nucl Mater 375:52–64. https://doi.org/10.1016/j.jnucmat.2007.10.017
2. Gwinner B, Auroy M, Balbaud CF, Fauvet P, Larabi GN, Laghoutarisa P, Robin R (2016) Towards a reliable determination of the intergranular corrosion rate of austenitic stainless steel in oxidizing media. Corros Sci 107:60–75. https://doi.org/10.1016/j.corsci.2016.02.020
3. Sedriks AJ (1996) Corrosion of stainless steels, 2nd edn. Wiley-Interscience, New York
4. ASTM Designation (2014) Standard practices for detecting susceptibility to intergranular attack in austenitic stainless steels, vol A-262, pp 1– 20
5. Mudali UK, Dayal RK, Gnanmoorthy JB (1993) Corrosion studies on materials of construction for spent nuclear fuel reprocessing plant equipment. J Nucl Mater 203:73–82. https://doi.org/10.1016/0022-3115(93)90432-X
6. Raj B, Mudali UK (2006) Materials development and corrosion problems in nuclear fuel reprocessing plants. Prog Nucl Energy 48:283–313. https://doi.org/10.1016/j.pnucene.2005.07.001

7. Pelle J, Gruet N, Gwinner B, Schlegel ML, Vivier V (2020) On the role of Fe(III) ions on the reduction mechanisms of concentrated nitric acid. Electrochim Acta 335:135578. https://doi.org/10.1016/j.electacta.2019.135578
8. Ningshen S, Mudali UK, Ramya S, Raj B (2011) Corrosion behaviour of AISI type 304L stainless steel in nitric acid media containing oxidizing species. Corros Sci 53:64–70. https://doi.org/10.1016/j.corsci.2010.09.023
9. Shit G, Ningshen S (2019) The corrosion behavior of compositional modified AISI type 304L stainless steel for nitric acid application. Anti-Corros Methods Mater 66:149–158. https://doi.org/10.1108/ACMM-02-2018-1906
10. Bhise S, Kain V (2012) Methodology based on potential measurement for predicting corrosion behaviour of SS 304L in boiling nitric acid containing oxidizing ions. Corros Eng Sci Technol 47:61–69. https://doi.org/10.1179/1743278211Y.0000000016
11. Hasegawa S, Kima SY, Ebina T, Tokuda H, Ito T, Nagano N, Hitomia K, Ishiia K (2016) Effect of nitrate on Corrosion of austenitic stainless steel in boiling nitric acid solution containing chromium ions. J Nucl Sci Technol 53:1332–1341. https://doi.org/10.1080/00223131.2015.1107514
12. Kolman DG, Ford DK, Butt DP, Nelson TO (1997) Corrosion of 304 stainless steel exposed to nitric acid-chloride environments. Corros Sci 39:2067–2093. https://doi.org/10.1016/S0010-938X(97)00092-9
13. Motooka T, Kato C, Yamamoto M (2010) Effect of re-oxidation rate of chromium and vanadium ions on corrosion rate of stainless steel in boiling nitric acid solutions. Zairyo-to-Kankyo 59:18–24 (Japanese). https://doi.org/10.3323/JCORR.59.18
14. Irisawa E, Yamamoto M, Kato C, Motooka T, Ban Y (2019) Effect of re-oxidation rate of additive cations on corrosion rate of stainless steel in boiling nitric acid solution. J Nucl Sci Technol 56:337–344. https://doi.org/10.1080/00223131.2019.1580624
15. Kain V, De PK (2005) Controlling corrosion in the back end of fuel cycle using nitric acid grade stainless steels. Int J Nucl Energy Sci Technol 1:220–231. https://doi.org/10.1504/IJNEST.2005.007146
16. Balbaud F, Sancheza G, Fauveta P, Santarinia G, Picard G (2000) Mechanism of corrosion of AISI 304L stainless steel in the presence of nitric acid condensates. Corros Sci 42:1685–1707. https://doi.org/10.1016/S0010-938X(00)00021-4
17. Laurent B, Gruet N, Gwinner B, Miserque F, Rousseau K, Ogle K (2017) The kinetics of transpassive dissolution chemistry of stainless steels in nitric acid: the impact of Si. Electrochim Acta 258:653–661. https://doi.org/10.1016/j.electacta.2017.11.110

Effects of Strontium Addition on Corrosion Properties of Al-12Si Alloy

Lalit Kumar Meena, Anney Kr. Sinha, Ravada Seshagiri, and Raghuvir Singh

1 Introduction

Currently, the materials designers in the industries are facing critical challenges due to the stringent requirements of energy-saving and reducing the emission of pollutants. In this perspective, the most economical approach to overcome these difficulties is the design and development of high-strength lightweight components [1]. The Al-Si alloys exhibit high strength to weight ratio due to which they are widely used in automotive, defense, space, aerospace, and washing machine industries [1–3]. Besides, they possess excellent castability and weldability, superior wear resistance, and recyclability. The Al-12Si alloys are the most suitable candidates for making engine pistons, auto hubs, crankcases, and washing machine spiders due to low thermal coefficient, superior impact resistance, high load-bearing capacity, and moderated corrosion resistance [4]. Silicon as a main alloying element aids fluidity to liquid aluminum and reduces shrinkage and porosity during solidification [5]. On the contrary, eutectic silicon crystallizes with coarse and lamellar morphology (needles or platelets) between ductile α-Al dendrites, which results in the deterioration of the mechanical properties, specifically ductility, and causes brittle failure [6]. There are several provisions proposed to improve the structural and mechanical properties of Al-Si alloys such as the addition of modifiers or alloying elements during the melting, heat treatment, and severe plastic deformation techniques [7–15]. Among all, eutectic modification techniques are deployed widely due to their

L. K. Meena · A. Kr. Sinha · R. Seshagiri · R. Singh (✉)
CSIR-National Metallurgical Laboratory, Jamshedpur, India
e-mail: rsr@nmlindia.org

L. K. Meena
e-mail: lalit@nmlindia.org

R. Seshagiri
e-mail: rsg@nmlindia.org

U. K. Mudali et al. (eds.), *Recent Trends in Electrochemical Science and Technology*, Springer Proceedings in Materials 15,
https://doi.org/10.1007/978-981-16-7554-6_3

simplicity and low capital cost. Generally, the addition of Sr, Na, or other rare-earth and alkali metals into Al-Si alloys results in modified eutectic structure, and thus improves the mechanical and physical properties [11, 15, 16]. Amongst all, Na and Sr are known to be the strongest eutectic modifiers and are frequently utilized in foundry industries. But, Sr addition has certain advantages over Na addition such as less volatility, high recovery rate (80–90%), easy to add, longevity in the melt during holding times, produces smoother castings, doesn't react with refractory, and is environment friendly [11]. The strontium modifier in the range of 100–400 ppm is assumed to retard the eutectic silicon growth by absorbing the re-entrant edge during solidification [17]. Thus Sr modifies the morphology of the eutectic phase from coarse brittle flakes to a finer and fibrous structure [18–22]. Further, Sr addition beyond this range is still able to modify the eutectic structure but produces coarse intermetallic (Al_4Si_2Sr) and porosity due to over modification which subsequently may influence the final properties of the alloys [23]. Further, Al-12Si alloys possess moderated resistance to uniform corrosion but often fail due to localized corrosion such as intergranular, pitting, and crevice corrosion depending upon the stability of the passive layer. The friction stir processing (FSP) is evidenced as an effective tool to improve the microstructure, mechanical, and corrosion properties through severe plastic deformation [14]. But, the complexity involved in processing, high initial or setup cost, and inability to process complex shapes restricted the industrial viability of the process. Additionally, heat treatment and selective laser melting (SLM) are other recognized processes to improve the corrosion resistance of Al-12Si alloys [24]. However, the heat treatment as an additional process step increases the initial cost of the material due to the high temperature involved and is restricted to the heat-treatable alloys. Further, the huge laser cost, support configuration requirement, and low building rates have restricted the application of the SLM process [24]. The eutectic modification process has certain advantages over the above-mentioned processes and is effective in altering the eutectic Si morphology, size, distribution of secondary phases, and intermetallic compounds which largely govern the passive layer stability and in turn the corrosion behavior [15, 25, 26]. Also, the eutectic modifiers influence the chemical composition of the oxide layer formed over the surface during exposure to the corrosive medium and thus alter the corrosion resistance of Al-Si alloys [27]. Lee et al. reported that the addition of 200 ppm of Sr to Al-Si alloy results in increased hardness, fracture toughness, and ultimate tensile strength (UTS) and reduced corrosion resistance as compared to the unmodified alloy [25]. The corrosion potential (E_{corr}) is reported to shift towards a more negative side from −557.7 mV to −748.1 mV while corrosion current (i_{corr}) increased from 0.51 to 1.76 $\mu A/cm^2$ in 0.5 M H_2SO_4 solution with Sr addition. The decrease in corrosion resistance was attributed to the increased grain boundary area between α-Al and eutectic silicon phases. Duygun et al. [28] also revealed that the addition of 250 ppm of Sr to Al-9Si alloy increased the area of eutectic phases from 26.1 to 33.0% and decreased its corrosion resistance due to galvanic coupling in 3.5% NaCl solution. Nevertheless, Ozturk et al. [22] reported the beneficial effect of the Sr addition on the corrosion behavior of A356 alloys in a 3.5% NaCl solution. The E_{corr} was shifted slightly towards the more negative side from −623 to −650 mV and i_{corr} decreased to one-sixth of 13.82 μA/cm^2 (2.21 μA/cm^2) after the addition of 250 ppm of Sr addition in die-cast alloy. The improved corrosion resistance was attributed to the stable passive oxide layer growth throughout the surface that was associated with the fine fibrous uniformly distributed eutectic phase. Apart from the metal/alloy and corrosive media, there are several other factors such as manufacturing process, fluid flow and temperature which affect the corrosion behavior [7].

Although, several investigations address the modification of the eutectic in Al-Si alloy using Sr but the literature on Sr addition beyond 0.2% and the resultant corrosion behavior is not explored to a greater extent. Besides, the effect of the addition of Sr to Al-Si alloy on the corrosion properties is contradictory. Further, there is no literature available that addresses the corrosion behavior of the Al-Si alloys in the flowing corrosive mediums despite the industrial relevance of the alloy. The present endeavor aims to investigate the effect of Sr addition (up to 0.35%) on the corrosion behavior of Al-12Si alloys in static and flowing alkaline chloride environments.

2 Material and Methods

The unmodified and Sr modified Al-12Si alloy with Sr wt% of 0.05, 0.1, 0.18, 0.25, and 0.35 were prepared by the typical melting and permanent die casting route. The commercial Al-12.6Si and Al-10Si master alloy ingots were utilized for this purpose. The preheated Al-12.6Si alloy ingots were melted in a 3 kg capacity high-frequency induction furnace using a graphite crucible under the cover flux C65/11 at the temperature 750–760 °C. After ensuring the complete melting, the desired weight of preheated Al-10Sr was added to achieve the composition with Sr ~0.05%, 0.1%, 0.18%, 0.25%, and 0.35% in the melt. The mixture was stirred with a graphite rod for 40–50 s and held for 20 min to ensure complete dissolution and homogenization. Afterward, the melt was degassed with argon and immediately poured into the preheated (for 1 h at 200 °C) cast-iron permanent mold at around 720 °C. The cast mold (size ~310 × 80 × 20 mm^3) was allowed to cool to room temperature in the open atmosphere. Thus cast plates were cut into ~10 × 10 × 5 mm^3 size specimens for microstructure and hardness determination. All the specimens were successively polished from 120, 400, 600 to 1000 grit SiC paper and then subjected to cloth polishing with alumina powder (particle size ~0.5 μm) to obtain a mirror finish. In order to reveal the microstructure, the specimens were etched for 20 s with Keller's reagent (2.5% HNO_3 + 1.5% HCl + 1% HF + 95% distilled water). The microstructural features were revealed through an optical microscope (Leica Microsystems, DM2500M) and electron probe micro-analyzer (EPMA)-energy-dispersive X-ray spectroscopy (EDS) (JEOL, AJXA-8230). The macro and microhardness measurements (in VHN) were carried out using Vickers hardness tester at 5 kgf and 50 gf load respectively, with 15 s dwell time. The corrosion behavior of unmodified and modified Al-12Si alloys was investigated by electrochemical polarization and immersion test methods (as per the standard ASTM-G31) in alkaline 3.5% NaCl solutions (pH

~11). All the specimens were finished up to 600 grit SiC paper and degreased thoroughly with acetone before each experiment. The specimens were cut into size ~20 × 20 × 5 mm^3 for the long-term immersion experiments (of four-month duration). The electrochemical studies include potentiodynamic polarization and electrochemical impedance spectroscopy (EIS), were carried out in both static and flowing conditions (fluid flow ~75 l/min) in Cl^- containing medium. These studies were performed using a potentiostat/galvanostat (Gamry Instruments, Ref 600) and an electrochemical cell comprised of three electrodes; a reference electrode (platinum disc), a working electrode (modified/unmodified Al-12Si alloy), and an auxiliary electrode (platinum disc). The flow-involved electrochemical experiments were conducted in a closed-loop corrosion rig setup where ~1 cm^2 of the working specimen was exposed to a flowing medium. All the flow-involved experiments were conducted in a long pipe section (4″ dia.) maintaining laminar flow conditions. Specimens were immersed in the solution for 3600 s and open circuit potential (OCP) was monitored before the start of the polarization test. The EIS experiments were carried out in the frequency range 0.01–100,000 Hz at open circuit potential with 10 mV AC perturbation. The potentiodynamic polarization experiments were performed in the potential range from −0.25 to 1.5 V w.r.t. OCP versus V_{Pt}, with a scan rate 1.67 mV/s. The EIS300 and DC105 techniques were utilized to analyze the Bode/Nyquist plots and polarization curves respectively. The corrosion products formed after the electrochemical measurements in the static and flowing conditions were analyzed using a scanning electron microscope (SEM)-EDS.

3 Results and Discussion

3.1 *Microstructure and Hardness*

The as-cast unmodified and Sr modified alloys designations and chemical compositions are illustrated in Table 1. The optical micrographs (Fig. 1) revealed a typical eutectic microstructure for all unmodified and modified alloys

Table 1 Chemical compositions of the prepared unmodified and modified Al-12Si alloys

Sample	Si	Cu	Mg	Mn	Fe	Zn	Pb	Sn	Sr	Al
UM	12.05	0.80	0.36	0.29	0.76	0.29	0.08	<0.03	**0**	Bal.
M 1	12.05	0.80	0.36	0.29	0.76	0.29	0.08	<0.03	**0.05**	Bal.
M 2	12.05	0.80	0.36	0.29	0.76	0.29	0.08	<0.03	**0.1**	Bal.
M 3	12.05	0.80	0.36	0.29	0.76	0.29	0.08	<0.03	**0.18**	Bal.
M 4	12.05	0.80	0.36	0.29	0.76	0.29	0.08	<0.03	**0.25**	Bal.
M 5	12.05	0.80	0.36	0.29	0.76	0.29	0.08	<0.03	**0.35**	Bal.

The unmodified alloys consist of α-Al dendritic phase and eutectic mixture in the interdendritic regions comprising of silicon crystallized with coarse and plate like morphologies (Fig. 1a). The microstructure consists of primary silicon rich phase (small grey and coarser grey color particles) with quadrilateral and irregular shapes inside the α-Al phase which evolves as a result of low solubility of silicon in aluminum [10].

The Sr addition yielded into much finer α-Al matrix and transformation of acicular to fine fibrous eutectic structure (Fig. 1b–f); these fine fibrous eutectics were noticed to grow into coarser and longer fibers when Sr was increased to 0.18% or higher, as displayed in Fig. 1e, f. The eutectic phases were found more uniformly distributed throughout the α-Al matrix when 0.25% Sr was added to Al-12Si as compared to other Sr modified alloys. Figures 2, 3 and 4, demonstrate the EPMA-EDS area maps

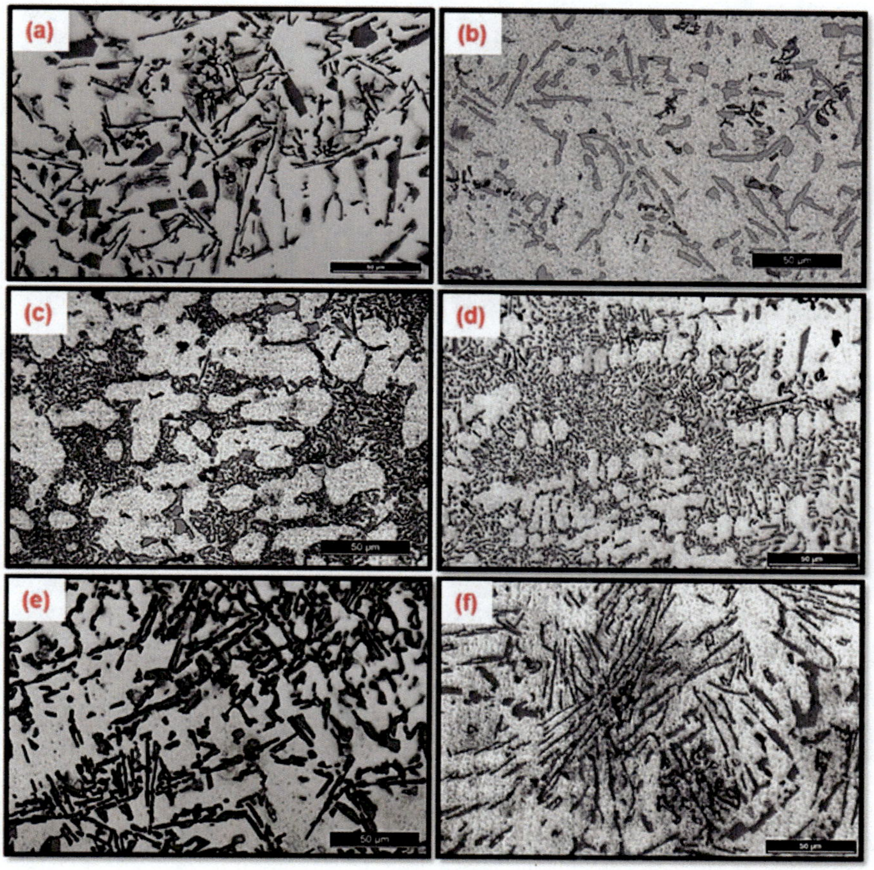

Fig. 1 Optical micrographs of etched unmodified and Sr unmodified alloys; **a** 0% Sr, **b** 0.05% Sr, **c** 0.1% Sr, **d** 0.18% Sr, **e** 0.25% Sr and **f** 0.35%

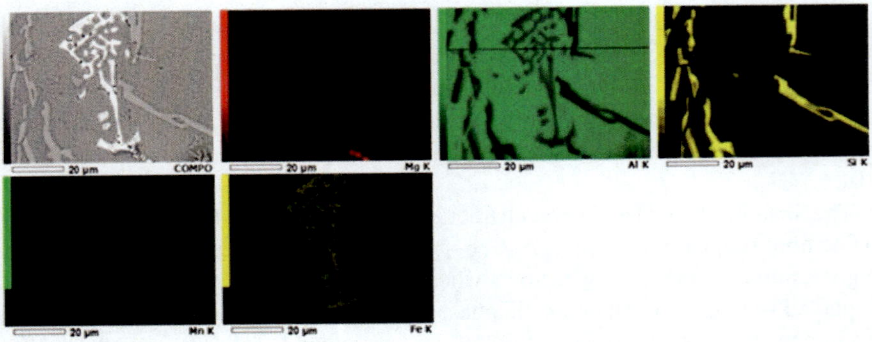

Fig. 2 EPMA-EDS area elemental map of etched unmodified alloys (0% Sr)

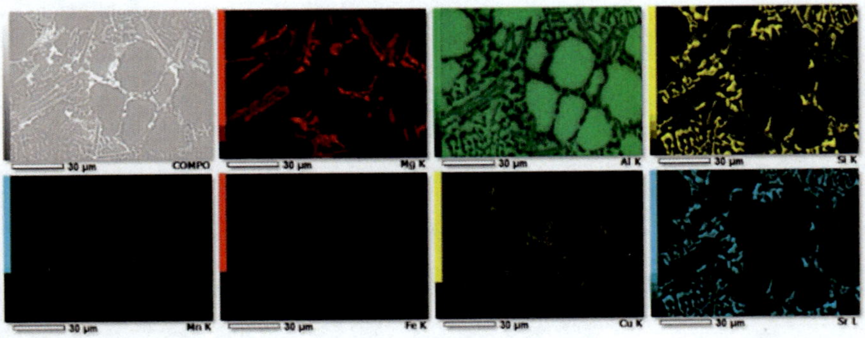

Fig. 3 EPMA-EDS area elemental map of etched modified alloys (0.1% Sr)

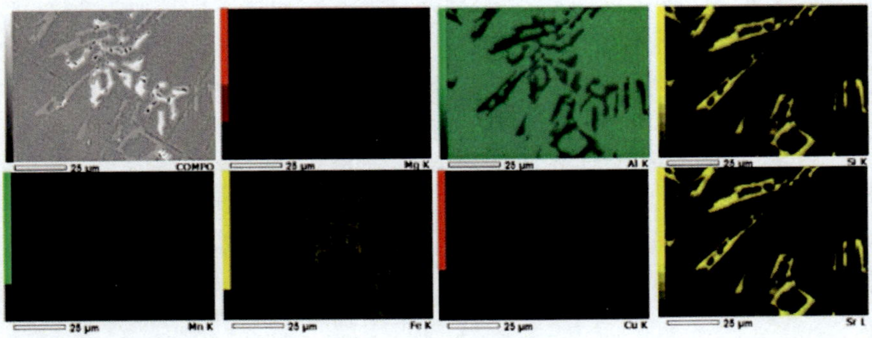

Fig. 4 EPMA-EDS area elemental map of etched modified alloys (0.25% Sr)

of unmodified and modified alloys and revealed the distribution of various elements including across the matrix.

The script-like structure was found to contain mainly Fe, Mn, Al, and Si elements associated with the unmodified α-Al matrix (Fig. 2). This phase is known as β-phase and precipitates due to the reaction of impurities (Fe & Mn) with the Al and Si prior to the eutectic reaction [16]. The strontium is mainly associated with the Al-Si eutectic regions and modified the secondary phase precipitation, as shown in Figs. 3 and 4. The fine fibrous microstructure mainly contained the Si, Sr, and Al and evolved due to the depression of eutectic temperature and retardation of eutectic silicon growth [19, 29]. Further, the coarsening of eutectic fibers and Sr enriched acicular Si lamella beyond 0.18% Sr are evident from the EPMA micrographs and EDS area map of 0.25% Sr modified alloys (Fig. 4). The eutectic silicon remained modified but the coarsening and precipitation due to the over modification resulted in the formation of Al_4Si_2Sr intermetallic compounds [23]. It is evident that the Sr modifies the shape, size, and distribution of eutectic structure which in turn may affect the physical, chemical, and mechanical properties of the Al-12 Si alloys. This effect is clearly manifested in all the strontium added Al-12Si alloys. Strontium is known to depress the eutectic temperature of Al-Si alloys and thus aids fluidity to the melt. Also, shrinkages found to be restricted to the local regions in the presence of strontium which further increases the feeding efficiency [22]. This resulted into much finer Si and α dendrites in the Sr modified alloys.

The hardness is generally influenced by the microstructural features of the materials. Figure 5 displays the effect of addition of Sr in Al-12Si alloy on the average macro and microhardness determined at 5 kgf and 50 gf loads respectively. The both macro and microhardness were found to increase up to 0.18% Sr content and then decreased slightly upon a further increase in Sr content, yet remained nearer to the unmodified alloy.

The overall microhardness is found to be slightly improved with increasing Sr content up to 0.18%. The change in microhardness is attributed to the morphology of α-Al, eutectic and secondary phases in addition to their shape, size, distribution, and phase fraction. Also, this is greatly influenced by the quality of casting and defects

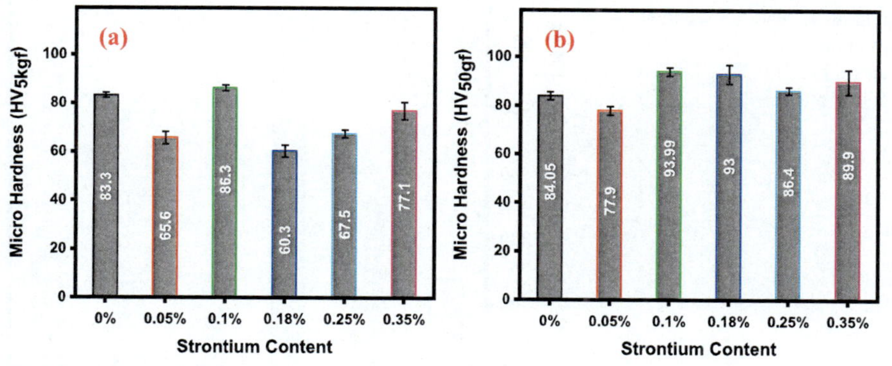

Fig. 5 **a** Macro and **b** microhardness relationship with Sr content in unmodified and modified alloys

such as pinholes, open holes, and subsurface blowholes, etc. It may thus be concluded that Sr content ~0.18% seems to be the upper limit for the modification of Al-12Si alloys as thereafter the eutectic structure begins to coarsen with finer α-Al dendrites and reduce the microhardness. The effect of Sr modification on corrosion resistance is discussed in the subsequent section.

3.2 Corrosion Behavior Characterization

The corrosion characteristics of unmodified and Sr modified alloys were determined by weight loss measurements during long-term exposures in alkaline 3.5% NaCl solutions (pH ~ 11) for four months. The corrosion rate was calculated using the equation mentioned below:

$$\text{Corrosion Rate(mpy)} = \frac{K * \text{Weight loss}}{\text{Density} * \text{Area} * \text{Exposure time}} \quad (1)$$

where K is conversion constant (3.45×10^6), weight loss is in grams, density is in g/cm^3 (~2.57 g/ cm^3), area is in cm^2, and exposure time is in hours.

The corrosion rate calculated for the unmodified and Sr modified alloys is shown in Fig. 6. The corrosion rate is observed to increase (5.04 mpy) with the Sr content in the alloy up to 0.1%, thereafter it is reduced to values lower than the unmodified Al-12Si alloy on increasing the Sr content up to 0.25%. The Sr addition ~0.1% increased the corrosion rate by ~261 and 0.25% Sr addition lowered it by about 5% as compared to the unmodified alloy (corrosion rate ~1.4 mpy).

The highest corrosion rate may be attributed to the fibrous and uniformly distributed eutectic phase equivalent to surrounding a large number of local anode

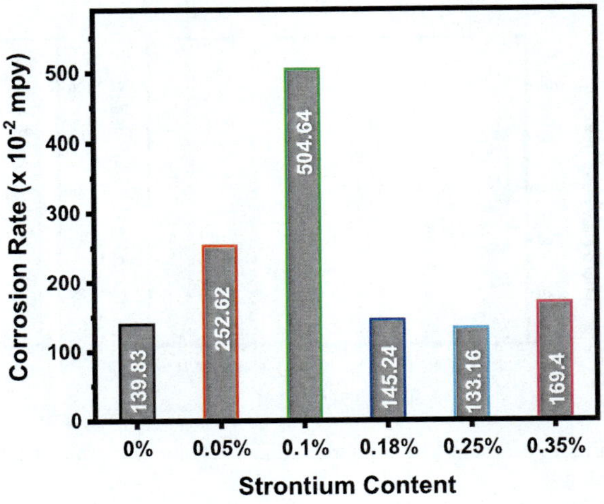

Fig. 6 Corrosion rate relationship with Sr content in unmodified and modified alloys in static 3.5% NaCl solution (pH ~11)

and cathode accelerating galvanic corrosion between the α-Al and secondary phases. While secondary phase in 0.25% Sr modified alloy, is larger (acicular Si-Sr) but relatively lesser local anode/cathode distribution and thus slightly improved the corrosion resistance of the alloy. No linear relation between the corrosion rate and Sr content is seen but 0.25% Sr content resulted in improved corrosion resistance.

The corrosion resistance was also evaluated by potentiodynamic polarization methods in alkaline 3.5% NaCl solution in the absence and presence of flow ~75 l/min. The potentiodynamic polarization curves obtained for the unmodified and modified alloys in static conditions are presented in Fig. 7 and electrochemical parameters determined from the polarization curves are listed in Table 2. Figure 7 and Table 2 demonstrate that the 0.25% Sr containing alloy possessed the highest corrosion resistance as evident by the lowest i_{corr} ~3.74 μA/cm^2 while 0.1% Sr alloys exhibited a lowest corrosion resistance (i_{corr} ~14.18 μA/cm^2) amongst all alloys. Further, unmodified and 0.25%Sr modified alloys exhibited pretty close current densities which are attributed to the quality of the castings. Despite having porosity, 0.25%Sr modified alloy exhibited slightly lower current density which clearly indicates improvement in the corrosion resistance with Sr addition.

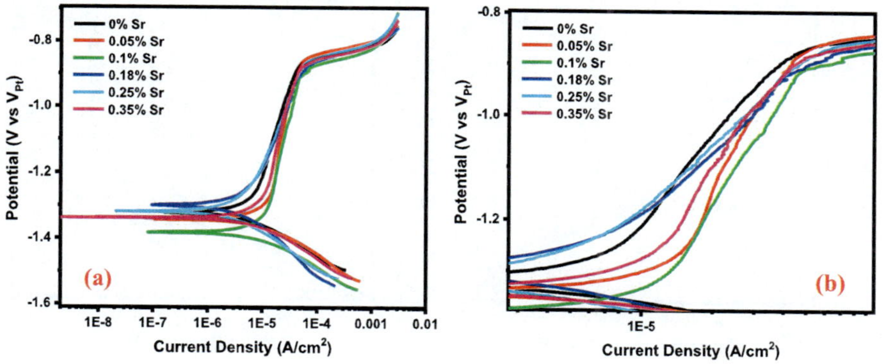

Fig. 7 Potentiodynamic polarization curves obtained for the unmodified and modified alloys in static 3.5% NaCl solutions (pH ~11)

Table 2 Electrochemical parameters determined from polarization curves in static 3.5% NaCl solution (pH ~11)

Sample	E_{corr} (mV)	i_{corr} (μA/cm^2)	E_{pit} (mV)	$E_{pit} - E_{corr}$ (mV)	$i_{passive}$ (μA/cm^2)
UM	−1322.0 ± 1.0	3.89 ± 0.1	−867.7 ± 7.4	454.3 ± 6.4	20.32 ± 2.91
M 1	−1346.5 ± 2.04	9.42 ± 1.12	−850.2 ± 2.08	496.35 ± 4.12	21.40 ± 0.71
M 2	**−1378.0 ± 5.0**	**14.18 ± 1.9**	**−879.4 ± 10.4**	**498.6 ± 5.4**	**27.70 ± 1.81**
M 3	−1306.5 ± 6.5	5.17 ± 0.46	−822.2 ± 0.05	484.25 ± 6.45	20.91 ± 2.46
M 4	**−1336.5 ± 18.5**	**3.74 ± 1.02**	**−857.7 ± 0.55**	**478.8 ± 17.95**	**19.23 ± 1.09**
M 5	−1336.5 ± 1.5	6.32 ± 0.45	−871.6 ± 0.7	464.9 ± 0.8	21.89 ± 3.35

The corrosion potential is found to shift towards slightly more negative values for both the alloys, as Sr is more electronegative than aluminum. Both, the unmodified and modified alloys exhibited passive-transpassive polarization behavior indicating a sharp pitting potential. While a wider passive region ($E_{pit} - E_{corr}$) of Sr modified alloy as compared to the unmodified alloys is observed, the pitting potential did not show specific trend with the Sr addition. The highest pitting potential (least negative) value was shown by the alloy with 0.18% Sr containing alloy.

The experiments in flow conditions were performed to simulate the dynamic fluid impact on corrosion of Al-Si alloy in several industrial environments. Figure 8 elucidates the corrosion behavior of the unmodified and modified Al-12Si alloy in flowing alkaline 3.5% NaCl solution at a flow rate of ~75 l/min and the electrochemical parameters obtained are listed in Table 3. The results demonstrate much higher corrosion current densities in flow conditions as compared to that of the static conditions. Amongst all, 0.25% Sr containing alloy exhibited the lowest i_{corr} ~47.29 μA/cm² while 0.1% Sr alloys showed the highest i_{corr} ~130.83 μA/cm², a trend similar to that observed in the static conditions. Also, the passive current density

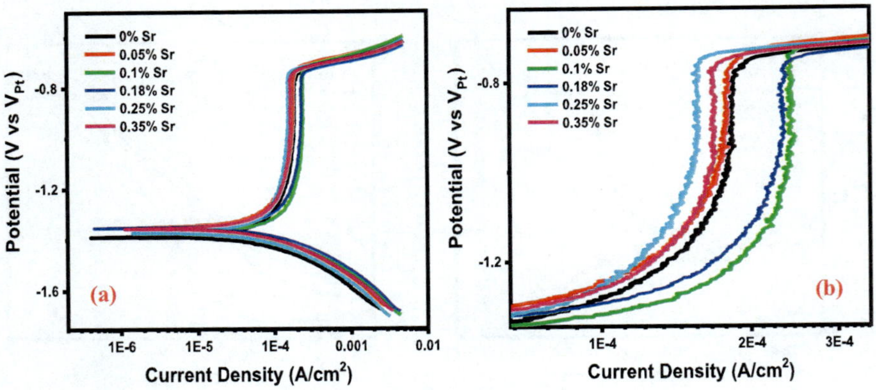

Fig. 8 Potentiodynamic polarization curves obtained for unmodified and modified alloys in flowing 3.5% NaCl solutions (pH ~11) at flow rate ~75 l/min

Table 3 Electrochemical parameters determined from polarization curves in flowing 3.5% NaCl solution (pH ~11) at flow rate ~75 l/m

Design	E_{corr} (mV)	i_{corr} (μA/cm²)	E_{pit} (mV)	$E_{pit} - E_{corr}$ (mV)	$i_{passive}$ (μA/cm²)
UM	−1388 ± 2.16	53.34 ± 5.29	−737.1 ± 6.25	650.87 ± 6.11	152.28 ± 16.59
M 1	−1358.5 ± 4.5	65.23 ± 6.51	−720.4 ± 1.35	638.05 ± 3.15	162.8 ± 8.4
M 2	**−1375.7 ± 2.87**	**130.83 ± 12.31**	**−721.4 ± 2.61**	**654.23 ± 3.28**	**259.59 ± 5.75**
M 3	−1356.3 ± 16.52	103.48 ± 6.12	−735.2 ± 2.70	621.13 ± 18.11	218.02 ± 18.32
M 4	**−1350.7 ± 2.06**	**47.29 ± 0.17**	**−727.9 ± 0.71**	**622.77 ± 1.44**	**118.67 ± 10.63**
M 5	−1351.7 ± 6.94	93.96 ± 10.80	−733.1 ± 0.29	618.57 ± 6.67	159.23 ± 6.52

follows a similar trend as corrosion current densities as $i_{passive}$ ~259.59 μA/cm^2 and 118.67 μA/cm^2 for 0.1% and 0.25% Sr containing alloys, respectively. Larger corrosion current density and pitting potential (E_{pit}) (nobler side) are found along with the wider passivation ranges for all the alloys as compared to the static conditions.

The corrosion of Al-Si alloys is a complex phenomenon and is considered to progress according to the following reactions [30]:

Anodic reaction

$$Al \rightarrow Al^{3+} + 3e^- \tag{2}$$

Cathodic reaction

$$O_2 + 2H_2O + 2e^- \rightarrow 4OH^- \tag{3}$$

As corrosion progresses, there is a local increase in pH as per Eq. 3 at the cathodic sites; this is especially important when the solution is stagnant and a continuous increase in pH is likely. This local increase in pH may be adequate to destabilize the passive/ pseudo passive aluminum oxide/hydroxide films at those weak locations [31]. Also, the stagnant conditions promote local crevice formation which may decrease the pitting resistance of the material. Thus, the lower pitting potentials reported in static conditions may be attributed to the locally destabilized aluminum oxide/hydroxide films due to pH increase. The corrosion current density which is governed by the mass transport at the alloy-solution interphase is higher in the flowing medium as compared to static conditions. The result substantiates the mass controlled corrosion rate as corrosion rate in the flowing medium is several fold higher than in the stagnant condition.

The immersion and potentidynamic polarization studies were complemented by electrochemical impedance spectroscopy (EIS) in order to reveal the corrosion resistance of the alloy in both static and flowing 3.5% NaCl solutions. The representative Nyquist curves for unmodified and modified alloys (0.1 & 0.25% Sr) obtained from EIS studies are plotted in Fig. 9. The Nyquist curves were fitted with the two-time constant electrochemical equivalent circuit (EEC) as shown in Fig. 10.

The electrochemical parameters determined from Nyquist and Bode plots acquired in static and flowing conditions are listed in Tables 4 and 5, respectively. The pore resistance (R_{pore}) and charge transfer resistance (R_{ct}) relationship with Sr content is plotted in Fig. 11a, b. In static conditions, both the pore/oxide layer resistance and charge transfer resistance are found to be highest for 0.25% Sr containing alloys amongst all, e.g., R_{pore} ~3.37 kohm cm^2 and R_{ct} ~7.6 kohm cm^2, while 0.1% Sr alloys exhibited the lowest R_{pore} ~1.5 kohm cm^2 and R_{ct} ~4.8 kohm cm^2 (Fig. 11a and Table 4).

The variation in pore and charge transfer resistance is attributed to the defects present in oxide film which directly influence the stability of oxide layer, i.e., lesser the defects higher the stability and pore resistance, and vice versa. Also, a similar trend was observed in the flowing conditions where 0.25% Sr containing alloys exhibited the highest pore resistance, R_{pore} ~ 206.17 Ω cm^2 which is in line with

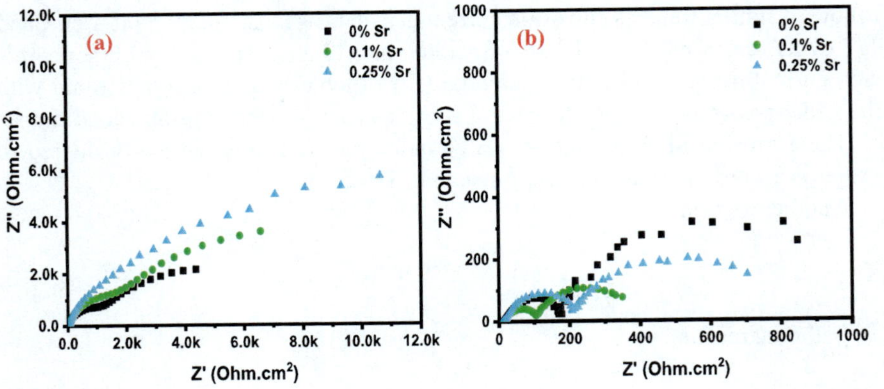

Fig. 9 The Nyquist curves for unmodified and modified alloys obtained from EIS studies in **a** static and **b** flowing 3.5% NaCl Solution

Fig. 10 The electrochemical equivalent circuit used for fitting the EIS curves of unmodified and modified alloys

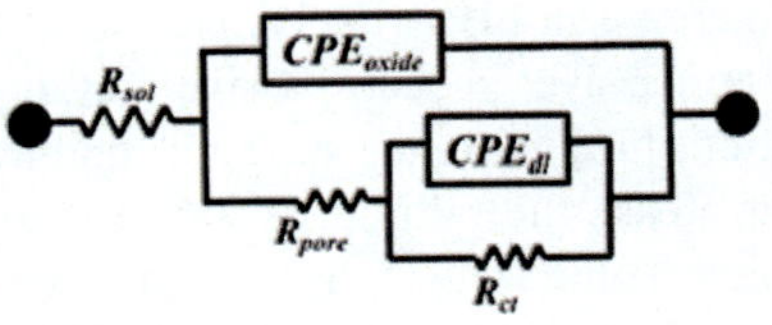

the results obtained from the anodic polarization curve (Fig. 11b, Table 5). Both the R_{pore} and R_{ct} are found to decrease in flowing conditions as compared to the static 3.5% NaCl solution, thus a much increased severity of uniform corrosion in flowing fluids is realized. The unmodified and modified alloys (0.1% & 0.25% Sr) were subjected to SEM–EDS observations after the electrochemical measurements in order to reveal the oxide layer characteristics and corroded morphology if any. The SEM micrographs and EDS measurements of the unmodified and modified alloys evaluated in static 3.5% NaCl solutions are depicted in Fig. 12.

The 0.25% Sr alloy was covered with a very thick oxide layer which was composed of Al oxide or Al-Si oxides as revealed by EDS analysis. The thick corrosion products were mainly comprised of Al-oxide while compact thin layers consisted of mixed, e.g., Al and Si oxides. The 0.1% Sr containing alloy was found to corrode near the eutectic and intermetallic regions, and exhibited a very thin layer of oxides. Figure 13 shows the SEM micrographs and EDS analysis of the unmodified and modified alloys evaluated in flowing 3.5% NaCl solutions. The 0.25% Sr containing alloy was found to contain Al-Si oxide layer compared to 0.1% Sr containing alloy which proves relatively higher oxygen contents in corrosion products. The unmodified alloy was surrounded by a cracked and relatively thicker Al-oxide layer.

In Al-12Si alloys, corrosion progresses through galvanic coupling in eutectic regions where Si acts as a local cathode to the eutectic Al phase. In such a scenario, the shape, size, and distribution of Si particles govern the severity of corrosion. As mentioned previously, the stability of the protective oxide layer also depends upon

Table 4 Electrochemical parameters determined from EIS curves in static 3.5% NaCl solution (pH ~11)

Sample	R_{pore} (Ohm cm^2)	R_{ct} (Ohm cm^2)	R_{sol} (Ohm cm^2)	Y_{oxide} ($S * s^{\eta}/cm^2$)	n_{oxide}	Y_{dl} ($S * s^{\eta}/cm^2$)	n_{dl}
UM	2370	6920	43.03	2.13E−04	0.846	1.04E−03	0.862
M 1	1897	5478	37.82	1.45E−04	0.897	1.47E−03	0.840
M 2	**1497**	**4798**	**49.23**	**1.72E−04**	**0.908**	**1.61E−03**	**0.823**
M 3	2172	5896	44.12	1.27E−04	0.897	1.26E−03	0.868
M 4	**3374**	**7603**	**44.10**	**1.77E−04**	**0.849**	**8.6E−04**	**0.855**
M 5	2078	5308	44.54	1.43E−04	0.887	1.33E−04	0.860

Table 5 Electrochemical parameters determined from EIS curves in flowing 3.5% NaCl solution (pH ~11) at flow rate ~75 l/min

Desgn	R_{pore} (Ohm cm^2)	R_{ct} (Ohm cm^2)	R_{sol} (Ohm cm^2)	Y_{oxide} ($S*s^{\eta}/cm^2$)	n_{oxide}	Y_{dl} ($S * s^{\eta}/cm^2$)	n_{dl}
UM	198.93	1081.6	16.83	1.1E−04	0.829	6.40E−03	0.918
M 1	167.8	734.27	15.37	1.61E−04	0.801	7.49E−03	0.916
M 2	110.23	259.4	14.48	3.86E−04	0.727	1.38E−02	0.895
M 3	105.4	270.17	15.14	4.65E−04	0.717	1.36E−04	0.899
M 4	**206.17**	427.13	15.15	1.75E−04	0.779	9.36E−03	0.903
M 5	155.57	376.13	14.90	2.99E−04	0.741	1.03E−02	0.907

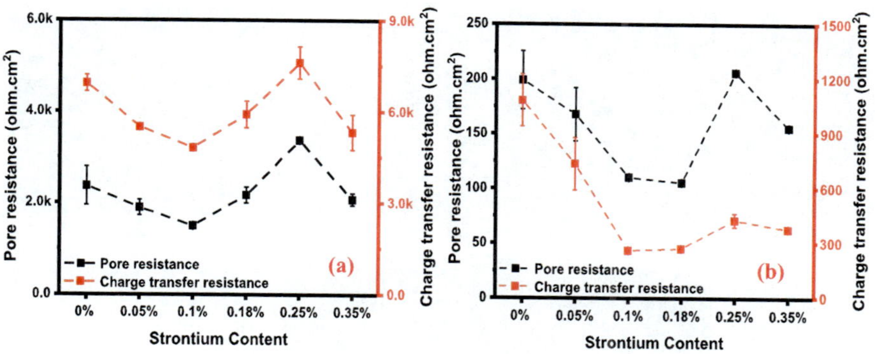

Fig. 11 Pore resistance (R_{pore}) and charge transfer resistance (R_{ct}) relationship with Sr content in 3.5% NaCl solutions **a** static conditions and **b** flow rate ~75 l/min

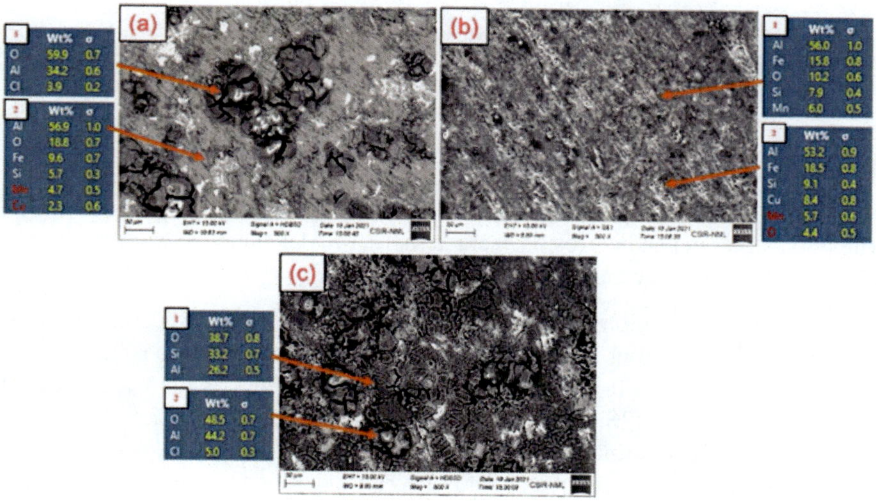

Fig. 12 SEM micrographs with EDS analysis of **(a)** unmodified (0% Sr), and **(b)** 0.1% Sr modified alloy, and **(c)** 0.25% Sr modified alloy in static 3.5% NaCl solutions

the shape and size of eutectic Si particles. In the presence of Sr, the much finer and uniformly distributed Si particles form a stable oxide layer and thus resulting in reduced corrosion severity. Being more electronegative than Al, Sr remains associated with eutectic Si particles, which narrow down the electrochemical potential difference between α-Al and Si particles.

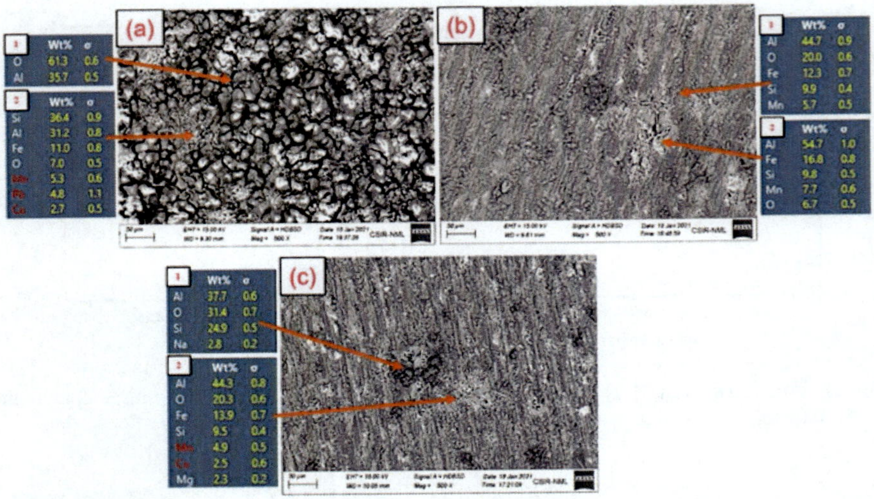

Fig. 13 SEM micrographs with EDS analysis of **(a)** unmodified (0% Sr) and modified alloys with, **(b)** 0.1% Sr and **(c)** 0.25% Sr addition in flowing 3.5% NaCl solutions (flow rate ~75 l/min)

4 Conclusions

The microstructural and corrosion characteristics of Al-12Si alloys with varying Sr contents are summarized as follows:

- The strontium additions to Al-12Si alloy resulted in the modification of Al-Si eutectic to fine fibrous structure up to 0.18% which started to coarsen beyond this Sr content. The 0.18% Sr content can be considered as the upper limit of modification in Al-12Si alloys from the mechanical properties point of view.
- The Sr modification influences the size and distribution of both α-Al dendrites and Al-Si eutectics. Also, the growth of script-like secondary phases was found to be restricted with Sr addition.
- The corrosion resistance is improved by Sr modification in both the flowing and static NaCl conditions. The 0.25% Sr containing alloys exhibited the lowest corrosion current densities and highest corrosion resistance in both the static and flowing conditions.
- Overall 0.18% Sr content can be considered as the optimized level from the mechanical and corrosion resistance properties point of view.

References

1. Hernandez FCR, Ramírez JMH, Mackay R (2017) Al-Si alloys: automotive, aeronautical, and aerospace applications. Springer International Publishing, Cham

2. Vijayan V, Prabhu KN (2016) The effect of simultaneous refinement and modification by cerium on microstructure and mechanical properties of Al-8% Si alloy. Int J Cast Met Res 29:345–349. https://doi.org/10.1080/13640461.2016.1144698
3. Biswas P, Patra S, Mondal MK (2018) Effects of Mn addition on microstructure and hardness of Al-12.6Si alloy. IOP Conf Ser: Mater Sci Eng 338. https://doi.org/10.1088/1757-899X/338/1/012043
4. Yang Y, Chen Y, Zhang J, Gu X, Qin P, Dai N, Li X, Kruth J-P, Zhang L-C (2018) Improved corrosion behavior of ultrafine-grained eutectic Al-12Si alloy produced by selective laser melting. Mater Des 146:239–248. https://doi.org/10.1016/j.matdes.2018.03.025
5. Olohunde ST, Hafizi ARM, Jamaliah I, Al-Bakoosh AA, Segun OO, Sadiq IO (2019) Corrosion resistance of aluminium–silicon hypereutectic alloy from scrap metal. J Bio-Tribo-Corrosion 5:41. https://doi.org/10.1007/s40735-019-0224-x
6. Chen Y, Zhang J, Gu X, Dai N, Qin P, Zhang L-C (2018) Distinction of corrosion resistance of selective laser melted Al-12Si alloy on different planes. J Alloys Compd 747:648–658. https://doi.org/10.1016/j.jallcom.2018.03.062
7. Kaiser MS (2019) Corrosion behaviour of Al-12Si-1Mg automotive alloy in acidic, alkaline and salt media containing Zr traces. J Chem Technol Metall 54:423–430
8. De Giovanni M, Kaduk JA, Srirangam P (2019) Modification of Al-Si Alloys by Ce or Ce with Sr. JOM 71:426–434. https://doi.org/10.1007/s11837-018-3192-6
9. Timpel M, Wanderka N, Schlesiger R, Yamamoto T, Lazarev N, Isheim D, Schmitz G, Matsumura S, Banhart J (2012) The role of strontium in modifying aluminium-silicon alloys. Acta Mater 60:3920–3928. https://doi.org/10.1016/j.actamat.2012.03.031
10. Biswas P, Patra S, Kumar Mondal M (2020) Structure-property correlation of eutectic Al-12.4 Si alloys with and without Zirconium (Zr) addition. Int J Cast Met Res 33:134–145. https://doi.org/10.1080/13640461.2020.1769319
11. Ebhota WS, Jen T-C (2017) Effects of modification techniques on mechanical properties of Al-Si cast alloys. In: Aluminium alloys—recent trends in processing, characterization, mechanical behavior and applications. InTechOpen, London
12. Liu Q, Liu M, Xu C, Xiao W, Yamagata H, Xie S, Ma C (2018) Effects of Sr, Ce and P on the microstructure and mechanical properties of rapidly solidified Al–7Si alloys. Mater Charact 140:290–298. https://doi.org/10.1016/j.matchar.2018.04.018
13. Prashanth KG, Scudino S, Klauss HJ, Surreddi KB, Löber L, Wang Z, Chaubey AK, Kühn U, Eckert J (2014) Microstructure and mechanical properties of Al-12Si produced by selective laser melting: effect of heat treatment. Mater Sci Eng A 590:153–160. https://doi.org/10.1016/j.msea.2013.10.023
14. Sun HH, Yang S, Jin D (2018) Improvement of microstructure, mechanical properties and corrosion resistance of cast Al–12Si alloy by friction stir processing. Trans Indian Inst Met 71:985–991. https://doi.org/10.1007/s12666-017-1232-5
15. Osório WR, Garcia LR, Goulart PR, Garcia A (2007) Effects of eutectic modification and T4 heat treatment on mechanical properties and corrosion resistance of an Al-9 wt%Si casting alloy. Mater Chem Phys 106:343–349. https://doi.org/10.1016/j.matchemphys.2007.06.011
16. Hekimoğlu AP, Çalış M, Ayata G (2019) Effect of strontium and magnesium additions on the microstructure and mechanical properties of Al–12Si alloys. Met Mater Int 25:1488–1499. https://doi.org/10.1007/s12540-019-00429-6
17. Timpel M, Wanderka N, Schlesiger R, Yamamoto T, Isheim D, Schmitz G, Matsumura S, Banhart J (2013) Sr–Al–Si co-segregated regions in eutectic Si phase of Sr-modified Al–10Si alloy. Ultramicroscopy 132:216–221. https://doi.org/10.1016/j.ultramic.2012.10.006
18. Jenkinson DC, Hogan LM (1975) The modification of aluminium-silicon alloys with strontium. J Cryst Growth 28:171–187. https://doi.org/10.1016/0022-0248(75)90233-X
19. Dahle AK, Nogita K, McDonald SD, Dinnis C, Lu L (2005) Eutectic modification and microstructure development in Al-Si Alloys. Mater Sci Eng A 413–414:243–248. https://doi.org/10.1016/j.msea.2005.09.055
20. Nogita K, McDonald SD, Dahle AK (2004) Solidification mechanisms of unmodified and strontium-modified hypereutectic aluminium-silicon alloys. Philos Mag 84:1683–1696. https://doi.org/10.1080/14786430310001659803

21. Szymczak T, Gumienny G, Pacyniak T (2015) Effect of Sr and Sb modification on the microstructure and mechanical properties of 226 silumin pressure casts. Arch Foundry Eng 15:105–108. https://doi.org/10.1515/afe-2015-0019
22. Öztürk İ, Hapçı Ağaoğlu G, Erzi E, Dispinar D, Orhan G (2018) Effects of strontium addition on the microstructure and corrosion behavior of A356 aluminum alloy. J Alloys Compd 763:384–391. https://doi.org/10.1016/j.jallcom.2018.05.341
23. Gruzleski JE, Closset BM (1990) The treatment of liquid aluminum-silicon alloys. American Foundry Society, Illinois
24. Chen H, Zhang C, Jia D, Wellmann D, Liu W (2020) Corrosion behaviors of selective laser melted aluminum alloys: a review. Metals (Basel) 10:102. https://doi.org/10.3390/met10010102
25. Lee SL, Cheng YC, Chen WC, Lee CK, Tan AH (2012) Effects of strontium and heat treatment on the wear-corrosion property of Al-7Si-0.3Mg alloy. Mater Chem Phys 135:503–509. https://doi.org/10.1016/j.matchemphys.2012.05.015
26. Farahany S, Ourdjini A, Bakhsheshi-Rad HR (2016) Microstructure, mechanical properties and corrosion behavior of Al-Si-Cu-Zn-X (X=Bi, Sb, Sr) die cast alloy. Trans Nonferrous Met Soc China (English Ed 26:28–38. https://doi.org/10.1016/S1003-6326(16)64085-2
27. Cardinale AM, Macciò D, Luciano G, Canepa E, Traverso P (2017) Thermal and corrosion behavior of as cast Al-Si alloys with rare earth elements. J Alloys Compd 695:2180–2189. https://doi.org/10.1016/j.jallcom.2016.11.066
28. Duygun İK, Hapçı Ağaoğlu G, Dispinar D, Orhan G (2019) Time-dependent corrosion properties of Sr-modified AlSi9 alloy analyzed by electrochemical techniques. J Alloys Compd 803:786–794. https://doi.org/10.1016/j.jallcom.2019.06.330
29. Dahle AK, Nogita K, Zindel JW, McDonald SD, Hogan LM (2001) Eutectic nucleation and growth in hypoeutectic Al-Si alloys at different strontium levels. Metall Mater Trans A Phys Metall Mater Sci 32:949–960. https://doi.org/10.1007/s11661-001-0352-y
30. Zheng Y, Luo B, He C, Ren Z, Wang S, Yin Y (2019) Corrosion behaviour of the Al-2.1–Mg-1.8–Si alloy in chloride solution. Bull Mater Sci 42:228. https://doi.org/10.1007/s12034-019-1923-0
31. Berlanga-Labari C, Biezma-Moraleda MV, Rivero PJ (2020) Corrosion of cast aluminum alloys: a review. Metals (Basel) 10:1384. https://doi.org/10.3390/met10101384

Evaluation of Corrosion Rate and Scratch Resistance in Chromium Alloyed Irons Influenced by Manganese Addition and Process Parameters

T. Pramod, P. Sampathkumaran, S. Seetharamu, Narendra Dube, Bincy Rose Vergis, R. K. Kumar, and C. Ranganathaiah

1 Introduction

It is very much desirable to develop ferrous systems, which can sustain chemical attacks, provide higher wear resistance and desirable surface features. They depend on bulk material conditions like composition, hardness, heat treatment, microstructure, and also the surface characteristics like scratch resistance and surface roughness of the constituent materials. To meet the industrial requirements, the alloying elements and processing conditions are chosen in such a way that the component integrity gets enhanced in terms of corrosion, wear, etc. The corrosion behavior of metallic materials is dictated by the surrounding medium, environmental conditions, etc. The corrosion properties are assessed based on the weight loss method as well

T. Pramod (✉)
The Energy and Resources Institute, Bangalore 560071, India

P. Sampathkumaran
Department of Mechanical Engineering, Sambhram Institute of Technology, Bangalore 560097, India

S. Seetharamu
Department of Mechanical Engineering, R.V College of Engineering, Bangalore 560064, India

N. Dube
Ducom Instruments (P) Ltd., Bangalore 560058, India

B. R. Vergis
Department of Chemistry, BMS Institute of Technology and Management, Bangalore 560064, India

R. K. Kumar
Materials Technology Division, Central Power Research Institute, Bangalore 560080, India
e-mail: rkkumar@cpri.in

C. Ranganathaiah
JSS University, Mysuru 570015, India

U. K. Mudali et al. (eds.), *Recent Trends in Electrochemical Science and Technology*, Springer Proceedings in Materials 15,
https://doi.org/10.1007/978-981-16-7554-6_4

as corrosion rate measurement as laid down in the standard procedure. Kaputkina et al. [1] studied corrosion resistance of high-strength austenitic steels. There was no pitting corrosion reported due to the presence of the alloys like nickel, manganese, and chromium. The assessment of scratch resistance and surface roughness parameter is generally done using specially designed scratch and surface roughness tester respectively. In this context, the scratch tester has emerged as a very useful tool to monitor the parameters such as adhesion of a coating to the underlying substrate, the effective frictional conditions of the coating, or the bulk material surface. Pohl et al. [2] have investigated the deformation behavior of AISI 1045 and AISI 304L materials with an increase in scratch load and reported that work hardening affects scratch parameters due to load application during scratching and also studied the micro mechanism involved during scratch indentation.

Many researchers have reported the scratch [3, 4] and corrosion [5, 6] behaviors of both ferrous and non-ferrous materials, but the information on these two parameters in high chromium (HiCr) alloyed irons is scanty. These two tests have been chosen as this material finds application in milling systems (grinding media) and pumps (impeller) in thermal power stations, wherein any scratch/crack arising out of repeated impacts of the grinding balls in liners and corrosion taking place during water lifting in impellers, would end up in damage as the HiCr irons (hard carbides) used are brittle in nature. Hence, the addition of manganese (austenite stabilizing element) to high chromium iron enhances the matrix toughness characteristics. Keeping in view the significance of corrosion resistance and scratch resistance in ferrous materials, especially high chromium manganese (HiCrMn) alloyed irons prepared in the grey cast iron mold by induction melting method, their responses with respect to metallurgical parameters have been looked into in both as-cast and heat-treated conditions with manganese at two levels (5 and 10% by weight) and adoption of cast section sizes of 12, 24 and 40 mm.

2 Materials and Methods

2.1 Materials

The high chromium iron castings measuring $150 \times 125 \times 25$ mm^3 were made by induction melting route with manganese at two levels (5 and 10% by weight) and adopting different cooling rates owing to employing varying section sizes (12, 24, and 40 mm). The test samples of size ($75 \times 25 \times 6$ mm^3) were prepared from the casting for evaluation of scratch resistance and corrosion rate. The sample details such as the designation, manganese weight percentage, section size, and the processing conditions are provided in Table 1.

Table 1 HiCrMn iron sample details and their designations

S. No.	Sample designation	Mn weight (%)	Section size (mm)	As-cast/heat-treated
1	5M12AC	5	12	As-cast
2	5M12HT	5	12	Heat-treated
3	5M24AC	5	24	As-cast
4	5M24HT	5	24	Heat-treated
5	5M40AC	5	40	As-cast
6	5M40HT	5	40	Heat-treated
7	10M12AC	10	12	As-cast
8	10M12HT	10	12	Heat-treated
9	10M24AC	10	24	As-cast
10	10M24HT	10	24	Heat-treated
11	10M40AC	10	40	As-cast
12	10M40HT	10	40	Heat-treated

2.2 Methods

The cooling rate of 5 and 10 M series having cast sizes of 12, 24, and 40 mm have been ascertained while pouring the molten liquid into the metal mold to get the specimen dimension of $150 \times 125 \times 25$ mm^3. The solidification rates have been obtained using a thermocouple dipped in a metal mold containing molten metal and is connected to a potential-metric strip chart recorder to get the temperature–time plots and was found to be 54, 37, and 28 s for 12, 24, and 40 mm section sizes, respectively. The heat treatment of HiCrMn iron samples consisted of austenitizing at 960 °C for four hours and then oil quenching to room temperature, subsequently tempering for half an hour at 200 °C. The austenitizing temperature of 960 °C and soaking time of four hours was optimized based on the preliminary studies carried out involving maximum dissolution of cementite in the matrix. Also, no attempt has been made in this work to consider the secondary carbide precipitation occurring during heat treatment. The chemical compositions are determined using an optical emission spectrophotometer (ASTM E415-99a Standard). Archimedes Principle was used to carry out density measurements. The retained austenite was measured using an X-ray diffractometer (Rigaku X-Ray Stress Analyzer) as per ASTM E975-03 Standard. The carbide volume percentages were calculated [7, 8] using the empirical relationship given below in Eq. (1) as well as using image analyzer software.

$$\text{Carbides} = 12.33\text{C} + 0.55\text{Cr} - 15.2 \quad (1)$$

The corrosion tests were carried out in a highly corrosive sodium chloride medium, being a strong electrolyte, results in the formation of metal ions and hydroxides. The corrosion rate was carried out at 25 °C for 720 h and 100 °C for four hours in reflux conditions as per ASTM G1-90 Standard. The corrosion rate was calculated using

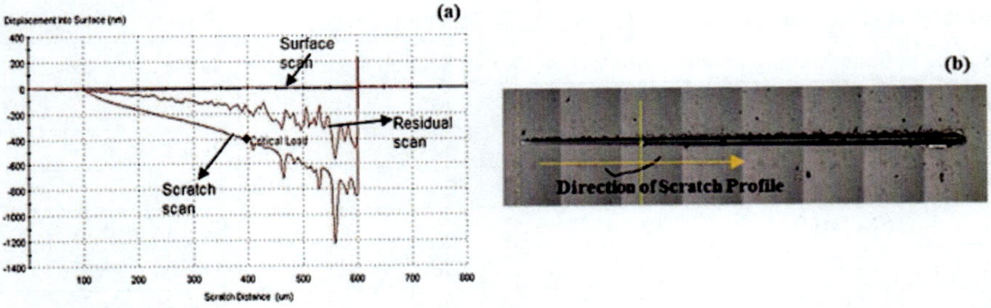

Fig. 1 A typical **a** scratch scan and **b** scratch profile

Eq. (2).

$$\text{Corrosion rate(mm/h)} = (8.76\text{W}/\text{DAT}) \tag{2}$$

where W is the difference between the final weight (g) and initial weight (g) of the sample, D is alloy density (g/mm^3), A is exposed area (mm^2) and T is the exposure time of the sample (h).

Rockwell C hardness (Struers) was measured at a load of 150 kg with diamond indenter having the cone angle of 120°, which was calibrated using a standard block having 60 ± 1HRC. The scratch adhesion test (Ducom Model TR101) was carried out as per ASTM G171-03 standard to measure the scratch width and depth involving scanning electron microscope (SEM, Hitachi TM4000) and digital depth gauge. The normalized scratch hardness (Hs) was calculated using Eq. (3)

$$H_{\rm s} = \frac{8L}{3.14(S_{\rm W})^2} \tag{3}$$

where $H_{\rm s}$ is normalized scratch hardness, L is load (N) and $S_{\rm W}$ is scratch width (μm). Figure 1 shows a typical scratch width monitored during the movement of the stylus and its scratch profile. Light and scanning electron microscopy was used to carry out the microstructural examination.

3 Results and Discussion

Table 2 shows the chemical composition of the samples analyzed. The Rockwell hardness, retained austenite, corrosion rate, density, and carbide volume (both experimental and empirical) are given in Table 3.

Table 2 Chemical composition of HiCrMn

Sample	C	Mn	Designation composition (weight %)					
			Si	Cr	Ni	Mo	S	P
5M12	2.66	4.5	2.69	18.58	0.91	1.87	0.046	0.080
5M24	2.61	4.6	2.78	18.62	0.92	1.82	0.044	0.090
5M40	2.58	4.8	2.10	17.90	0.81	1.90	0.057	0.075
10M12	2.53	9.7	1.90	17.65	0.80	1.85	0.040	0.090
10M24	2.48	9.7	1.85	17.60	0.82	1.80	0.040	0.085
10M40	2.46	9.9	1.92	17.85	0.83	1.90	0.042	0.087

Table 3 Rockwell hardness, retained austenite, corrosion rate, density, and carbide volume

Sample designation	Rockwell C hardness	Retained austenite (%)	Corrosion rate (%)		Density (g/cc)	Carbide volume (%)	
			Room temperature (25 °C for 720 h)	Reflux condition (100 °C for 4 h)		Experimental	Empirical
5M12AC	56	54	1.51	0.51	7.70	29.0	27.8
5M12HT	63	47	1.59	0.54			
5M24AC	55	57	1.32	0.43	7.58	27.5	27.2
5M24HT	61	50	1.40	0.49			
5M40AC	52	61	1.00	0.35	7.54	26.5	26.5
5M40HT	56	57	1.33	0.41			
10M12AC	52	60	1.29	0.32	7.45	26.0	25.7
10M12HT	55	55	1.69	0.37			
10M24AC	50	63	1.04	0.25	7.42	25.5	25.1
10M24HT	53	60	1.21	0.30			
10M40AC	48	65	0.95	0.19	7.31	24.5	24.9
10M40HT	50	62	1.08	0.23			

3.1 Corrosion Rate

In HiCrMn iron samples, Fe undergoes oxidation in the presence of oxygen and chlorine to form ferrous hydroxide and ferrous chloride. The pH values of the immersion medium are observed to be 7.5 before the corrosion test and increase to 9.0–9.5 after the completion of the corrosion experiments. Thus the pH measurements substantiate the basic nature of the corrosion products.

It is seen from Fig. 2 that the higher Mn content sample (10 M series) in both the condition (as-cast and heat-treated) have yielded a lower corrosion rate compared to 5 M series sample. In fact, 10M40 shows a lower corrosion rate than 5M40, which may be traced to the higher austenite content. In the as-cast condition, the austenite

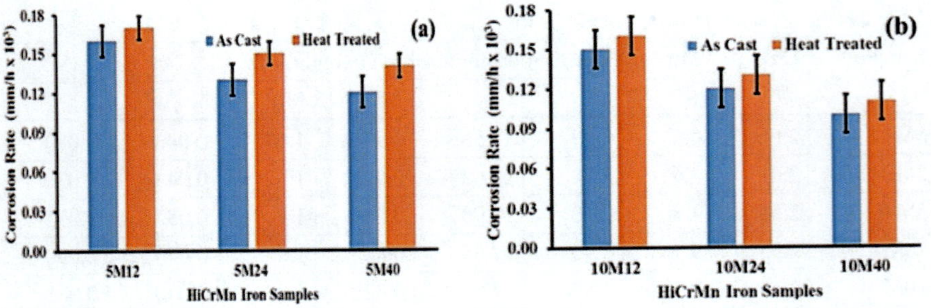

Fig. 2 Corrosion rate of 5 and 10 M series of as-cast and heat-treated samples at room temperature

phase is predominantly present along with the formation of carbides of chromium, iron, and molybdenum in 5M12. Following the heat treatment, the austenite content has decreased and there is a good possibility that some of the secondary phases would have formed. The phase transformation in 5M12HT is responsible for the recording of higher hardness.

The corrosion rate in the reflux condition is higher for 5 and 10 M series compared to samples immersed in the solution at room temperature (Fig. 3). It is quite understandable that when the samples are taken to the boiling temperature, the thermal kinetics favors the formation of corrosion products at a shorter time than at room temperature yielding a higher corrosion rate. This particular observation is substantiated from the literature wherein the corrosion rate has been reported [9] to be much higher in reflux conditions than in room temperature as the oxygen concentration at higher temperatures reacts with the metals to form corrosive products such as hydroxides at a faster rate. The same analogy holds well in the present case also.

It is observed from Table 3 that the hardness showed a decreasing trend with the increase in section size due to the increase in austenite content in the casting. Thus based on these experimental observations, it is observed that 10M40, which is exhibiting higher austenite content is less prone to corrosion than either 5M40/5M12 in both the conditions (as-cast and heat-treated). The major portion of corrosion in both 5 and 10 M series comes from the iron phase, i.e., austenite as it is presumed

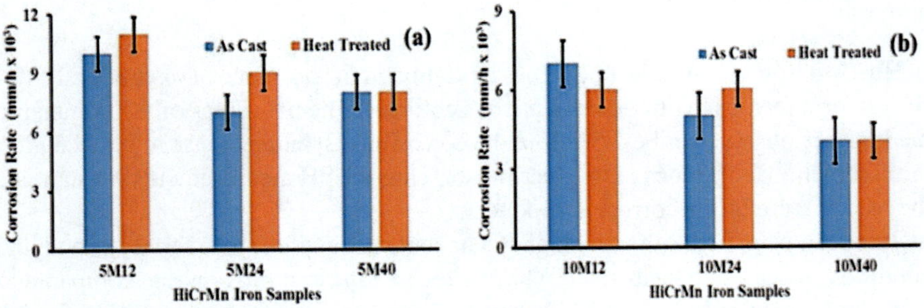

Fig. 3 Corrosion rate of 5 and 10 M series of as-cast and heat-treated samples at 100 °C

that the carbides generally do not get attacked from the corrosion media. Further, the austenite phase originated from the addition of Mn element, and hence one could observe fewer corrosion products getting formed due to the higher level of Cr and Mn additions in HiCrMn irons. However, the degree of corrosion rate in HiCrMn samples has shown variations in the corrosion rate as they are dependent on the austenite content as well as carbide volume. This clearly shows that the incorporation of Mn in Cr irons plays a major role in reducing the corrosion rate. It is also reported that the addition of Mn and Ni to chromium favors the formation of austenite at room temperature [10].

The corrosion rate of 5 and 10 M series have been arrived at, by taking Eq. (2) into consideration. It is seen that irrespective of the section size, the corrosion rate reduces with a rise in the Mn content from 5 to 10% for both room and high temperatures respectively. These trends are substantiated by the light microscopic features, wherein the corrosion rate obtained is very much dependent upon the austenite levels present in matrix. The corrosion rate trends obtained at room temperature of 5 and 10 M series are now discussed in terms of percentage reduction. The reduction in the corrosion rates of as-cast condition of 10M40 over 5M40 and 5M12 is 76 and 63% respectively. Whereas, the corrosion rate in heat-treated conditions of 10M40 reduces to the tune of 82% and 65% compared to 5M40 and 5M12 respectively. From these trends, it is noticed that the percentage reductions in the corrosion rate of heat-treated samples are much less compared to the as-cast counterparts. This is quite understandable and logical due to the higher level of austenite content present in the as-cast samples. In summary, the least corrosion rate in the reflux state is exhibited by 10M40 whereas 5M12 shows the highest irrespective of whether it is as-cast or heat-treated.

Handoko et al. [11] studied the effect of austenitizing temperatures namely 950, 1050 and 1150 °C affecting the corrosion rate in dual-phase stainless steels. It is inferred that the higher the temperature, the lower is the corrosion rate. Further, it is reported that raise in austenitizing temperature, increases the retained austenite content and the grain size. In the current study, the HiCrMn samples have been subjected to an austenitizing temperature of 950 °C but varied the Mn content as well as the section thickness of the casting to assess the corrosion behavior. With an increase in austenite content, the corrosion rate decreased due to an increase in manganese level as well as-cast section thickness. Jiang et al. [12] studied high nitrogen stainless steel and focused on finding out the relationship between corrosion behavior (immersion method) and microstructure subjected to variation in austenitizing temperature. They reported that higher austenitizing temperature yielded higher corrosion resistance on account of great stability of passive film, larger passivation tendency, and decrease in pit growth rate. A similar argument holds good in the present case also as the sample is having a higher amount of austenite, the corrosion behavior has vastly improved. Kim et al. [13] have studied the corrosion properties of high-strength steel and showed that the key elements Cr and Mn are mainly responsible for reducing the corrosion rate which contributes to the grain refinement in the matrix as well formation of a corrosion product which is a non-adherent protective oxide layer. In the present study, the presence of higher austenite levels and eutectic carbides of Cr have contributed to the lower corrosion rate in the 10% Mn bearing sample compared to a 5% Mn counterpart.

3.2 Scratch Parameters

Figures 4 and 5 show the scratch width and depth in as-cast and heat-treated conditions of 5 M and 10 M series at three different loads of 10, 45, and 80 respectively. With the increase in manganese content from 5 to 10%, irrespective of the conditions, both the scratch width and depth show an increase in trend. Further, it is noticed that the scratch width, as well as depth of 5 and 10 M series in as-cast condition (Figs. 4a, c and 5a, c), are higher compared to the corresponding heat-treated (Figs. 4b, d and 5b, d) samples.

Also, it is seen that the average scratch widths of the heat-treated samples 5M12HT and 5M24HT are smaller than those of the corresponding as-cast samples namely, 5M12AC and 5M24AC which is in full agreement with the increase in hardness reported in Table 3. The samples 5M40AC and 5M40HT follow the trend of other samples as far as the Rockwell C hardness is concerned. The scratch width in heat-treated sample 5M40HT is consistently lower than that seen in the as-cast sample for all three loads. In the case of 10M12HT and 10M24HT at lower load (10 N), the scratch width is smaller than that of 10M12AC and 10M24AC and the trend repeats at higher load as well. It is reported that the higher the hardness, the lower is the

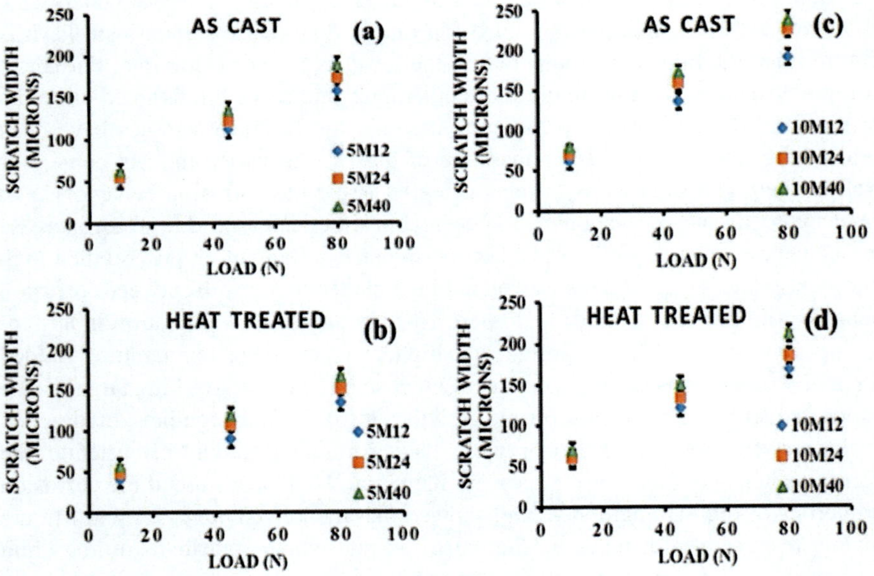

Fig. 4 Scratch width versus normal load for 5 and 10 M series in both as-cast and heat-treated conditions

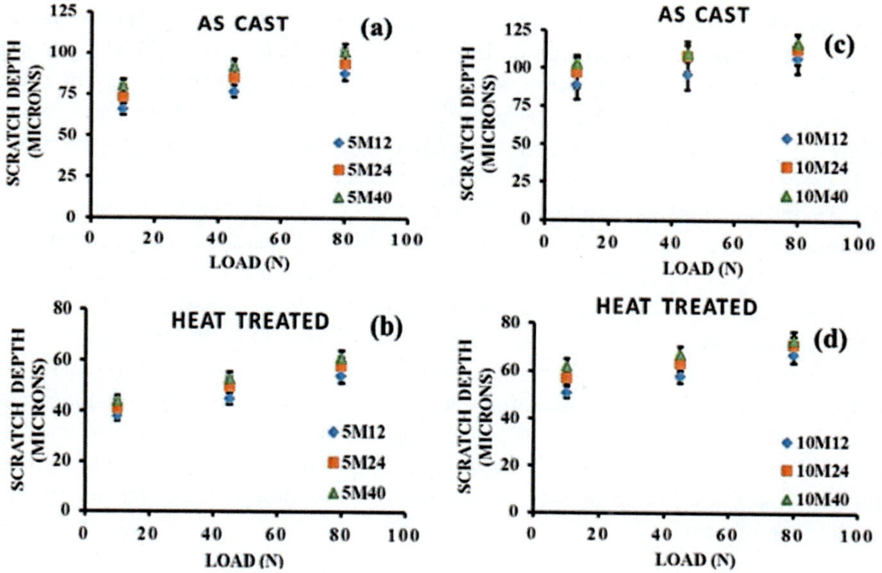

Fig. 5 Scratch depth versus normal load for 5 and 10 M series in both as-cast and heat-treated conditions

scratch width as well as the depth and the same trends are observed in the present work [2]. Based on the above discussion, it may be inferred that the least section size (12 mm) samples possess the least scratch width in both 5 and 10% Mn series compared to 24 and 40 mm section size samples in both the conditions (as-cast or heat-treated).

The reduction in the scratch width of 5M12 over 5M40 and 10M40 for all the loads (10, 45, and 80 N) employed are in the range of 15–17% and 34–36% respectively in the as-cast samples. Similarly, the reduction in terms of the scratch width in the heat-treated conditions for the sample 5M12 compared to either 5M40 or 10M40 are in the range of 19–26% and 37–41% for the loads 10, 45, and 80 N respectively. It is quite evident from the results that the scratch width of heat-treated samples reduces to a large extent as compared to the as-cast ones. The percentage decrease obtained in the scratch depth of 5M12 compared to 5M40 and 10M40 for the three loads (10, 45, and 80 N) in the as-cast condition is to the tune of 13–18% and 25–36%, respectively. As regards the heat-treated samples, the reduction in the percentage of the scratch depth noticed in 5M12 compared to 5M40 or 10M40 for the same load range is 11–14% and 26–29%, respectively.

Poh et al. [2] conducted scratch resistance on iron and AISI grades (steel) 1045 and 304L and reported that the work hardening potential of the samples are very much responsible for piling up of the deformed layer, where it is said that massive piling up of the material is more in iron than steel. The SEM and atomic force microscope (AFM) topography of the scratch in pure iron in relation to pile up and deformed volume next to the scratch groove have been stressed. In the present case also the scratch adhesion characteristics involving different degrees of scratch width and piling up of debris are noticed on account of an increase in Mn content in HiCr irons with various section sizes. The deformation pattern of the HiCrMn iron samples of 10 M series shows deeper scratches compared to 5 M series because of the larger pile up of material during scratching involving ploughing mechanism. It is reported [2] that lower scratch width is assisted with lesser tangential force, smaller frictional levels, and higher hardness, and this argument may hold good in the present work also as 10 M series have yielded lower hardness levels compared to 5 M series. These variations may be due to the variations in the degree of work hardening characteristics as well as the generation of pile up of particles of different degrees along the groove length in a single pass while conducting the scratch test. The piled up of materials are pushed aside or maybe chipped off close to the edges during scratching due to the prevailing operating mechanisms such as ploughing and cutting action. Mezlini et al. [14] have carried out a detailed investigation of scratch abrasive testing on aluminum alloys which involved mainly ploughing and wedge formation and resulted in increase in scratch cross section with an increase in the application of load.

The purpose of determining the normalized scratch hardness is to find out the material behavior under scratch load where half the projected area of the loading is taken into account. Figure 6 shows the plot of normalized scratch hardness (H_S) versus indentation hardness (H) at different loads during the scratch adhesion test (5 and 10 M series in both conditions). The repeatability of the test run in terms of coefficient of variation determination is done as per the ASTM G 65-16e1 Standard.

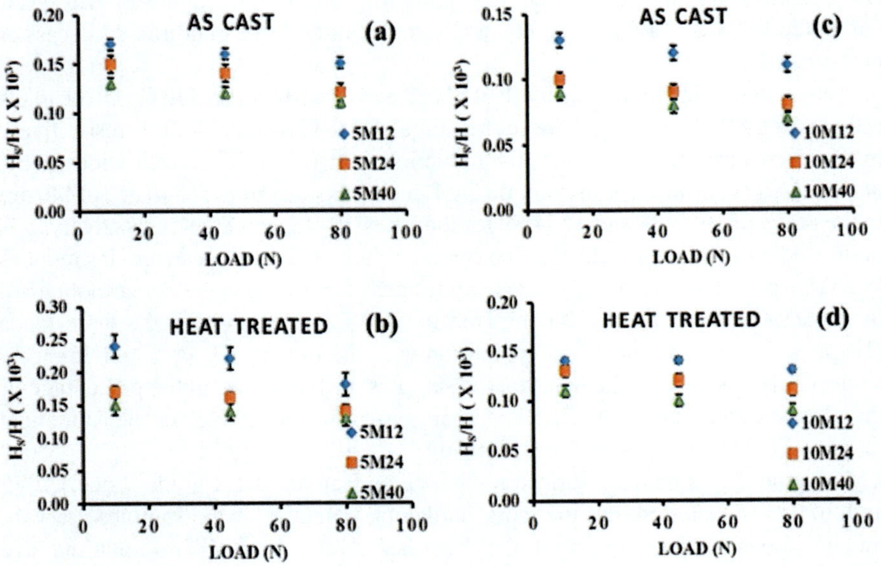

Fig. 6 Normalized scratch hardness versus load for both 5 and 10 M series in as-cast and heat-treated conditions

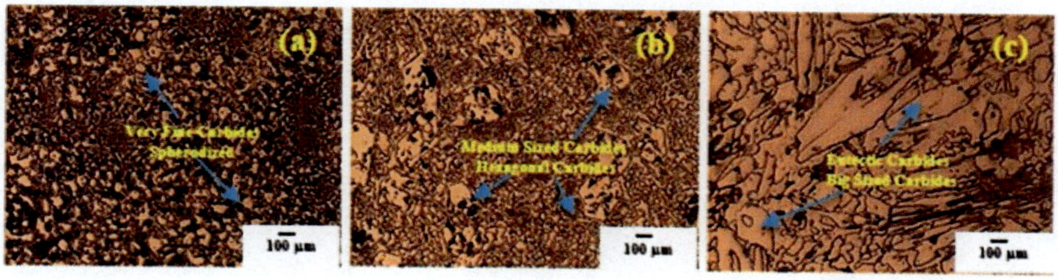

Fig. 7 Surface microstructures of **a** 5M12HT, **b** 5M24HT and **c** 5M40HT Samples at 500X

It is observed that the normalized scratch hardness of samples 5 and 10 M series shows a decreasing trend with respect to the increase in load application. There are variations noticed with respect to microstructures of 5 and 10 M series having varied percentages of austenite, martensitic, and carbides wherein higher normalized scratch hardness is obtained for higher austenite content samples compared to samples having lower austenite levels at a given load.

Balaji et al. [15] reported the abrasive wear resistance of ferrous microstructures with different hardness levels evaluated by a scratch tester and reported that bainite and pearlite displayed different responses compared to martensitic and tempered martensitic structures and could also characterize the groove with microstructures. The present observations noticed are very much similar to the trends reported with respect to the normalized scratch hardness versus load of the steel samples having different microstructures. A good correlation of the average scratch width and depth with the changes in alloy hardness, varying manganese levels, cooling rates, and heat treatment is observed.

3.3 *Light Optical Micrographs*

Figure 7 shows the light optical micrographs of heat-treated 5M12, 5M24, and 5M40 samples. 5M40HT reveals bigger sized primary eutectic carbides (Fig. 7c) randomly distributed in a matrix of austenite (57%) with carbide volume (26.5%) compared to 5M24HT which shows medium-sized carbides and small-sized hexagonal carbides (Fig. 7b) with retained austenite of 50% and carbide volume of 27.5%. 5M12HT has yielded a very fine carbide network in an austenite matrix of 47% and carbide volume of 29%. Further, the carbide distribution in 5M12HT (Fig. 7a) matrix is quite uniform and spherodized. These microstructural features support the scratch resistance data wherein 5M40HT has shown the highest scratch width because of the lowest hardness and carbide volume with the highest retained austenite. On the other hand, 5M12HT reveals the least scratch width due to the highest hardness and carbide volume with the least retained austenite content. The carbides irrespective of their size and shape; whether they are spherical or hexagonal, big or small, primary or secondary, as-cast or heat-treated are randomly distributed in the matrix in the case of 5M40 compared

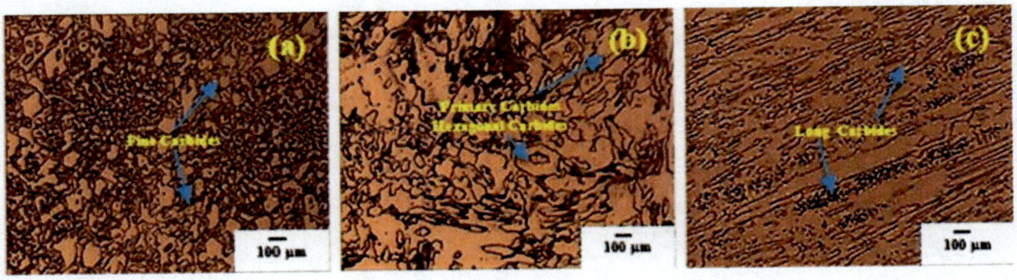

Fig. 8 Surface microstructures of **a** 10M12HT, **b** 10M24HT and **c** 10M40HT Samples at 500X

to 5M12 and 5M24 samples. 5M12 has uniform carbides dispersion by solidification process in a metal mold having higher thermal conductivity resulting in better heat transfer characteristics [16, 17]. Further, these carbides are very fine in size on the surface due to the prevailing higher cooling rate and grow bigger in size towards the core of the material. Therefore the surface properties when evaluated would be much superior compared to the mid-portion of the casting. In the present case, the section thickness of the metal mold being minimum (12 mm) tries to cool faster from all the sides resulting in the formation of a very fine grain structure. This has yielded the least scratch width for 5M12 when compared to 5M24 and 5M40 samples.

The microstructures of the heat-treated samples of 10 M series are shown in Fig. 8. 10M12 yielded fine carbides in an austenitic matrix of 55% with carbide volume of 26%, whereas 10M24 shows the lamellar type of primary carbides along with medium-sized hexagonal carbides in the matrix of austenite (60%) with carbide volume of 25.5%. 10M40 exhibits long carbides with an austenite content of 62% with a carbide volume of 24.5%. The fracture properties of these long carbides are load direction-dependent. The carbides irrespective of the direction of the load application are firmly supported by the matrix and are not easily broken. Whereas the long carbides are liable to get fractured easily on the application of load in the perpendicular direction, as they are more brittle and bigger in size compared to globular/spherical type of carbides along its length (axial). Thus in the present case, 10M40HT sample possessing long carbides during scratch loading is more prone to damage than 5M12HT samples as it contains a very fine carbide network. The actual situation in the samples would be more complex due to the variation in austenite content in the matrix, different carbide morphological features as well as change in the effective hardness on the surface observed.

The SEM morphology of 5M12HT and 10M12H are shown in Fig. 9. The evidence of spherodization of carbides (globular shape) in heat-treated 5M12 is very clearly seen in Fig. 9a which is not noticed in 10M12. The carbide size is found to be in the range of 2–5 µm. Hence, the SEM observations give a very good correlation between the scratch and corrosion results obtained in this work.

Thus, the significant point that emerges from this work, is to choose 10% manganese content with 24 mm section size in high chromium manganese iron alloy for achieving better properties to produce parts such as liners, impellers, etc. where resistance to wear and corrosion are important.

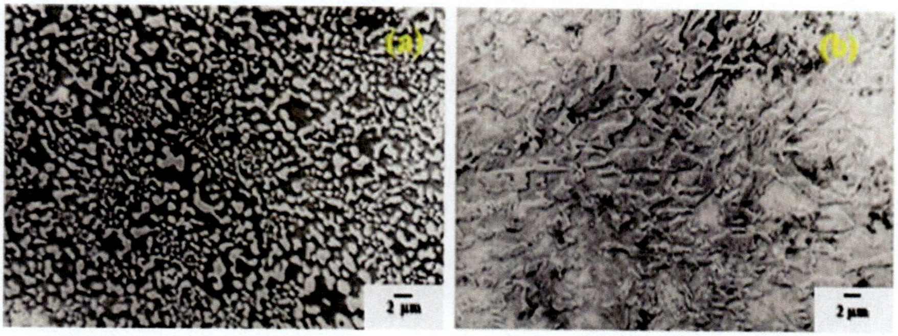

Fig. 9 SEM morphology of **a** 5M12HT and **b** 10M12HT

4 Conclusions

The following important points emerge from this investigation.

- Good correlations have emerged among the composition, corrosion rate, hardness, scratch width, and depth as well as metallurgical features of Mn-doped chromium iron alloys irrespective of cooling rate and heat treatment.
- The corrosion results indicate the higher the manganese addition, the lower is the weight loss due to seawater corrosion at room temperature and reflux condition.
- In 5 and 10% Mn bearing samples the corrosion rate decreases with an increase in section size. The sample with higher section thickness and lower hardness has shown a lower corrosion rate due to the presence of higher austenite content and vice versa.
- The corrosion rate shows higher in reflux conditions over the room temperature on account of reaction kinetics.
- The scratch parameters are very much dependent on the composition. With an increase in manganese content, the scratch width and depth increase.
- The scratch width and depth are found to be higher for the higher section thickness sample in 5 and 10% Mn bearing samples both in as-cast and heat-treated conditions. Further, 10M40 shows higher scratch width and depth, whereas 5M12 shows the least.
- The normalized scratch hardness like in the case of scratch width/depth is highest for 5M12HT and least for 10M40HT.
- The above data have been substantiated with the hardness, retained austenite, density, carbide volume, and light microscopic features.
- Among the systems studied, 10% Mn with 24 mm section size in high Cr Mn Fe alloy exhibited improved properties compared to others.

Acknowledgements The authors wish to acknowledge with thanks the management of TERI, SaIT, RVCE, DUCOM, BMSIT, CPRI, and JSS University to present and publish this paper.

References

1. Kaputkina LM, Svyazhin AG, Smarygina IV, Bobkov TV (2016) Corrosion resistance of high strength austenitic chromium nickel manganese steel containing nitrogen. Steel Transl 46:644–650. https://doi.org/10.3103/S0967091216090023
2. Poh F, Hardes C, Theisen W (2016) Scratch behavior of soft metallic materials. AIMS Mater Sci 3:390–403. https://doi.org/10.3934/matersci.2016.2.390
3. Deuis RL, Subramanian C, Yellup JM (1996) Abrasive wear of aluminum composites. Wear 201:132–144. https://doi.org/10.1016/S0043-1648(96)07228-6
4. Zhang ZF, Zhang LC, Mai YW (1995) Particle effects on friction and wear of aluminum matrix composites. J Mater Sci 30:5999–6004. https://doi.org/10.1007/BF01151519
5. Natesan M, Selvaraj S, Manickam T, Venkatachari G (2008) Corrosion behavior of metals and alloys in marine industrial environment. Sci Technol Adv 9:1–7. https://doi.org/10.1088/1468-6996/9/4/045002
6. Marimuthu V, Kannoorpatti K (2016) Corrosion behaviour of high chromium white iron hard facing alloys in acidic and neutral solutions. J Bio-Tribo-Corros 2:1–12. https://doi.org/10.1007/s40735-016-0056-x
7. Pearce JTH (1984) Structure and wear performance of abrasion resistant chromium white cast irons. AFS Trans 126:599–621
8. Hebbar BM, Seshan S (1994) Fracture toughness of high chromium cast irons. AFS Trans 118:349–356
9. Melchers RE (2002) Effect of temperature on the marine immersion corrosion of carbon steels. Corrosion 58:768–782. https://doi.org/10.5006/1.3277660
10. ASM Hand Book 3 (2016) Alloy phase diagram introduction to phase diagrams. ASM International, Philadelphia, USA
11. Handoko W, Pahlevani F, Sahajwalla V (2018) Effect of retained austenite stability in corrosion mechanism of dual phase high carbon steel. Int J Mater Metall Eng 12:1–5. https://doi.org/10.5281/zenodo.1314650
12. Jiang Z, Feng H, Li H, Zhu H, Zhang S, Zhang B, Han Y, Zhang T, Xu D (2017) Relationship between microstructure and corrosion behavior of martensitic high nitrogen stainless steel 30Cr15Mo1n at different austenitizing temperatures. Mater 10:861–880. https://doi.org/10.3390/ma10080861
13. Kim B, Kim S, Kim H (2018) Effects of alloying elements (Cr, Mn) on corrosion properties of the high strength steel in 3.5% NaCl solution. Adv Mater Sci Eng 1–13. https://doi.org/10.1155/2018/7638274
14. Mezlini S, Kapsa Ph, Henon C, Guillemenet J (2004) Abrasion of aluminium alloy: effect of subsurface hardness and scratch interaction simulation. Wear 257:892–900. https://doi.org/10.1016/j.wear.2004.05.004
15. Narayanaswamy B, Ghaderi A, Hodgson P, Cizek P, Chao Q, Safi M, Beladi H (2019) Abrasive wear resistance of ferrous microstructures with similar bulk hardness levels evaluated by a scratch tester method. Metall Mater Trans A: 1–12. https://doi.org/10.1007/s11661-019-05354-2
16. Junyl S, Yuding J (1987) The effect of orientation and thickness of carbides on abrasive wear resistance of high chromium iron. In: International conference on wear of materials, ASME, New York, pp 661–671
17. Seetharamu S, Sampathkumaran P, Kumar RK (1995) Erosion resistance of permanent moulded high chromium irons. Wear Part I 186–187:159–167. https://doi.org/10.1016/0043-1648(95)07173-3

Stress Corrosion Cracking Behavior of Spring Steel in Aggressive Corrosion Environment

H. K. Basukumar and K. V. Arun

1 Introduction

Spring steel is vastly used in major manufacturing industries such as automotive and aerospace for its resilience and elastic properties with high yield strength. The majority of the spring steel requirements are met with carbon steel, alloy steel, and corrosion-resistant steel. Some of the desired mechanical properties are yield strength, toughness, and fatigue strength to withstand high stress, vibrational load, shock, or impact loading. Steel as a material with desired mechanical properties like high tensile strength, ductility, and high toughness is used in automotive industries [1–4]. The alloy spring steels have a definite place in the field of spring materials, particularly for afore-stated service conditions. Like high carbon steels, alloy spring steels also can endure high and low-temperature environments. The action of service load and the environment will be enforced on the components, regardless of the nature of the loads. The failure of the component can be attributed to crack initiation and its propagation originated from the combined effect of these conditions. The initiation is supposed to include some induction of microcrack during service environment and their intensification, due to applied working load, this ends up in stress corrosion cracking. In an engineering application where materials are often exposed to the action of the basic and acidic service environment can be severe to the properties of the material, subsequently may affect adversely the performance of the components, system as a whole may be adversely affected. These engineering material failures in services can appear in form of the big engineering disasters. Such kinds of failures lead to a huge loss of life, property and adversely affect society as a whole. One of the common forms of environmental degradation is corrosion. Due to corrosion,

H. K. Basukumar (✉)
KLE Institute of Technology, Hubballi, Karnataka 580027, India

K. V. Arun
Government Engineering College, Haveri, Karnataka 581110, India

U. K. Mudali et al. (eds.), *Recent Trends in Electrochemical Science and Technology*, Springer Proceedings in Materials 15,
https://doi.org/10.1007/978-981-16-7554-6_5

there is variation in the material properties and loss in the form of capital in the industrial sector. Corrosion cannot be eliminated but can be minimized with the proper selection of material to suit particular service conditions. Despite all the precautionary measurements taken for such environmental service conditions, the materials fail. The conjoint action of these factors in combination leads to sudden and catastrophic failure. The synergic effect of the high rate of loading, environmental conditions, and material is termed stress corrosion cracking (SCC). The study of failure because of SCC in carbon steel was highlighted in the early nineteenth century with the catastrophic collapse of suspension bridge across the Ohio river killing 46 people in vehicles on the bridge at the time. Rust in the eye section had caused a stress corrosion crack, which went critical as a result of high bridge loading and low temperature. The failure was exacerbated by a high level of residual stress in the eye section. One of the causes for the failure of the suspension springs in automobile vehicles, driven over different service environments was the SCC failure mechanism. The interaction of service environment and mechanical working load is the most general case. Almost all corroding structures are stressed and all stressed structures are exploited in one or the other corrosive environment [5, 6]. The service failures in engineering components occur when a component is subjected to applied stress and a slow environmentally induced crack growth until a critical crack length is reached. The susceptibility of failure of spring steel is closely related to its performance under the prevalent service environment which can be catastrophic [7–11]. Corrosion leads to loss of material, lowering the structural efficiency in terms of strength, and ductility. The corrosion resistance of spring steel must therefore be studied and the mechanism of corrosion understood. In the present research, the overall corrosion mechanism in spring steel was analyzed to examine the SCC of spring steel under the influence of the pre-stress state.

2 Materials and Methods

One of the widely used low alloy springs steel is chrome vanadium steel, commercially known as EN 47. Globally, it is also designated with other common trade names as SAE/AISI6150, SUP10. It is suitable for overall engineering purposes with a higher strength than mild steel. EN47 spring steel is one of the prominently used materials in automotive and industrial suspension systems such as spring clamps, vehicle flat, and coil spring which are subjected to static or variable dynamic stresses and even for impact loads. The procured material has been chemically tested for its composition as shown in Table 1.

Table 1 Chemical composition of test specimen material EN47

Composition	Carbon	Silicon	Manganese	Phosphorus	Sulphur	Chromium	Vanadium
wt%	0.451	0.247	0.660	0.018	0.010	1.0	0.155

SCC is a process of cracking, arising from the cumulative effect of tensile stress and a corrosive environment. The requisite tensile stresses may be in the form of micromechanically induced residual stresses or directly applied stresses or may already occur during primary mechanical processes. SCC tests were performed on pre-stressed chemically treated and untreated tensile cylindrical specimens. These tests are preferred for a relative assessment of the susceptibility of alloys to SCC. However, the application of data obtained from these tests is to postulate that stress corrosion cracking has occurred in corrosive media. General corrosion of the specimen was simulated in the laboratory by immersion technique [12–15]. The specimens were corroded with HNO_3 of 35% concentration solution by immersion for about 5 min of exposure. The nitric acid exhibits the most aggressive corrosive nature in most of the metal, especially in steel. Therefore, nitric acid is considered for the simulated laboratory condition. The nitric acid with 20–22% concentration is used for passivation, below and beyond which acts as a corrosive media [1]. This typical solution demonstrates corrosion behavior through anodic dissolution, which is one of the mechanisms for stress corrosion cracking. The samples were subjected to a mechanical tensile test to measure their mechanical strength and other material properties under both untreated and treated to corrosive environmental conditions and to evaluate their performance. A typical experimental setup used in the laboratory is shown in Fig. 1.

One of the most commonly used mechanical tests conducted on the material is the tensile test. In this test, measuring the deformation caused due to applied load till the fracture point will lead to the understanding of the mechanical behavior of the

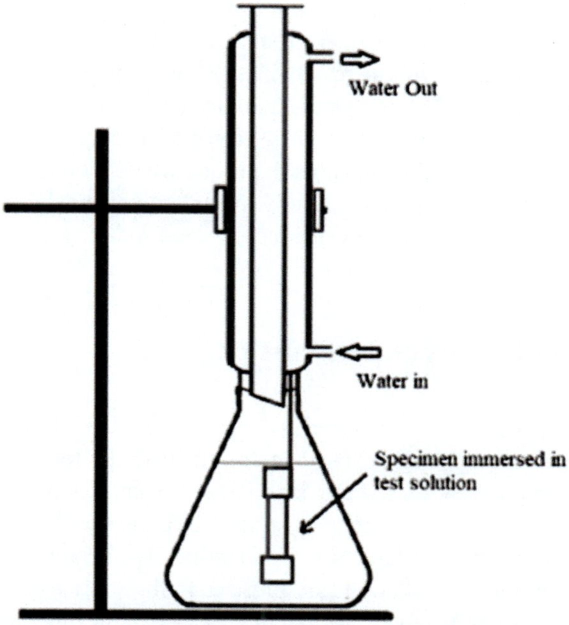

Fig. 1 A typical test setup used in the study

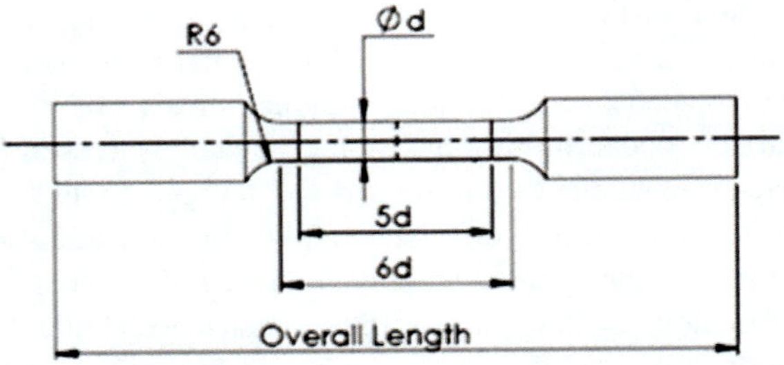

Fig. 2 Typical tensile Specimen drawing

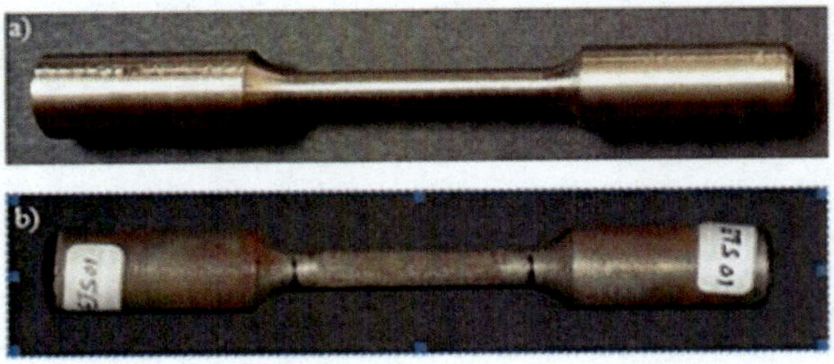

Fig. 3 Standard tensile **a** untreated and **b** treated specimens

material. This will enable us to anticipate how the material behaves in its intended applications. Cylindrical tensile specimens (110 mm overall length, 30 mm gauge length, and 6 mm gauge diameter) were machined from round bars of alloy EN47 in such a way that the gauge length to the diameter ($1/d$) ratio of these specimens was maintained at five, according to the ASTM E8-04. All samples were machined to have the same dimensions. Figure 2 presents the configuration and the dimensions of the cylindrical bar specimens used in tensile testing. The specimens as shown in Fig. 3, were cleaned for the residual imperfection present on the surface by degreasing with the ethanol. The samples were polished well to achieve the necessary metallographic finish before the tests were performed. The tensile tests are performed on cylindrical specimens before and after exposure to a defined environment. All the tensile testing has been conducted in a fully integrated universal testing machine (50 kN capacity) at room temperature using a cross-head velocity of 0.3 mm/minute.

3 Results and Discussion

The fracture has occurred in the gauge length of the test sample. The fractured surface observations were carried out for the specimen tested in low and high magnifications. Figure 4 shows fracture surfaces of the untreated and stress corroded specimen. As portrayed by the nature of spring steel, the specimen failed in a brittle fashion characterized by little plastic deformation. The center of the radial-fan markings is near the fracture initiation site, illustrating the surface cracks were induced by the virtue of the applied load. The fracture surface of the untreated specimen (Fig. 4a) reveals a

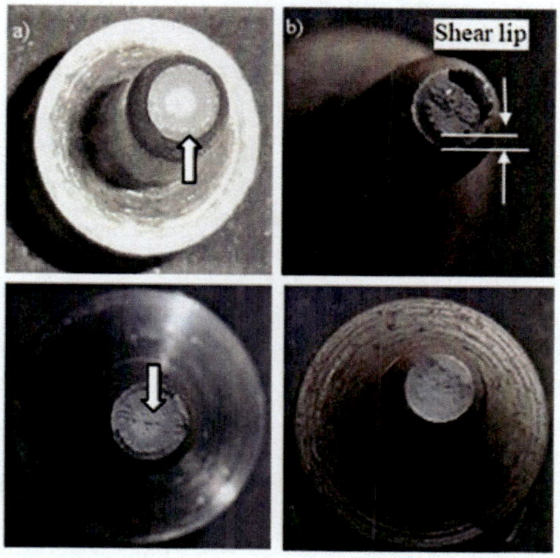

Fig. 4 Fracture surfaces of **a** untreated and **b** treated specimens

flat fracture surface with shallow radial marks as shown by the arrow associated with the cleavage mechanism, while the bright fibrous core area has dimples, typically associated with a nucleation-growth-coalescence mechanism. Dull and fibrous is the characteristic of the surface of a ductile fracture. A close look at the fractured surface Fig. 4b of the corroded specimen shows a narrow zone of intense shear. This shear region reveals a localized deformation due to pre-existing cracks. Room temperature tensile stress/strain curves of the investigated steel samples are presented in Fig. 5, from which tensile properties were obtained graphically. The elongation and reduction in the area are the indicators of the ductility of the material. The graphical representation of mechanical tensile properties obtained from the test is represented in Fig. 6.

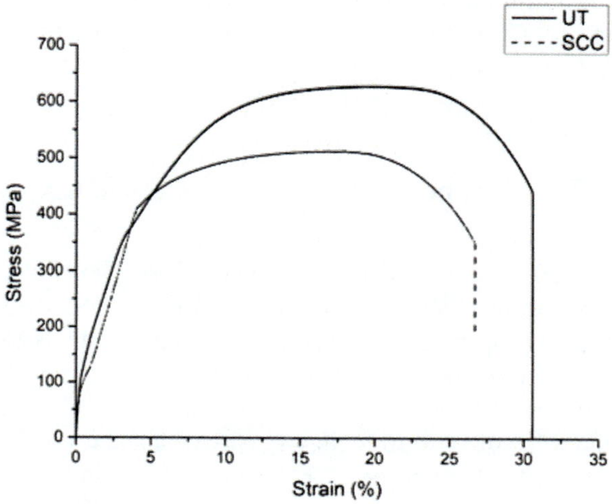

Fig. 5 Stress/Strain behavior of untreated and treated specimen

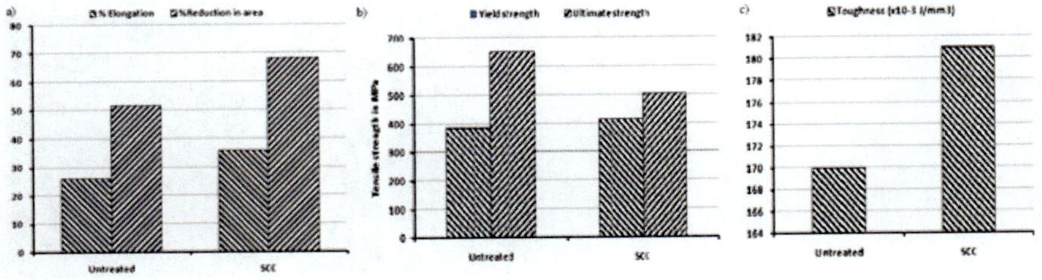

Fig. 6 Mechanical properties of untreated and treated specimens of EN 47 subjected to tensile test **a** % elongation and reduction in area, **b** yield strength and ultimate strength (MPa), and **c** toughness (10^{-3} J/mm^3)

3.1 Stress/Strain Behavior of Treated and Untreated Material

The stress/strain curve is one of the reliable and most common sources for the evaluation of mechanical properties of engineering materials. Stress/ strain curves of EN47 grade steel were obtained to evaluate their tensile behaviors. The stress-strain curve is generated by plotting the applied stress and the deformation in terms of strain caused due to it. The stress/strain curves for the tension test samples were shown in Fig. 5. The treated material has exhibited a stiffer mode of deformation with delayed fracture. The linear nature of the stress-strain curve depicts the elastic deformation region, indicates that the load-bearing capacity is more compared to the SCC specimen. The stress corroded material cannot sustain more loads even in elastic limit and similarly yield of the untreated specimen is comparatively high. The treated specimen exhibited brittle behavior than the untreated specimens and showed much smaller elongation at fracture location during the tension test; however, for the treated specimen, the necking could not be ignored. Its fracture surfaces show ductile failure. Even though the yield strength of SCC specimen is less compared to the untreated, the useful life will be higher than that of the untreated. This is very clearly depicted in Fig. 4b. The loss of material due to anodic dissolution and hardening is thought to be the cause.

3.2 Influence of SCC on Volumetric Changes and Toughness

The effect of stress corrosion is noticed in terms of change in yield strength, ultimate strength, and ductility of the untreated and treated tensile test specimens. The uniaxial tensile properties of the treated and untreated steels are summarized in Fig. 6. It is a compilation of the tensile properties of the samples that were untreated and treated with HNO_3. The yield stress is calculated as 0.2% strain. For the untreated specimen the value of yield strength was 385 MPa and for the SCC sample, a higher value of 416 MPa was recorded. The value increased by almost 8% when the SCC sample is compared with the untreated counterpart. The SCC sample test result reveals a noticeable detrimental influence in UTS compared to the untreated sample when deformed in uniaxial tension. The UTS for the untreated sample was 650 MPa, while strength was 506 Mpa for the counterpart. The yield strength of the SCC has been increased whereas the ultimate tensile strength has decreased significantly and it is about 22%. The elongation of the untreated specimen is about 26%, while the elongation of the SCC specimen is about 36%. From the test results, it can be inferred that the SCC tested specimen increased the ductility but reduced the ultimate strength. Deformation on the fracture surface is caused will be discussed in the section of micro-mechanism failure analysis. The anodic dissolution of material led to the loss in cross-section resulting in varied tensile behavior of stress corroded specimen. The effect of SCC is witnessed in terms of loss material as a reduction of area. There is a decrease in the strain with an increase in the load-carrying capacity in the elastic range indicates the loss of ductility.

3.3 Micro-mechanism Failure Analysis

The fracture surface of the failed samples was examined using a scanning electron microscope (SEM) to evaluate fracture modes. The macroscopic fracture and the characteristics of the tensile fracture surface were examined. This examination also provided information on the microscopic mechanisms prevailing tensile fracture surface. During slow strain rate tensile load failure, the formation of the voids, coalescence of voids, initiation of crack, crack growth, and rapid failure has occurred in the untreated specimen. The formation of multiple cracks at the circumference penetrates the material at a rapid speed; finally, fails as bulk shearing of the material. The formation of multiple cracks acts as triggering portion originated by the virtue of the stress corrosion, are denser on the circumference of the material compared to untreated, but the depth of the dense cluster cracks is very less. The fractured facets of untreated specimens exhibit dimples and tears ridges around the periphery of the facets as illustrated in Fig. 7. The fracture facets of the untreated specimens reveal an elongated fracture, whereas pre-stressed brittle failure for treated specimen surfaces. Similar sets of facets were witnessed [16–20]. It consists of many small dimples and several deep cracks, indicating a ductile fracture mechanism as indicated in Fig. 8b. The microstructure of the stress corroded specimen (Fig. 8) shows pre-existing cracks which led to the mixed mode of brittle and ductile cracking inferred due to the presence of a mixture of cleavage and micro void coalescence at the micro-scale and the same can be witnessed by the presence of shear lips. The image analysis reveals a quasi-brittle fracture. The failure indicates that the mechanism of localized plastic deformation, induced by anodic dissolution plays a dominant role in the SCC and is a consequence of a decohesive process. In essence, the SCC is a corrosive state. Corrosion is a continuous electrochemical mechanism in which both crack initiation and propagation are regulated by dissolution. Both mechanical fracturing and dissolution have been combined in this experiment. When exposed to corrosive media, the

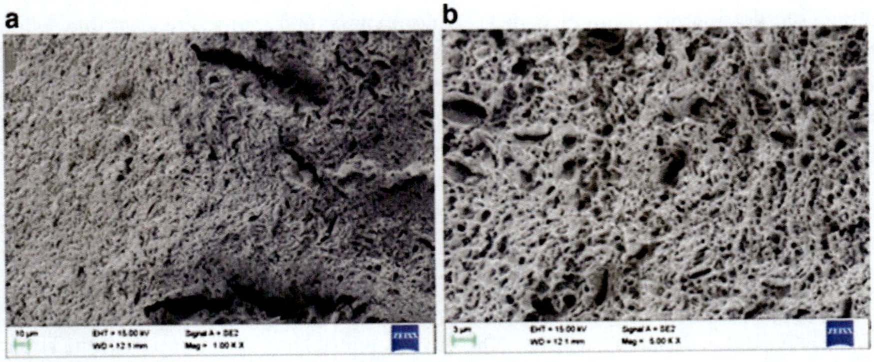

Fig. 7 SEM images of untreated specimen. **a** Initiation of multi-cracks due to tensile loading and **b** overload fracture surface showing microvoids and microcracks

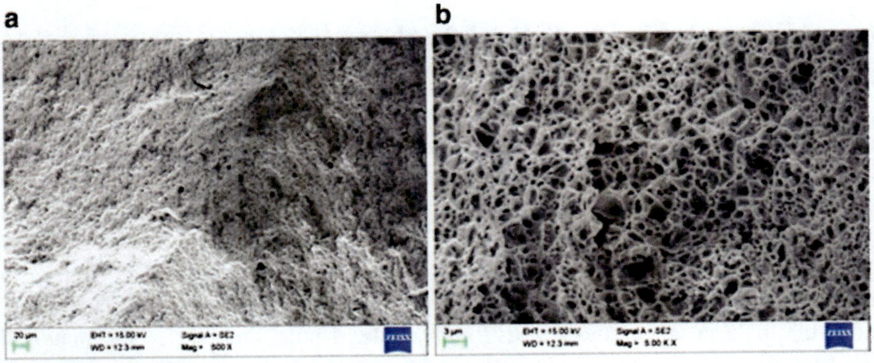

Fig. 8 SEM images of treated fracture surface showing **a** stress corrosion dominant region showing subcritical cracks and **b** coalescence of voids and fracture at the overload zone

pre-existing microcracks caused by prestressing formed a fine series of small corrosion subways. These subways expand in diameter and length until the remaining fragment is stressed to the point of ductile deformation and fracturing. As a result of the grooved fracture surface, microvoid coalescence at the fractured fragment was observed experimentally.

4 Conclusions

SCC is one of the key sources of an unforeseen and disastrous breakdown of industrial plants. The applied load, material condition, and corrosive environment are three aspects that are important for this form of failure to occur. The cracking condition for each metal or alloy is unique and not all environmental conditions endorse the

SCC failures. The influence of the pre-stress condition on spring steel was studied to investigate the stress corrosion cracking of spring steel.

- The stress corroded material cannot sustain more loads even in the elastic limit. The yield strength of the untreated sample is comparatively high. The treated specimen exhibited brittle behavior compared to the untreated specimens and showed much smaller elongation at fracture.
- The fracture surface of the untreated specimen reveals a flat fracture surface with shallow radial marks associated with a cleavage mechanism while dimples, typically associated with a nucleation-growth-coalescence mechanism, are present with the bright fibers at the core region. The stress corroded specimen microstructure reveals pre-existing cracks, which contributed to the mixed mode of brittle and ductile cracking inferred due to the existence of a mixture of cleavage and microvoid coalescence. The same can be interpreted with the appearance of shear lips.
- The fractography analysis shows a quasi-brittle fracture, which suggests that the mechanism of plastic deformation localization caused by anodic dissolution plays a dominant role in stress corrosion cracking.
- The corrosion mechanism is related to the influence of anodic dissolution of the steel, which can be considered significant. Cracking morphology indicated stress corrosion cracking of carbon steel.

References

1. Fontana MG (2010) Corrosion engineering, 5th edn. Tata McGraw Hill Education Private Limited, NewDelhi
2. Types of Spring Materials (2020) https://www.thomasnet.com/articles/metals-metal-products/types-of-spring-materials/
3. Suzuki T, Kubot M, Kozawa S, Miyamoto H (2019) Low-alloyed high-strength suspension spring steel. Nippon steel technical report, No. 122
4. Basukumar HK, Arun KV (2017) Study on the combined effect of corrosive environment and impact loading on the behaviour of low carbon steel. In: International conference on advances of mechanical engineering sciences. ISBN 978-81-932966-3-9D
5. Ashida Y, Daigo Y, Sugahara K (2017) An industrial perspective on environmentally assisted cracking of some commercially used carbon steels and corrosion-resistant alloys. JOM 69:1381–1388. https://doi.org/10.1007/s11837-017-2403-x
6. Chatterjee UK (1995) Stress corrosion cracking and component failure: causes and prevention. 20(February):165–184
7. Spring steels. Available from: https://www.ispatguru.com/spring-steels. Accessed May 2020
8. Eliaz N, Shachar A, Tal B, Eliezer D (2002) Characteristics of hydrogen embrittlement, stress corrosion cracking and tempered martensitic embrittlement in high-strength steels. Eng Fail Anal 9:167–184
9. Woodtli J, Kieselbach R (2000) Damage due to hydrogen embrittlement and stress corrosion cracking. Eng Fail Anal 7:427–450. https://doi.org/10.1016/S1350-6307(99)00033-3
10. Khalifeh A (2019) Stress corrosion cracking damages, failure analysis. IntechOpen. https://doi.org/10.5772/intechopen.80826

11. Wang X, Chen YL, Niu G (2017) The study on corrosion resistance of high-strength spring steel. Corr Eng Sci Technol. https://doi.org/10.1080/1478422X.2017.1384628
12. Gariboldi E, Nicodemi W, Silva G, Vedani M (1994) Mechanical properties of spring steels at room and low temperatures. Metall Sci Technol 12:11–21
13. Logan H (1962) Studies of the stress-corrosion cracking of low-carbon steels. J Res Natl Bur Stand-C Eng Instrum 66(4):347–356
14. Prakash TL, Hara JO, Malik AU (1999) Studies on stress corrosion cracking (SCC) behavior of various metals and alloys used in the desalination and power plants. Desalination 123:215–221. https://doi.org/10.1016/S0011-9164(99)00079-X
15. Sivaprasad S, Tarafder S (2007) Corrosion of stressed components in: training programme on industrial corrosion: evaluation and mitigation. NML, Jamshedpur, pp 68–91
16. Roy AK (2005) Environment-induced degradation and crack-growth studies of candidate target materials. 1–11. https://digitalscholarship.unlv.edu/hrc_trp_sciences_materials/52
17. Becker WT, Lampman S (2002) Fracture appearance and mechanisms of deformation and fracture, vol 11, ASM Handbook. Failure analysis and prevention. ASM International, pp 1162–1178
18. Becker WT (2002) Lesson 3 principles of failure analysis; ductile and brittle fracture. ASM International
19. Townsend HE (2001) Effects of alloying elements on the corrosion of steel in industrial atmospheres. Corrosion 57(6):497–501. https://doi.org/10.5006/1.3290374
20. Suárez F, Gálvez JC, Cendón DA, Atienza JM (2016) Fracture of eutectoid steel bars under tensile loading: Experimental results and numerical simulation. Eng Fract Mech 158:87–105. https://doi.org/10.1016/j.engfracmech.2016.02.044

Effect of pH on Electrodeposition of Ni–Cd Alloy Coatings and Their Anticorrosion Performance

Cindrella N. Gonsalves, I. M. Sneha, and A. Chitharanjan Hegde

1 Introduction

Cadmium (Cd) is a versatile element known for its durability, excellent electrical conductivity, lubricity, high ductility, and solderability, and hence it is considered a key component in various technologically important applications especially in the battery industry. In accordance with the data provided in the U.S Geological survey report of 2009, about 8% of cadmium is used in the coating industry of mild steel, aluminum, and other non-ferrous metals, like copper, zinc, tin, etc. due to its anticorrosive property in saline and alkaline environments [1]. Besides this, electroplated Cd has been used as a brightener since 1910 [2], as it imparts a glossy appearance to the base metal, and thus giving them an aesthetic value. However, due to its sacrificial property, similar to zinc coating they tend to corrode before the base metal [3]. As a result of which it is known to indirectly contribute towards cadmium poisoning, thus obstructing its widespread use. Therefore, they are preferably used in combination with other metals, like Ni [4–7], Fe [8], Co [9], Cu [10], Zn [11], and Ag [2] by the technique of its alloy coating. When Cd is alloyed with iron group elements, like Ni [12–14] they exhibit a peculiar codeposition behavior termed anomalous, where Cd though less noble than Ni deposits more readily leading to high Cd content Ni–Cd deposits, than in the bath [2]. Thus, by knowing the principles of Ni-based alloy plating, it is possible to alter the codeposition behavior by changing the operating parameters, like cathode current density (c.d.), electrolyte pH, temperature, metal salt concentration [15], additives [9, 16], ion mass transport [4, 17], etc. The experimental results revealed that Ni-based alloy coatings having low Cd content tend to be brittle and undergo spontaneous exfoliation [2]. In recent years, glycine (NH_2CH_2COOH) the simplest of amino acids is popularly used as an additive in various binary and

C. N. Gonsalves · I. M. Sneha · A. C. Hegde (✉)
Department of Chemistry, Electrochemistry Research Laboratory, National Institute of Technology Karnataka, Surathkal, Mangaluru 575025, India

U. K. Mudali et al. (eds.), *Recent Trends in Electrochemical Science and Technology*, Springer Proceedings in Materials 15,
https://doi.org/10.1007/978-981-16-7554-6_6

ternary Zn-based alloy coatings of Fe group elements [18–23]. As the behavior of Cd is analogous to that of Zn, the present work is focused on developing glossy, adherent, anticorrosive Ni–Cd alloy coatings from a low concentration bath of Ni–Cd, using citrate salt as a complexing agent, glycine as an additive under different conditions of bath pH. It is pertinent to mention here that the purpose of the development of any electrodeposited alloy coatings is to use it as the most prevalent corrosion-resistant material on the surface of the most widely used engineering material in the industry. In this direction, the present Ni–Cd alloy coating bath has been proposed by keeping $CdCl_2$ content in the bath to the minimal (only 5 g L^{-1}). The purpose of this study is, to establish the effect of pH on the surface morphology of Ni–Cd alloy coatings, and to minimize the use of environmentally hazardous Cd in the bath.

2 Experimental

2.1 Preparation of Optimal Bimetallic Ni–Cd Bath

The preliminary studies were focused on the formulation of optimal Ni–Cd bath through the standard Hull cell method described elsewhere [24]. The bath was prepared by dissolving the $NiCl_2.6H_2O$ salt and $CdCl_2$ salt. All chemicals (LR grade) including trisodium citrate ($Na_3C_6H_5O_7.2H_2O$) as a complexing agent, NH_4Cl as conducting/supporting electrolyte, and glycine (NH_2CH_2COOH) as additive were homogeneously dissolved in double distilled water. The electrolyte is later filtered to remove undissolved impurities. Hull cell analysis was performed in a trapezoidal PVC container of 267 cm^3 electrolyte holding capacity, containing cathode as mechanically buffed pre-treated copper panels (10.0×7.0 cm^2) with an exposed area of 10.0×3.0 cm^2, and anode as graphite having the same exposed area. Solution pH was initially noted, and the cathode was immersed in the electrolyte in an inclined position for the galvanostatic Hull cell test. 1.0 A cell current was applied, using a DC power source (DC power Analyzer, Agilent Technologies, Model: N6705C) for 300 s. The binary Ni–Cd alloy coating was deposited on the cathode and then rinsed with double distilled water. The coated Hull cell panels, after drying were overlaid on the Hull cell ruler to assess the current density (c.d.) range to get the satisfactory coating. The uniformity of Ni–Cd alloy coating was improved by adding 1.0 g L^{-1} glycine into the electrolyte. The composition and operating variables of the bath required to get satisfactory coatings of Ni–Cd alloy on copper were arrived and are given in Table 1.

Table 1 Optimal Ni–Cd bath composition and operating parameters used for the solution pH variation study

Bath content	Composition ($g\ L^{-1}$)	Operating parameters
$NiCl_2.6H_2O$	66.0	Cathode: Copper panel
$CdCl_2$	5.0	Anode: Graphite
$Na_3C_6H_5O_7.2H_2O$	100.0	Current density: 2.0 A dm^{-2}
NH_4Cl	33.0	Duration: 600 s
NH_2CH_2COOH	1.0	Electrolyte pH: 2.0, 4.0, 7.0, 8.0, 10.0

2.2 *Effect of pH on Electrodeposition*

To understand the effect of electrolyte pH on the coatings formed, conventional galvanostatic electrodepositions were carried out at different pH with a fixed c.d. of 2.0 A dm^{-2}. Copper substrate (7.5 × 3.0 cm^2 dimension) with an exposed plating/deposition area of 3.0 × 3.0 cm^2 was used as the cathode. Before electrodeposition, the mechanically buffed copper panels were subjected to solvent degreasing using trichloroethylene (TCE) as a solvent and pickled with 1:1 HNO_3. The solution pH was altered in a sequence of 2.0, 4.0, 7.0, 8.0, and 10.0, and accordingly, the Ni–Cd alloy coatings were carried out for 600 s in a custom-made electrolytic cell. At the end of the deposition, the panel was removed from the electrolytic cell, rinsed with double distilled water, air-dried, and placed in a desiccator to avoid moisture attack. As the pH of the bath increased to > 10.0 it resulted in the precipitation of $Ni(OH)_2$. Therefore, electrodepositions were confined to the value of pH up to 10.0.

2.3 *Electrochemical Corrosion Study*

The anticorrosion performance of binary Ni–Cd coatings electrodeposited at different pH was evaluated in 3.5% NaCl medium at 303 K using a computer-controlled Potentiostat/Galvanostat (PAR VersaSTAT3-400, Ametek Instruments). The electrochemical corrosion studies were performed in a customized three-electrode cell, as shown in Fig. 1, using platinized platinum as the counter electrode, saturated calomel electrode (SCE) as the reference electrode, and Ni–Cd electrodeposited coatings as working electrode (WE), with an exposed surface area of 1.0 cm^2.

Initially, the open circuit potential (OCP) was identified by subjecting it to a steady-state condition for 5 min. Further, electrochemical impedance spectroscopy (EIS) study was carried out by applying an alternate current (AC) of a small amplitude of 10 mV, in the frequency range of 100 kHz–10 MHz. Nyquist plots corresponding to different Ni–Cd coatings were drawn. The corrosion rates (CR's) of Ni–Cd alloy

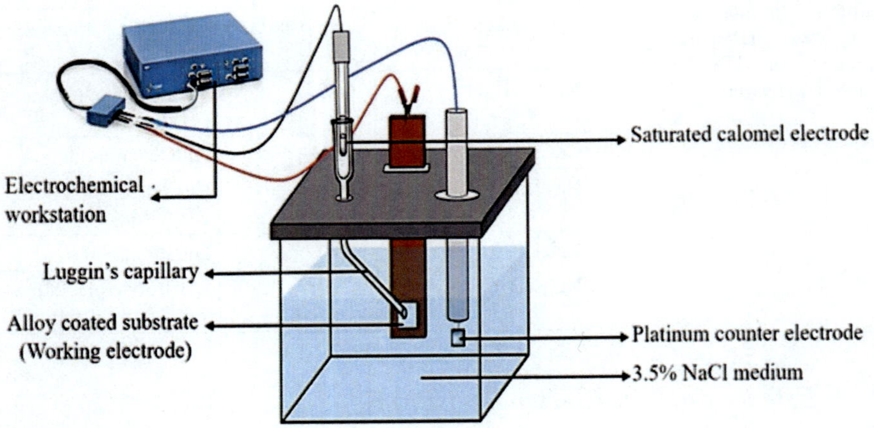

Fig. 1 The experimental set up used for the electrochemical corrosion study of electrodeposited Ni–Cd alloy coatings

coatings, corresponding to different c.d.'s was evaluated by potentiodynamic polarization approach. The polarization of WE was carried out at a scan rate of 1.0 mV s^{-1} with a potential ramp of ±250 mV from OCP. CR's were calculated by Tafel's method of extrapolation. The corrosion current (i_{corr}) and corrosion potential (E_{corr}) values were obtained and the corrosion rate (CR) was calculated using the standard Eq. (1),

$$\mathrm{CR}(\mathrm{mm\,y^{-1}}) = \frac{K \times \mathrm{EW}_{\mathrm{alloy}} \times i_{\mathrm{corr}}}{D_{\mathrm{alloy}}} \tag{1}$$

where K is the conversion factor (0.00327), i_{corr} is corrosion current density in μA cm^{-2}, $E.W_{alloy}$ is the equivalent weight of the alloy, and D_{alloy} is the density of the alloy in gcm^{-3}.

2.4 Surface Analyses

Changes in the surface morphology of binary Ni–Cd coatings electrodeposited at 2.0 A dm^{-2} at varied pH were studied using a Scanning Electron Microscope (SEM, Model Zeiss, Germany), interfaced with Energy Dispersive X-ray Spectroscopy (Oxford Instruments). To confirm the presence of constituent elements (Ni and Cd), within the coatings, and approximate wt.% of metals at which they have been alloyed in the coatings, Energy Dispersive X-ray Analysis (EDAX) was carried out. Besides, the crystallographic phase structure determination was carried out through X-ray Diffraction studies (JEOL, JDX-8P) using Cu K_α ($\lambda = 1.5406$ Å) radiation, in continuous scan mode with a scan rate of 2°min^{-1}.

3 Results and Discussion

3.1 Influence of pH on Electrodeposits

Among many other deposition parameters, the pH of the electrolyte is considered to bring in a remarkable change in the nature of electrodeposits. Figure 2 shows the photographs of the Ni–Cd alloy coatings obtained at varied pH from the Hull cell optimized Ni–Cd bath at 2.0 A dm^{-2}. It was observed that at acidic pH the coatings were dull and powdery. This is because, during electrolytic deposition, there is a discharge of H^+ ions along with the metal ions being deposited. At acidic pH, the concentration of H^+ ions is more, or in other words, there is excessive diffusion of H^+ ions towards the cathode surface. Moreover, in aqueous electrolytes, Cd tends to precipitate out as its hydroxide at lower pH than Ni [2] and this causes a detrimental effect on the plating rate, thereby leading to burnt deposit as shown in Fig. 2. The coatings obtained at pH 8.0 were visually bright and adherent. Usually, Ni–Cd coatings though appear bright, are brittle, and tend to exfoliate. But this has been drastically reduced by the combined effect of trisodium citrate along with a small amount of glycine (1.0 g L^{-1}), thereby intensifying the brightness and homogeneity of the coating at pH 8.0. This could be because between pH 4.0 and 9.0, glycine exists in its zwitterionic form $(NH_2CH_2COO)^{\pm}$, thus acts as a buffer and prevents the precipitation of metal hydroxides from electrolytic bath [25]. However, with further increase of electrolyte pH to 10.0, the deposit showed a non-uniform appearance, *i.e.,* having both bright and dull areas that could be ascribed to the local alkalinity caused by the increased H_2 evolution at the vicinity of the cathode surface [2], at the same time the existence of glycine in its anionic form $(NH_2CH_2COO)^-$ above pH 9.0 [25] leads to the formation of $[Ni(NH_2CH_2COO)_3]^-$ complex which accelerates the electroreduction of Ni^{2+} ions [26, 27].

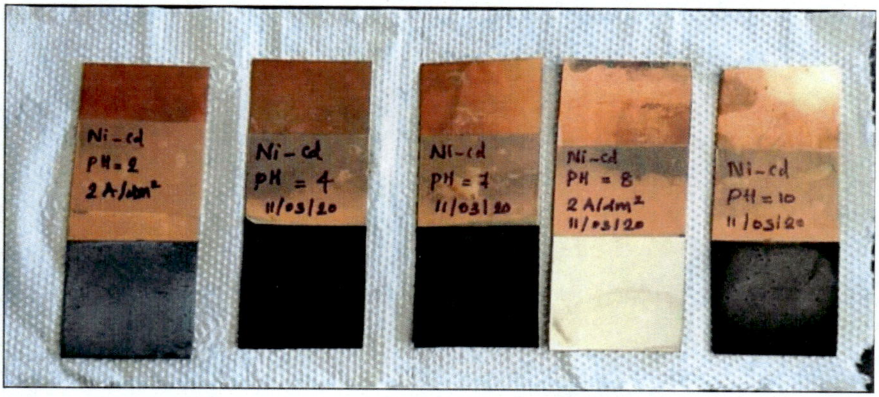

Fig. 2 Nature of binary Ni–Cd alloy coatings electrodeposited at pH 2.0, 4.0, 7.0, 8.0, and 10.0 using c.d. = 2.0 A dm^{-2}

3.2 SEM-EDAX Study

The SEM images of binary Ni–Cd alloy coatings developed at pH 2.0, 8.0, and 10.0 (only representative) are shown in Fig. 3. It is observed that at pH = 2.0, a large number of Ni–Cd micro-spheroid-like structures were observed which decreased with an increase of pH up to 8.0. It may be explained by the fact that on passing the direct current (DC) through an electrolyte, alongside the deposition of Ni–Cd alloy coating on the surface of the cathode, rigorous discharge of H_2 gas do take place on its surface, which is responsible for the formation of pits/gunnies on its surface. But at pH 10.0, a drastic change in the surface morphology was found with continued unevenness on the surface, but with no pores. This change in the nature of the deposit could be due to the excessive alkalinity of the electrolyte [4]. Thus, it may be concluded that the pH of the electrolyte influences the surface morphology of the Ni–Cd alloy coatings.

To know the composition of Ni–Cd alloy coatings deposited at different pH, Energy Dispersive X-ray Analysis (EDAX) was carried out. The composition of the binary Ni–Cd alloy coatings, in terms of wt.% of Ni and Cd, developed at different electrolyte pH are shown graphically in Fig. 4. It may be seen that at the acidic conditions of the bath, i.e., pH = 2.0, the wt% of Cd (less noble metal) in the deposit is much higher (53.1%) than in the bath (7.6%), shown by the horizontal line in Fig. 4. At low pH, the wt% of more readily depositable Cd is much higher in the deposit than noble Ni. It may be attributed to the fact that less noble Cd suppresses the codeposition of Ni [2]. But with a gradual increase of pH, wt % of Cd in the deposit has decreased and reached a minimum value of 11.6%. This increase of noble metal (Ni) content with an increase of pH indicates that the mode of deposition of Ni–Cd alloy coatings tends to vary from anomalous to normal type. It may be attributed to the fact that Ni–Cd alloy coating can follow both anomalous and normal types of codeposition depending on the alkalinity of the bath, as envisaged by Brenner [2]. At lower limits of pH, Ni–Cd bath follows a peculiar anomalous type of codeposition., i.e., less noble Cd deposits preferentially than nobler Ni; and at higher limits of pH,

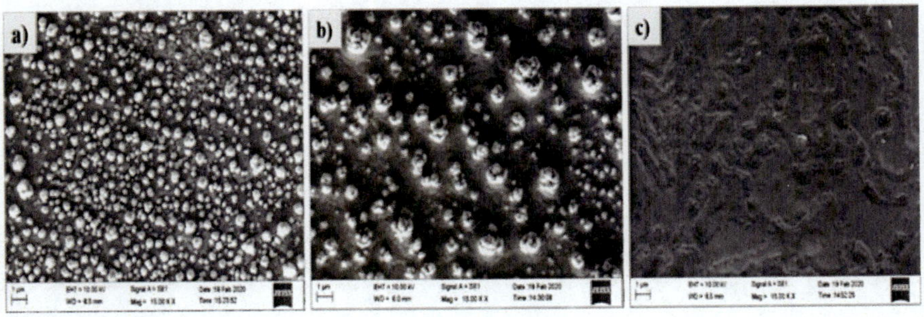

Fig. 3 Microstructure of Ni–Cd alloy electrodeposited alloy coatings at different pH: **a** 2.0, **b** 8.0 and **c** pH 10.0

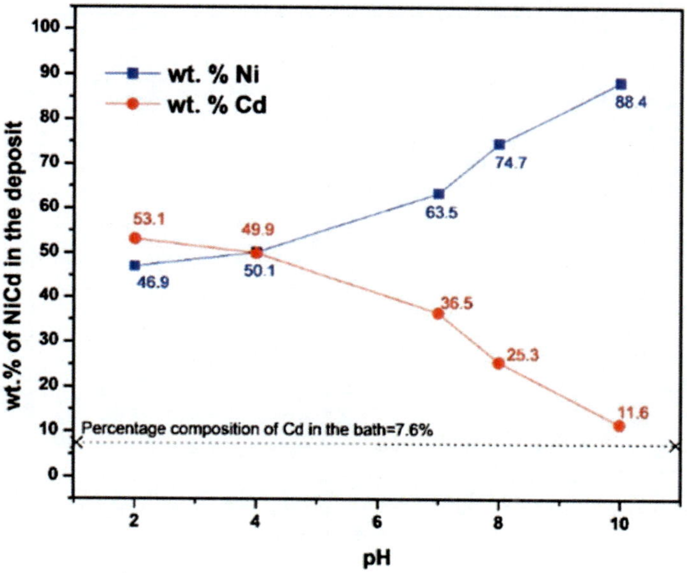

Fig. 4 Graphical representation showing the change of Ni and Cd content in the Ni–Cd deposit, with the change of pH

it follows a normal type of codeposition., i.e., nobler Ni deposits preferentially than less noble Cd. It is evident from the plot of change of metals contents in the deposit with the change of pH, shown in Fig. 4.

3.3 *Corrosion Performance Analysis*

3.3.1 Electrochemical Impedance Spectroscopy (EIS) Study

Electrochemical impedance spectroscopy (EIS) is an invaluable AC perturbation technique that measures the response of an electroplated specimen to a low amplitude AC signal applied over a range of frequencies at OCP. Generally, EIS data is evaluated utilizing Nyquist plots obtained by drawing imaginary (Z_{im}) impedance against real (Z_{re}) impedance. Figure 5 represents the Nyquist response obtained for electrodeposited binary Ni–Cd coatings developed at 2.0 A dm^{-2}, at different pH of the bath. It is observed that all the coatings exhibit only one capacitive loop or a single incomplete semicircle irrespective of varied pH. This implies that a single charge transfer process is involved in the corrosion process of all coatings. However, there is a significant increase in the value of polarization resistance (R_p) with an increase in electrolyte pH. The maximum axial radius observed for Ni–Cd coating corresponding to pH = 8.0 indicates that it is the most corrosion-resistant, compared to all other coatings. To understand the corrosion mechanism electrochemical simulation

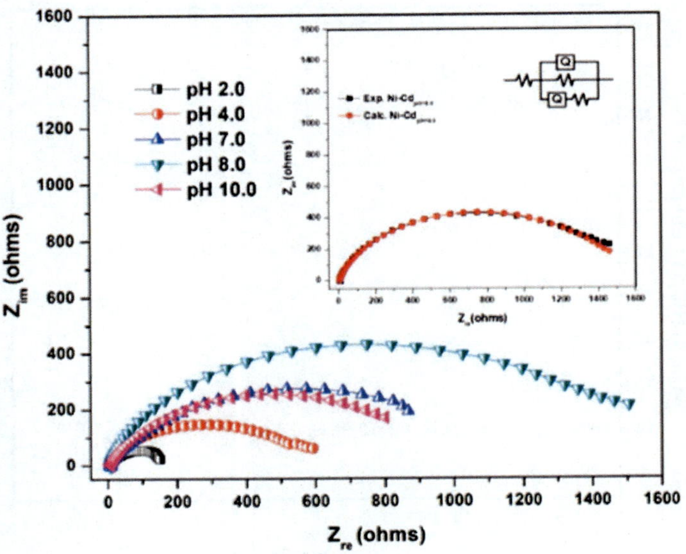

Fig. 5 Nyquist plots of Ni–Cd alloy coatings deposited at constant c.d. = 2.0 A dm^{-2} and different pH. The electrochemical simulation of the impedance response, with equivalent circuit, is shown in the inset

of Nyquist plot corresponding to Ni–Cd alloy coatings, showing the least corrosion rate was made and the data were fitted with an equivalent circuit. It is shown in the inset of Fig. 5.

3.3.2 Potentiodynamic Polarization Study

The corrosion behavior of binary Ni–Cd coatings electrodeposited at 2.0 A dm^{-2} under different electrolyte pH was quantified through potentiodynamic polarization study [28] recorded vs OCP. All corrosion rates (CR) were determined by extrapolating the Tafel response and the Tafel plots are as shown in Fig. 6. The observed E_{corr} and i_{corr} values corresponding to each coating obtained at varied pH are tabulated in Table 2. It is seen that the E_{corr} shifted to the nobler end up to pH 8.0, along with the decrease in i_{corr} value. From Eq. (1), it may be noted that i_{corr} is directly proportional to CR. Therefore, it can be inferred that the Ni–Cd alloy coating deposited at pH 8.0 shows a maximum corrosion resistance.

The composition and corrosion data of Ni–Cd alloy coatings corresponding to different c.d.'s are reported in Table 2. From the data, it may be observed that CR decreased with an increase of pH up to 8.0, and then decreased. This may be accounted for by the fact that at pH = 2.0, less noble Cd (wt% 53.1) deposited more readily than Ni, thus favoring anomalous type of codeposition. However, when the pH was raised to 8.0, the wt% of Ni in the alloy coating is found to be increased to 74.7%,

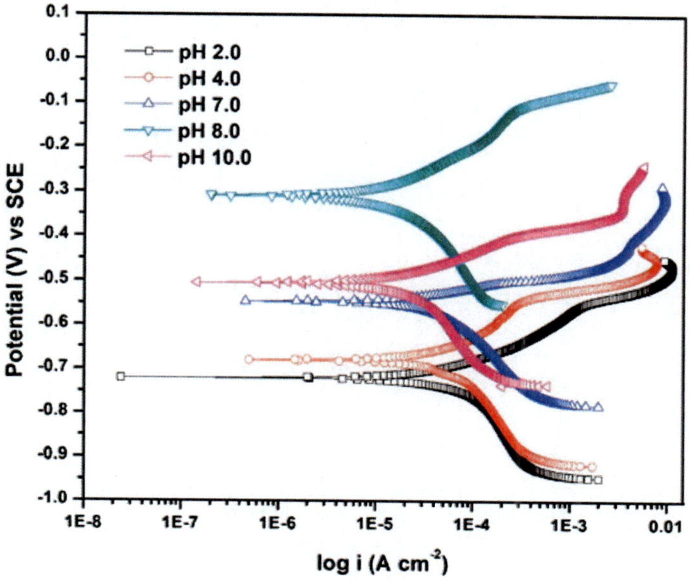

Fig. 6 Tafel plots obtained for binary Ni–Cd alloy coatings deposited at different electrolyte pH when subjected to potentiodynamic polarization study

supported by EDAX compositional analysis. On the further increase of pH to 10.0, the corrosion rate increased despite the increase of Ni content in the deposit. This may be attributed to the increased porosity of the coatings due to the formation of metal hydroxides [4]. Therefore, from the composition and CR data, it may be concluded that the pH of the bath has a significant role in increasing the Ni content in the deposit which facilitated the corrosion resistance of the binary Ni–Cd alloy coating. In other words, the inherent anomalous type of codeposition of Ni–Cd alloy coating has shown a tendency to follow a normal type of codeposition as the pH of the bath has increased.

3.4 XRD Study

XRD patterns of Ni–Cd alloy coatings deposited at the 2.0 Adm^{-2} but at different pH are shown in Fig. 7. The X-ray diffractograms showed the presence of both Cd and Ni in the deposit by exhibiting the XRD peaks at $2\theta = 43.2$ and $74.1°$ corresponding to Cd (102) (200) phases, and $2\theta = 44.5$ and $50.8°$ representing Ni (111) (200) phases respectively, arrived by comparing those with representative JCPDS data files. The additional Cd (101) peak corresponding to pH = 2.0 at $2\theta = 38.7°$ was found to disappear with an increase of pH as shown in Fig. 7. It supports the fact that an increase of pH favored the deposition of coatings having higher wt.% of Ni.

Table 2 Corrosion parameters of binary Ni–Cd coatings deposited at 2.0 A dm^{-2} and different pH

pH	wt% Ni	wt% Cd	$E.W_{alloy}$	D_{alloy} (g cm^{-3})	E_{corr} (V) versus SCE	i_{corr} (μA cm^{-2})	C.R. (×10^{-2} mm y^{-1})	Coating appearance
2.0	46.9	53.1	39.3	8.79	0.721	97.05	142.1	Grey dull
4.0	50.1	49.9	38.5	8.80	0.681	76.88	110.25	Black powdery
7.0	63.5	36.5	35.5	8.83	0.547	48.91	64.49	Black dull
8.0	74.7	25.3	33.4	8.85	0.309	18.25	25.55	Glossy metallic
10.0	88.4	11.6	31.1	9.35	0.506	29.30	31.89	Grey dull

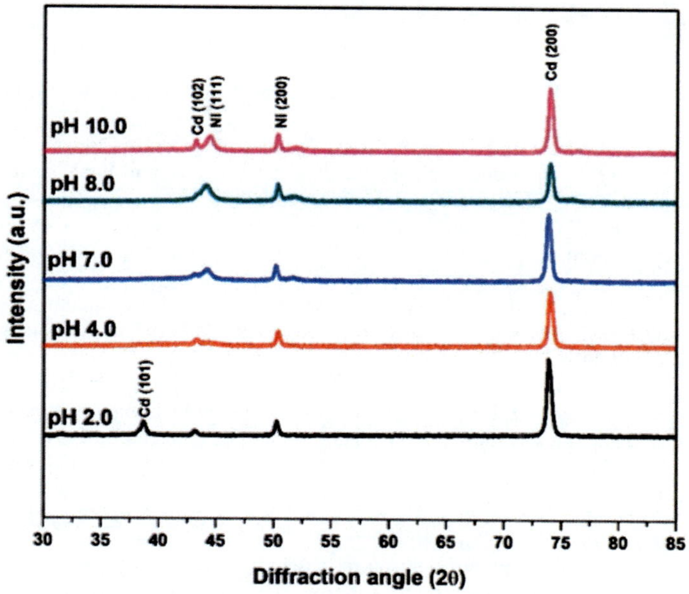

Fig. 7 XRD patterns of Ni–Cd alloy coatings deposited at 2.0 Ad m^{-2} and various pH

4 Conclusions

Based on the experimental investigations of this study, the following conclusions are made:

- A new citrate bath has been developed for obtaining smooth and uniform coatings of binary Ni–Cd alloy using glycine as the additive.
- Corrosion study of Ni–Cd alloy coatings deposited at different pH and constant c.d. of 2.0 A dm^{-2} revealed that pH of the bath was detrimental in controlling the composition and in turn the corrosion resistance property.
- A bright and adherent Ni–Cd alloy coating deposited at pH = 8.0, having about 74.7 wt% Ni is found to be the most corrosion-resistant compared to all other coatings.
- A drastic increase of corrosion resistance of Ni–Cd alloy coatings with increasing pH is attributed to the increased Ni content of the alloy. It may be attributed to the change in the inherent anomalous type of codeposition of Ni–Cd alloy coating to the normal type of codeposition.
- The least CR of Ni–Cd alloy coating deposited at pH 8.0 is ascribed to its improved surface morphology, composition, and phase structure supported by SEM-EDAX and XRD analysis, respectively.

Acknowledgements The authors are thankful to NITK Surathkal for the research facilities.

References

1. National Center for Biotechnology Information. https://www.ncbi.nlm.nih.gov/books/NBK304372/
2. Brenner A (1963) Electrodeposition of alloys: principles and practice. Elsevier
3. Wernick S (1932) The electrodeposition of cadmium from cadmium sulfate solutions. Trans Electrochem Soc 62:27. https://doi.org/10.1149/1.3493786
4. Rao VR, Hegde AC (2014) Magnetically induced codeposition of Ni–Cd alloy coatings for better corrosion protection. Ind Eng Chem Res 53:5490–5497. https://doi.org/10.1021/ie403639z
5. Sivakumar R, Rama Char TL (1973) Electrodeposition of nickel-cadmium alloys from a pyrophosphate bath. Electrodeposition Surface Treatment 1:227–237. https://doi.org/10.1016/0300-9416(73)90016-3
6. Wang P-N, Chang S-H (2018) Effect of the mechanical properties and corrosion behaviors of nickel-cadmium duplex electroplated AISI 4340 steel by using various solid solution treatments. Mater Trans 59:406–411. https://doi.org/10.2320/matertrans.M2017282
7. Smith C A (1992) Investigation of nickel-cadmium-diffused and cadmium plating replacement coatings. p 920943
8. Singh VB (1979) Electrodeposition of ternary nickel-iron-cadmium alloys. J Appl Electrochem 9:285–289. https://doi.org/10.1007/BF01112481
9. Singh VB, Tikoo PK (1978) Electrodeposition of ternary nickel—cobalt—cadmium alloys from acetate bath. Electrochim Acta 23:393–396. https://doi.org/10.1016/0013-4686(78)80081-4
10. Budniok A, Gala J, Pluta I (1979) Electrodeposition of copper in the presence of cadmium. Electrochim Acta 24:1247–1252. https://doi.org/10.1016/0013-4686(79)87080-2
11. Ganesan P, Kumaraguru SP, Popov BN (2006) Development of Zn–Ni–Cd coatings by pulse electrodeposition process. Surf Coat Technol 201:3658–3669. https://doi.org/10.1016/j.surfcoat.2006.08.143
12. Rao VR, Hegde AC (2014) Role of cadmium on corrosion resistance of Zn–Ni alloy coatings. Surf Eng Appl Electrochem 50:63–71. https://doi.org/10.3103/S1068375514010141
13. Addi Y, Khouider A (2011) Zinc-nickel codeposition in sulfate solution combined effect of cadmium and boric acid. Int J Electrochem 2011:1–7. https://doi.org/10.4061/2011/742191
14. Durairajan A, Haran BS, White RE, Popov BN (2000) Development of a new electrodeposition process for plating of Zn–Ni–X (X=Cd, P) alloys: I. Corrosion characteristics of Zn–Ni–Cd ternary alloys. J Electrochem Soc 147:1781. https://doi.org/10.1149/1.1393434
15. Rehim SDSAE, Wahaab SMAE, Abdella OM (2007) Electroplating of nickel–cadmium alloys from alkaline solutions. J Chem Tech Biotechnol 35:217–222. https://doi.org/10.1002/jctb.5040350503
16. Mohanty US, Tripathy BC, Singh P, Das SC (2002) Effect of Cd^{2+} on the electrodeposition of nickel from sulfate solutions. Part I: current efficiency, surface morphology and crystal orientations. J Electroanal Chem 526:63–68. https://doi.org/10.1016/S0022-0728(02)00771-4
17. Shetty AR, Hegde AC (2021) Magnetoelectrodeposition of Ni–Mo–Cd alloy coating for improved corrosion resistance. Chem Data Collect 32:100639. https://doi.org/10.1016/j.cdc.2020.100639
18. Kashyap R, Srivastava SK, Srivastava SC (1986) The role of addition agents in the electrodeposition of Ni–Mn–Zn alloys from a sulphate bath. Surf Coat Technol 28:129–137. https://doi.org/10.1016/0257-8972(86)90052-6
19. Karahan IH (2013) Effects of pH value of the electrolyte and glycine additive on formation and properties of electrodeposited Zn–Fe coatings. Sci World J 2013:1–7. https://doi.org/10.1155/2013/273953
20. Thangaraj V, Hegde AC (2007) Electrodeposition and compositional behaviour of Zn–Ni alloy. Indian J Chem Technol 14:246–252

21. Rashwan SM, Mohamed AE, Abdel-Wahaab SM, Kamel MM (2003) Electrodeposition and characterization of thin layers of Zn–Co alloys obtained from glycinate baths. J Appl Electrochem 33:1035–1042. https://doi.org/10.1023/A:1026280109296
22. Rashwan SM (2005) Study on the behaviour of Zn–Co–Cu alloy electroplating. Mater Chem Phys 89:192–204. https://doi.org/10.1016/j.matchemphys.2003.10.019
23. Chentsova EV, Solov'eva ND, Pochkina SY, Terin EV (2020) An effect of parameters of the reverse electrolysis mode on composition and properties of zinc–nickel–cobalt alloy. Russ J Appl Chem 93:369–374
24. Kanani N (2004) Electroplating: basic principles, processes and practice. Elsevier
25. Kovalska N, Pfaffeneder-Kmen M, Tsyntsaru N, Mann R, Cesiulis H, Hansal W, Kautek W (2019) The role of glycine in the iron-phosphorous alloy electrodeposition. Electrochim Acta 309:450–459. https://doi.org/10.1016/j.electacta.2019.03.203
26. Sotskaya NV, Dolgikh OV (2008) Nickel electroplating from glycine containing baths with different pH. Prot Met 44:479–486. https://doi.org/10.1134/S0033173208050123
27. Ibrahim MAM, Al Radadi RM (2015) Role of glycine as a complexing agent in nickel electrodeposition from acidic sulphate bath. Int J Electrochem Sci 10:4946–4971
28. Parthasaradhy NV (1989) Practical electroplating handbook. Prentice-Hall Inc., USA

A Study of Electrodeposited Gold Thin Films Using a Confocal Laser Scanning Microscope

Vaishali Umrania and Sharad Shukla

1 Introduction

Gold plating is used for microwave modules mainly to impart engineering properties. Gold plating is carried out on Aluminum alloys, Kovar, Invar, and SS 304 for imparting engineering properties namely electrical conductivity, corrosion resistance, solderability, electromagnetic interference (EMI)/ electromagnetic conductance (EMC), etc. The pulsed reverse current (PRC) plating is a multifaceted technique for formulating nano-ordered metals, alloys, and composite film formation [1–4]. Theoretical considerations of the PRC deposition process are stated in many reports [5–12]. There are a lot of studies on silver pulse plating [2, 13–16] but for reverse pulse gold plating, very less literature is found. The quality of gold plating is controlled by gold concentration, pH, temperature, current density, current waveform, mechanical agitation, anode–cathode ratio, etc. [17].

In this work, a process series is established for gold plating on Al 6061 T6 [18]. Acidic gold cyanide plating chemistry $K[Au(CN)_2]$ is used owing to its better plating rate, high current densities, uniform deposit distribution, and stability. The present study is aimed at investigating the effect of current density parameters on plating surface finish. The electrodeposition mechanism is better understood by using pulse current as a tool [19]. This method of deposition can improve the deposit properties. PC and PRC play a major role in determining the surface roughness and structure of gold deposits. PRC deposits demonstrated superior performance than DC for certain electronic applications [20]. Composite materials, semiconductors, metals, and alloys can be deposited by various combinations of the forward and reverse pulse

V. Umrania (✉) · S. Shukla
Space Applications Centre (ISRO), Ahmedabad, Gujarat, India
e-mail: vaishali@sac.isro.gov.in

S. Shukla
e-mail: sharadshukla@sac.isro.gov.in

U. K. Mudali et al. (eds.), *Recent Trends in Electrochemical Science and Technology*, Springer Proceedings in Materials 15,
https://doi.org/10.1007/978-981-16-7554-6_7

parameters. Enhanced characteristic materials can be obtained by careful control of mass transport and reaction kinetics [8]. Gold thin film is deposited from cyanide bath with pH below 7. Direct and square wave pulse currents have been experimented with. Pulse (PC and PRC) electrodeposition is applied to produce a deposit with a finer grain size in comparison with direct current (DC) electrodeposition.

Confocal Laser Microscope (CLM) is superior with respect to conventional optical microscopes. It can perform diversified measurements such as surface roughness measurement, height gap, and shape measurement [21]. In a confocal beam path, CLM uses laser light to capture small optical segments of samples and are compiled to form 3D images [22]. The surface morphology, the surface roughness of the gold thin films deposited by DC, PC, and PRC were analyzed by CLM.

2 Experimental

Al alloy 6061-T6 specimens (dimension: 100 mm × 25 mm × 1 mm) having three delta surface finish were used for gold plating. The process sequence was soap solution cleaning, acid cleaning, alkaline zincating, autocatalytic nickel plating, and top layer gold plating. Aluminum comes in the category of difficult-to-plate material because it has a natural oxide film in its stable state. It is very difficult to remove it and specific pre-treatment is required prior to electroplating. Standard guidelines as per ASTM B 253–87 were followed for pre-treatment of aluminum alloy 6061 T6 [23]. All the samples which were to be plated by DC, PC, and PRC method were electroless nickel plated together in one batch to keep similarity in the surface finish by electroless nickel plating. The coatings were produced in an electrolyte based on $K[Au(CN)_2]$ and citric acid/sodium citrate buffer. Plating was done for 30 min at an average current density of 2 A per square feet to get gold plating thickness of 2 ± 0.2 μm. Bath parameters such as gold concentration, temperature, and pH were studied in earlier work [17]. The plating bath parameters are given in Table 1.

The present investigation was focused on the relative study of surface finish of DC, PC, and PRC gold thin films on Al alloy 6061 T6. CLM microphotographs of gold thin films on aluminum specimens with various power supply regimes (DC, PC, and PRC) were recorded. Various parameters were monitored, such as pulse parameters, current graph, plating thickness, CLM image, and particle size. As per the theory described in the literature [5], sample parameters were decided such that the average current remains constant. Various pulse duty cycles of 1–100%, at frequencies 10–100 Hz with a variable current density of 10–100 A per square feet (ASF) were applied, and

Table 1 Plating bath parameters

S. No.	Parameters	Values
1	Gold concentration	10 ± 1 g/l
2	pH of the bath	4.9 ± 0.5
3	Temperature of the bath	65 ± 1 °C

its effect on particle size was studied. Approximately 80 samples were prepared using different PRC parameters to get desirable grain size and surface roughness. Pulse reverse power supply offers flexibility to choose forward and reverse current density, time, and duration separately. During PRC plating, current is swapped quickly amid different values, so pulses series of equal duration, polarity, and amplitude is formed. Each pulse is having zero current in between two successive pulses. During ON–time current is applied and during OFF—time current is not applied. So small grains are formed which can pack together in a better way resulting in less porous and fine-grained deposits [24].

In the case of DC, only one parameter can be selected, which is the current density of 1–3 ASF. In the case of PC total of seven parameters can be selected. In PRC, there is flexibility to choose 17 parameters. The range of parameters employed in this study for PC and PRC are given in Table 2. The various parameters involved in the electrodeposition have been studied for revealing the relation with the microstructure in order to improve the process of synthesizing nanostructured gold. The rate of metal deposition in PC might be the same as DC for identical current parameters. At a low duty cycle, a high current was required to obtain the same average deposition rate as DC and a minimum of 10–50% duty cycles was required to get PC. At high duty cycles, PC gives almost the same characteristic deposition as DC. During ON—time fully charged double layer was not formed, if frequency is high. In case of less OFF–time, double layer didn't get enough time to discharge [5]. So, lower duty cycles, lower frequency, and higher off times were applied. Thin film with grain size down to

Table 2 Parameters for PC and PRC

S. No	Parameters	PC	PRC
1	Forward on time	0.1–1 ms	0.1–1 ms
2	Forward off time	0–9 ms	0–9 ms
3	No. of forward cycle	1–30	1–30
4	Forward duration	1–300 ms	1–300 ms
5	Reverse on time	–	0.1–1 ms
6	Reverse off time	–	0–9 ms
7	No. of reverse cycle	–	1–10
8	Reverse duration	–	0–60 ms
9	Frequency	–	10–100 Hz
10	Forward current density	10–100 ASF	10–100 ASF
11	Reverse current density	–	0–250 ASF
12	Average current density	1–8 ASF	1–8 ASF
13	Current ratio	–	0–3
14	Positive duty cycle	–	1–100
15	Negative duty cycle	–	0–50
16	On duty cycle	1–100	1–100
17	Cycle time	1–100 ms	1–100 ms

a few nanometers was produced. The effects of frequency and duty cycle on coating thickness, morphology, and particle size were studied. After finalizing parameters for DC, PC, and PRC plating, six samples each designated as D1, D2, D3, D4, D5, and D6 were plated by DC, P1, P2, P3, P4, P5, and P6 were plated by PC and R1, R2, R3, R4, R5, and R6 were plated by PRC to get the average reading of surface roughness measurement. The topography, surface roughness, and surface morphology of DC, PC, and PRC gold thin films were studied by CLM analysis for the bare sample and plated samples. The samples were observed under a laser, and 3D photographs were taken at 17000X magnification.

3 Results and Discussion

The surface roughness of the specimen had a significant effect on coating surface appearance. Grain size reduction and smoother surfaces lead to a reduction in porosity. All 18 specimens were subjected to surface roughness measurement before and after DC, PC, and PRC gold plating. In an electroplating bath, oppositely charged ions are aligned near anode and cathode, forming an electrical double layer. For plating, positive ions must move through the solution to reach the cathode. During DC, double layer gets charged to a limited extent which creates a hindrance to upcoming ions to the substrate. When pulse current is applied, the positive ions get enough energy to break up electrical double layer and reach the cathode. Grain nuclei are initiated during off time in PC. So in unit area, more grains can be accumulated. At high current density, nucleation rate is also higher which results in a fine-grained thin film. It was evident that surface roughness was more after DC plating and decreased after PRC plating. Details of measurement are given in Table 3.

Table 3 Surface roughness data for DC, PC, and PRC plated samples

Specimen	Surface roughness Ra μm		Specimen	Surface roughness Ra μm		Specimen	Surface roughness Ra μm	
	DC plating process			PC plating process			PRC plating process	
	Before plating	After plating		Before plating	After plating		Before plating	After plating
D1	0.155	0.214	P1	0.200	0.195	R1	0.173	0.167
D2	0.172	0.203	P2	0.162	0.155	R2	0.187	0.157
D3	0.235	0.265	P3	0.235	0.229	R3	0.158	0.135
D4	0.202	0.186	P4	0.202	0.196	R4	0.209	0.184
D5	0.172	0.235	P5	0.192	0.188	R5	0.183	0.159
D6	0.180	0.251	P6	0.195	0.189	R6	0.207	0.185

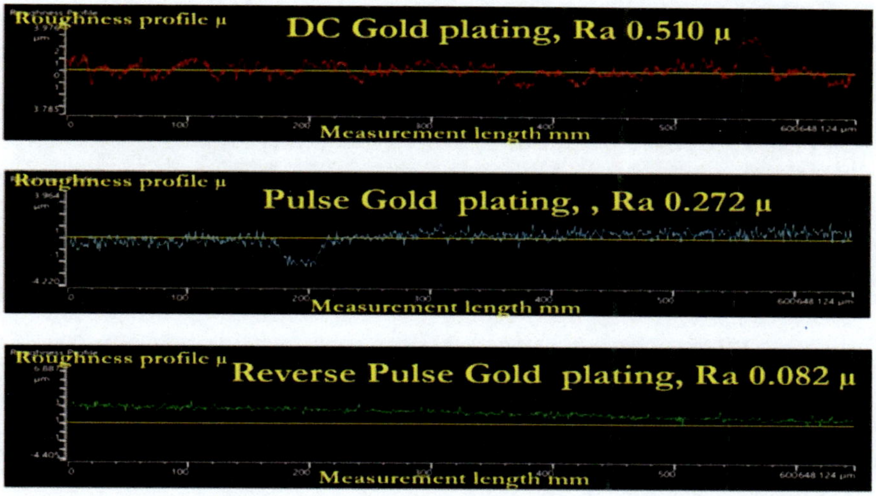

Fig. 1 Surface roughness analysis by CLM

Surface roughness profiles after DC, PC, and PRC gold plating are given in Fig. 1. The profile of reverse pulse plated coating was very smooth indicating lower roughness. The profile of pulse plating coating showed only one peak and the DC plated coating profile shows more peaks which is evident from its higher surface roughness.

No pores were visible in the CLM photographs of the gold thin film plated by PRC, and DC coating showed pores. The CLM photograph of the PRC plated gold thin film was denser than the DC. So, porosity is almost diminished in PRC plated thin film as evident in Figs. 2 and 3. It is clear that PRC gold plated sample showed smaller grain size and less porosity as compared to DC gold plated films.

3D Morphology indicated that PRC plated samples were more uniform than DC plated ones. The surface finish was better in PRC plating than DC plating (Fig. 4).

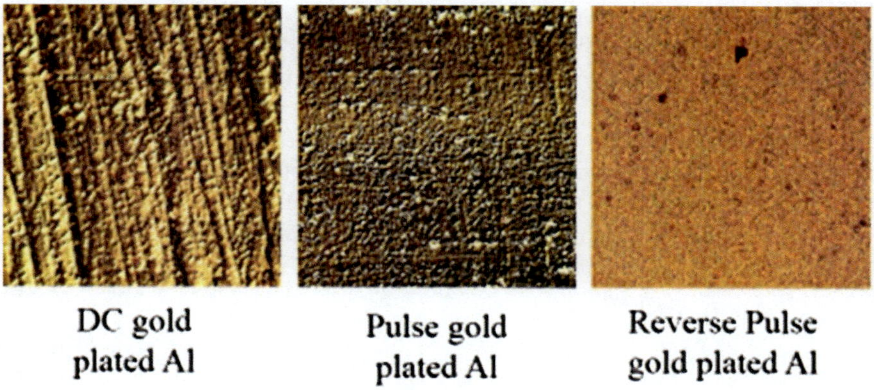

Fig. 2 Color images of gold plated specimens obtained by CLM

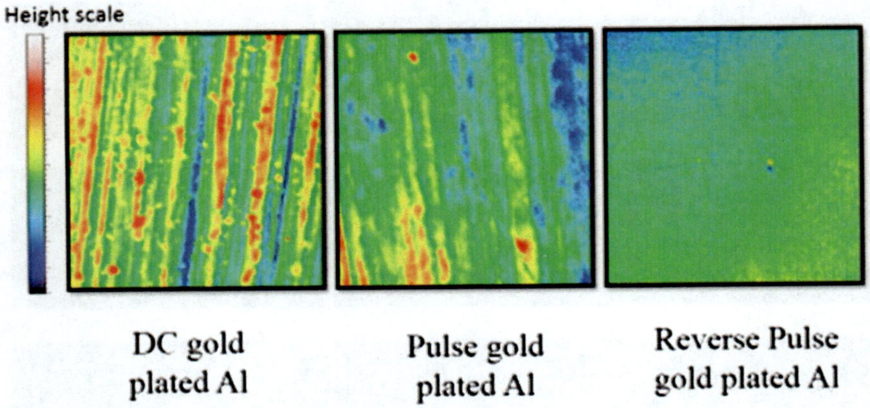

Fig. 3 Height analysis by CLM

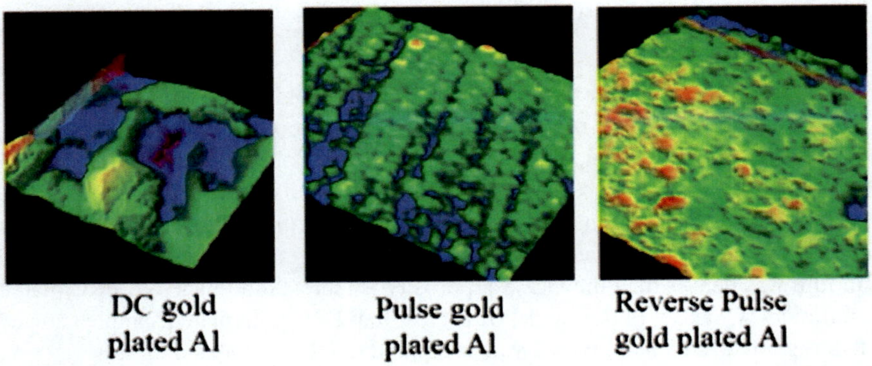

Fig. 4 3D Morphology images of DC, PC, and PRC gold plated Al

The grain size was measured from 3D morphology images. It was efficiently reduced to 25–50 nm for PRC film from 1300 to 2500 nm observed for DC film and 500–700 nm grain size was observed for PC films. In DC plating the voltage and current distributions are not always uniform, even if the work surface is planar. The concentration of the gold ions near the interface will also deplete over time. Thus the deposit may become non-uniform. In a pulsed deposition, between the deposition pulses, some time is offered to build the reactant concentration near the interface. In PRC, during anodic pulse, protruding material will preferentially etch, leading to a smooth surface in the final product.

4 Conclusions

An efficient plating method was developed and used for electroplating gold thin film on aluminum. Pulse reverse current (PRC) plating proved to be more advantageous than direct current (DC) plating for getting smoother and fine-grained deposits. The parameters were optimized by carrying out various experiments. From the experimental data, it has been concluded that less porous and fine-grained good quality deposits were obtained at lower duty cycles and lower frequency. Surface roughness was found to be significantly less in PRC film compared to DC and it was in between for PC film. Results indicated that PRC plated gold samples were more uniform and proved to be superior as compared to DC plated ones. This method of electroplating yields an improved finished appearance of metal surfaces.

Acknowledgements The authors appreciate the technical help received from colleagues during the course of experiments reported in this paper. The authors gratefully acknowledge the inspiration received from Shri Jaimin Desai, Group Director, EnTSG. SRA/PMQD-M team, for testing support and regards to reviewers for spending their valuable time in the evaluation. Sincere thanks to Shri A. Bhattacharya, DD ESSA, and Director-SAC for their kind approval of this paper.

References

1. Hassan MA, Hamid ZA, Nassif N (2014) Electrodeposition of silver nano particles on carbon sphere surfaces by pulse current. IJTRA 2:154–158
2. Dadvand N, Dadvand M (2014) Pulse electrodeposition of nanostructured silver-tungsten-COBALT oxide composite from a non-cyanide plating bath. J Electrochem Soc 161:D730–D735. https://doi.org/10.1149/2.0371414JES
3. Karami H, Babaei H (2012) Electroplating nanostructured hard gold through the pulse galvanostatic method. Int J Electrochem Sci 7:14
4. Chaparro A, Arnulfo W, Lopez EV (2007) Electrodeposition of nickel plates on copper substrates using PC y PRC. Matéria (Rio de Janeiro) 12:583–588. https://doi.org/10.1590/S1517-70762007000400006
5. Chandrasekar MS, Pushpavanam M (2008) Pulse and pulse reverse plating—conceptual, advantages and applications. Electrochim Acta 53:3313–3322. https://doi.org/10.1016/j.electacta.2007.11.054
6. Anqiang H, Barbara D, Akhlaghi S et al (2002) Pulse plating of gold-tin alloys for mieroeleetronie and optoelectronic applications. Plat Surf Finish 89:48–531
7. Devaraj G, Guruviah S, Seshadri SK (1990) Pulse plating. Mater Chem Phy 25:439–461. https://doi.org/10.1016/0254-0584(90)90111-M
8. Green TA, Roy S (2017) Application of a duplex diffusion layer model to pulse reverse plating. Trans IMF 95:46–51. https://doi.org/10.1080/00202967.2016.1214354
9. Hager HE (1986) Current levelling behaviour under pulse current electroplating conditions: The role of reaction pseudocapacitance. J Appl Electrochem 16:189–195. https://doi.org/10.1007/BF01093350
10. Bojkov C, Davis K, Krumnow M (2003) System and method for pulse current plating. Patent Application Publication, US, Pub No (US20040140219A1)
11. Cao Q-D, Fang L, Lv J-M, Zhang X-P, Dat NT (2018) Effects of pulse reverse electroforming parameters on the thickness uniformity of electroformed copper foil. Trans IMF 96:108–112. https://doi.org/10.1080/00202967.2018.1423736

12. Mohan S, Raj V (2005) A comparative study of DC and pulse gold electrodeposits. Trans IMF 83:72–76. https://doi.org/10.1179/002029605X29393
13. Shanthi C, Barathan S, Jaiswal R, Arunachalam RM, Mohan S (2008) The effect of pulse parameters in electro deposition of silver alloy. Mater Lett 62:4519–4521. https://doi.org/10.1016/j.matlet.2008.08.032
14. Cheng D-H, Su Y-T, Li K-J, Cao T-H, Zhang W (2005) Two-direction pulse plating process for cyanide-free silver plating. J Mater Protect 7:009
15. Jayapoorani S, Ghosh DK (2012) Effect of Pulse Reverse Plating Using Silver On Printed Circuit Boards. In: International Conference & Workshop on Recent Trends in Technology (TCET). pp 25–27
16. Vaghela V, Shah DNK, Limaye VV (2017) Enhancement in surface morphology by Pulsed electro deposition of pure Silver on Al alloy. Int J Sci Res Educ 5
17. Umrania V, Shah NK (2017) Optimization of gold deposition by pulse currents for optoelectronic devices. Indian journal of technology and education, Birla Vishvakarma Mahavidyalaya
18. Shukla S, Gomathi N, George R (2014) Autocatalytic silver-plating of aluminium radio frequency waveguides with autocatalytic nickel as the undercoat for space applications. Surf Topography: Metro Prop 2:045004. https://doi.org/10.1088/2051-672X/2/4/045004
19. Ch JR (1977) The Electrodeposition of gold by pulse review by Raub and Knödler of the application of the technique to the plating of gold. Gold Bulletin 10:38–44
20. Sondhi P, Stine K (2020) Electrodeposition of nanoporous gold thin films. IntechOpen. https://doi.org/10.5772/intechopen.94604
21. Lasertec Corporation (2018) Principle of confocal optics. About confocal microscope Lasertec Corporation.htm. In: www.lasertec.co.jp. Accessed 1 Mar 2020
22. Overview of Confocal Laser Scanning Microscopes. https://www.zeiss.com/microscopy/int/products/confocal-microscopes.html. Accessed 10 Mar 2018
23. ASTM B253-87(2010) Standard Guide for Preparation of Aluminum Alloys for Electroplating. https://www.astm.org/DATABASE.CART/HISTORICAL/B253-87R10.htm. Accessed 7 May 2018
24. Mordechay S, Milan P (2010) Modern electroplating, 5th edn. Wiley Inter Science

One-Pot Solvothermal Synthesis of Spinel $MgFe_2O_4$ Nanoparticles as a Promising Cathode Material for Rechargeable Mg-ion Battery

Vinay Gangaraju, Kunal Roy, Mahesh Shastri, Navya Rani Marilingaiah, Manjunath Shetty, Hiroaki Kobayashi, Takaaki Tomai, Prasanna D. Shivaramu, and Dinesh Rangappa

1 Introduction

In the last few years, a major part of the research in the chemical and electrochemical field has shown substantial interest and attention towards the development of novel rechargeable batteries as alternative energy storage systems to the lithium-ion battery technology [1–6]. Beyond the rechargeable Li-ion battery, currently, Mg-ion rechargeable batteries are getting much attention to alleviate energy storage needs due to their high natural abundance, safety, high specific capacity (2205 Ah kg^{-1}), and especially high volumetric energy density (3833 mAh cm^{-3}) [7–9]. Due to the divalency of Mg-ion, it can phenomenally lead to a large theoretical capacity by the redox reaction of Mg/Mg^{2+} [10]. Compared to Li-ion battery systems, the development of multivalent Mg-ion battery system could not make significant advancement previously because of two reasons: (a) kinetically baffled intercalation and Mg-ions diffusion within the lattice of host materials and (b) discordance of a high electrolytic window with the metallic anode and high-voltage cathodes [11–13].

V. Gangaraju · K. Roy · M. Shastri · M. Shetty · P. D. Shivaramu · D. Rangappa (✉)
Department of Applied Sciences (Nanotechnology), Centre for Post Graduate Studies, Visvesvaraya Technological University, Muddenahalli, Chikkaballapura 562101, India
e-mail: dinesh.rangappa@vtu.ac.in

N. R. Marilingaiah
Department of Basic & Applied Science, Dayananda Sagar University, Kumaraswamy Layout, Bengaluru 560078, India

H. Kobayashi · T. Tomai
Institute of Multidisciplinary Research for Advanced Materials, Tohoku University, Sendai 980-8577, Japan
e-mail: h.kobayashi@tohoku.ac.jp

T. Tomai
e-mail: tomai@tagen.tohoku.ac.jp

U. K. Mudali et al. (eds.), *Recent Trends in Electrochemical Science and Technology*, Springer Proceedings in Materials 15,
https://doi.org/10.1007/978-981-16-7554-6_8

So far, comparatively low capacities and cell potentials have been used in the case of intercalation cathodes besides the significant capacity declining over few cycles. The main factors behind these problems include slow diffusion of Mg^{2+}- ions by increasing the electrode polarization, which results in ion-trapping on the electrode, and co-intercalation of solvent molecules causing the severe structural distortion and decomposition of cathode material and instability of intercalates. Another problem is the formation of passivation layer solid-electrolyte interface (SEI) with Mg^{2+} transportation, which hinders the cell performance efficiency. Therefore, future research directions mainly involve the structural modifications of intercalation as well as conversion electrodes to improve cell efficiency [5]. Thus, $MgFe_2O_4$ spinel structured oxide can be an attractive electrode material for the application of Mg-ion batteries having a theoretical capacity of 804 mAh g^{-1} because the weak MgO does not get reduced during the discharge process of the cell [14–20].

Among various compounds that have been extensively studied as cathode materials for Mg-ion battery, metal sulfides with Chevral phase such as Mo_6S_8 are one of them [7]. But their low terminal voltage (1.0–1.2 V) limits their applicability as a future generation electrode material. Despite the potentially high energy density of some metal oxides (like V_2O_5, MoO_3, MnO_2, etc.), slow Mg^{2+} ion diffusivity makes their rate capability also low [21]. While discussing ion diffusivity, spinel metal oxides (AB_2O_4) have attracted growing interest because of their high diffusivity [22, 23]. To date, extensive work has been carried out on spinel metal oxides for Li-ion batteries, Mg-ion batteries, Na-ion batteries, etc., but very few of them were explored as cathode materials. Recently, Yokozaki et al. [24] have reported the spinel $MgMn_2O_4$ that can exhibit a theoretical capacity up to 540 mAh g^{-1} for both Mn^{3+}/Mn^{4+} and Mn^{3+}/Mn^{2+} redox transformation at the electrolyte window of < 3.5 V (vs. Mg/Mg^{2+}), while a discharge capacity of 60 mAh g^{-1} with 80% capacity retention have been found. Thackeray et al. [25] synthesized $LiMn_2O_4$ for the first time that exhibited outstanding electrochemical performance as a cathode during lithium intercalation with a terminal voltage of 3.0–4.0 V (vs. Li-metal) [26]. Moreover, it has been observed that the properties of spinels can be tuned by the partial replacement or substitution of Mn with Ni, Co, or Cr to increase the voltage up to 5.0 V [27–30].

In the past two decades, several investigations have been carried out on Mg-ion batteries for the development of new electrode material. Various spinel metal oxides including MgM_2O_4 (M = Mn, Fe, Ni, Co, etc.) are studied for the advancement of the electrode material. Several researchers have extensively studied the design and electrochemical performance analysis of spinel $MgMn_2O_4$ (capacity up to 270 mAh g^{-1}) [2, 24, 31], $MgCo_2O_4$ (capacity of 124 mAh g^{-1}) [32], and $Mg_{2/3}Ni_{4/3}O_2$ (capacity of 69.8 mAh g^{-1}) [32] as cathode materials for Mg-ion battery. Various spinel ferrites have been studied from their molecular and theoretical point-of-view including Fe_3O_4 [33–35], $ZnFe_2O_4$, and $MgFe_2O_4$, and reported that there would be a significant change in delivered capacity and structural deformation due to different divalent cation occupancy in the spinel.

Thus, the above-mentioned material has advantages and structural similarities with $MgCo_2O_4$ and $MgMn_2O_4$ (efficient cathode material for Mg-ion battery, as reported by Truong et al. [24, 32]). This has motivated us to design and analyze the performance of spinel $MgFe_2O_4$ as an emerging cathode material for Mg-ion batteries. One of the most effective methods to improvise the cathodic performance in an electrochemical cell is by controlling the size of the nanoparticles for boosting ion transportation into the host lattice [36]. Thus the hydrothermal and solvothermal processes are of great interest on account of their simplicity and efficiency, with good control over target particle sizes and morphology [37, 38]. Moreover, there are no reports on the one-pot solvothermal synthesis of $MgFe_2O_4$ as a cathode material for the Mg-ion battery available to date. Hence, in this investigation, the one-pot facile solvothermal synthesis of $MgFe_2O_4$ nanoparticles has been explored. Simultaneously, the cathode performance is analyzed by conversion reaction phenomena for the first time that can pave way for the application of $MgFe_2O_4$ as a cathode in a rechargeable Mg-ion battery.

2 Experimental Section

2.1 Materials

Required materials for the synthesis of $MgFe_2O_4$ nanoparticles (NPs) are $FeCl_3.6H_2O$ and $Mg(NO_3)_2{\cdot}6H_2O$ (Lobachemie Laboratory Reagents & Fine Chemicals) were used as major precursor salts. NaOH (ChemLabs) (3.5 M) and sodium dodecyl sulphate (SDS, Merck) (21 mM) were used as reducing agents and capping agents, respectively with ethylene glycol (SD Fine Chemicals Ltd.) (40 mL) as a solvent. All the chemicals used for this one-pot solvothermal synthesis were of analytical grade.

2.2 Preparation of $MgFe_2O_4$ Nanoparticles by One-Pot Solvothermal Method

During the synthesis of $MgFe_2O_4$, initially, the above-mentioned precursor salts were taken in a stoichiometric ratio of 1:2 (Mg:Fe) in different molar concentrations. Then, the precursor materials were added individually in 40 mL of ethylene glycol solvent. After adding both precursors (hydrated Mg-nitrate and Fe-chloride), the mixture was stirred for 1 h. Later, NaOH and SDS were added at conc. of 3.5 M and 21 mM, respectively, and stirred for another 1 h. The SDS was added as a capping agent in that solution. After 1 h, the whole homogeneous mixture was transferred into a Teflon-lined stainless steel autoclave by keeping the autoclave in an oven at 200 °C for 12 h [15]. After 12 h of reaction, the precipitate was magnetically separated from the upper solvent layer and washed several times with DI water and ethanol to remove the impurities. Then the washed precipitate was kept in a vacuum drying oven at 60 °C for 6 h.

2.3 Electrode Fabrication

The electrochemical performance of $MgFe_2O_4$ was analyzed using CR2032 type coin cell. The working electrode, i.e., cathode consisted of cathode material ($MgFe_2O_4$), commercial polyvinylidene-fluoride (PVDF as a binder), and acetylene black (as conductive material) in the ratio of 8:1:1 (based on wt %). These three materials were first mechanically ground using traditional mortar-pestle to prepare electrode paste by adding N-methyl 2-pyrrolidone (NMP)) as solvent as per the requirement. Then the prepared slurry was coated uniformly on the aluminium sheet and kept inside the oven for drying at 80 °C overnight. The coated cathode sheet was then cut into circular discs. Similarly, the anode was also prepared in the same manner consisting of Maxsorb (highly activated carbon), the polytetrafluoro ethylene (PTFE) (as a binder), and acetylene black (as conductive material) in the ratio of 8:1:1 without using solvent and rolled into anode sheet, kept for drying in an oven. After electrode preparation, the cell was assembled in the Ar-filled glove box. Polypropylene film (thickness of 20–25 μm) was used as a separator in between cathode and anode and 0.5 M of $Mg(OCl_4)_2$ in acetonitrile ($Mg(OCl_4)_2$/AN) was used as an electrolyte. After that, the fabricated cell was kept for charge–discharge cycling performance measurement on a multichannel battery tester (Arbin Work-Station).

2.4 Instrumentation and Characterization

The structural information of the synthesized powder was analyzed by powder X-ray diffractometry (XRD) (Rigaku Ultima 4 at 40 kV and 30 mA) using CuKα irradiation having a wavelength of 1.54 Å. Data were collected at a diffraction angle of 2θ with a step size of 0.02° and the average crystallite size (D_{avg}) was calculated using the following equation:

$$D_{avg} = 0.9\lambda/\beta \cos\theta \tag{1}$$

where λ is the wavelength of X-rays used (1.54060 Å), β is the full width at half maximum (FWHM), and θ is the angle of diffraction. To determine the thermal stability of $MgFe_2O_4$ in the range of 50–800 °C, thermogravimetric-differential thermal analysis (TG–DTA, PerkinElmer STA 8000) of $MgFe_2O_4$ was performed at a heating rate of 10 °C/min in N_2-flow. The morphology of the synthesized powder has been obtained using a scanning electron microscope (SEM, Hitachi 150). The compositional analysis of the synthesized powder has been obtained from Energy-dispersive spectroscopic (EDS) analysis using EDS analyzer (Thermax) attached with SEM. Transmission electron microscopy (TEM, Hitachi) and X-ray photo-electron spectroscopy (XPS, ULVAC-PHI5600)) data of the as-prepared samples were also obtained. Cyclic voltammetry (CV) and electrochemical impedance spectra (EIS) were obtained using an electrochemical system (Origa-Lys).

3 Results and Discussions

Based on the cation distribution in the crystal lattice, spinels (AB_2O_4) are categorized into two different models, i.e., normal spinels and inverse spinels. Usually, cubic close-packed systems are observed in normal spinels. Whereas, in inverse spinels, octahedral sites are occupied by all the A-cations and half of the B-cations because of the strong preference of A-cations towards octahedral sites according to Crystal Field Stabilization Theory (CFST) [39, 40]. In light of these recent studies, the XRD pattern of as-synthesized $MgFe_2O_4$ via the solvothermal method was recorded and displayed in Fig. 1. The XRD pattern was identified as the inverse cubic spinel structure with the space group of Fd-3 m. The cell parameters obtained from the crystallographic information from the XRD pattern characterization software were: $a = b = c = 8.388$ Å and $\alpha = \beta = \gamma = 90°$. These cell parameters are in good agreement with the previous report [41]. The major characteristic peak of the ferrite spinel is centered at $2\theta = 35.45°$ and the corresponding lattice plane is (3 1 1) providing evidence for the formation of ferrite spinel structure [42]. This observation is similar to the earlier reports [43, 44]. No residual α-Fe_2O_3 formation or other impurity phases were present in the sample. The average crystallite size of $MgFe_2O_4$ is 59.886 nm, obtained using Scherrer's formula (Eq. 1), corresponding to the lattice planes (2 0 2), (3 1 1), (4 0 0), (5 1 1), and (4 0 4).

Thermal stability of spinel $MgFe_2O_4$ has been evaluated by TG and DTA method measured in N_2-atmosphere as shown in Fig. 2. The gradual (initially) loss of weight up to 13.56% was observed below 300 °C, which can be attributed to the total removal of physiosorbed and inter-molecular water content, interlamellar anions, and

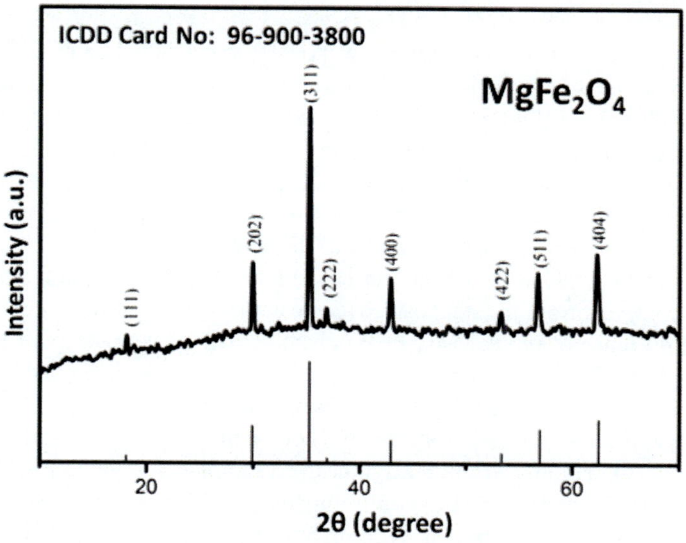

Fig. 1 XRD pattern of $MgFe_2O_4$ synthesized by one-pot solvothermal technique

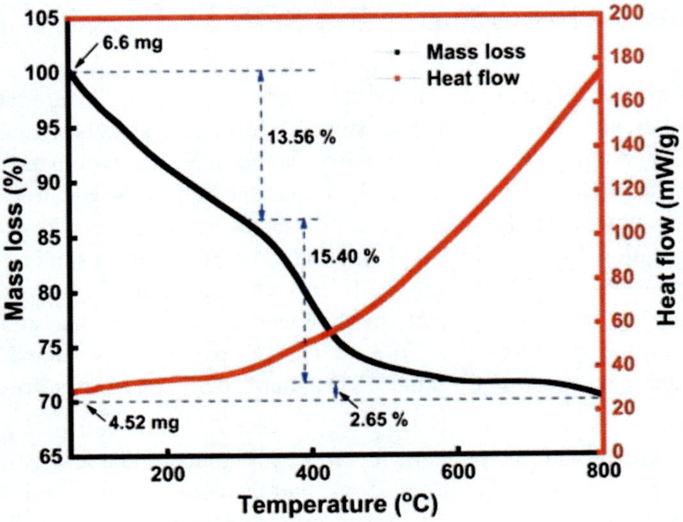

Fig. 2 TGA/DTA curves of as-synthesized $MgFe_2O_4$ synthesized at 200 °C for 12 h by ethylene glycol assisted solvothermal method

hydroxyl-molecules [45]. The small exothermic peak near 400 °C observed in the DTA curve and the next sudden weight loss of 15.4% in the TG curve, corresponding to the decomposition of existing solvent molecules (ethylene glycol) and continuous dehydroxylation [45, 46]. After 400 °C, there was a very small percentage of weight loss (2.65%) and continued to be stable almost throughout the temperature up to 800 °C because of pure $MgFe_2O_4$ phase formation by the conversion of unwanted metal oxides.

The morphology of the nano-$MgFe_2O_4$ powder was studied using SEM micrographs, which are shown in Fig. 3a, b. The SEM image shows the agglomerated particles with spherical shapes. The individual particle morphology was studied using TEM and the result is displayed in Fig. 3c, d. The TEM images clearly reveal the presence of very fine particles with a size below 50 nm with spherical shapes which are agglomerated. A close view of TEM images reveals that the particles are having almost uniform morphology and size distribution.

EDS spectral analysis has been further studied for the determination of the elemental composition of the synthesized ferrite with its phase purity. Figure 4 clearly shows the presence of regular Mg, Fe, O characteristic peaks with pure $MgFe_2O_4$ phase. Also, the EDS analysis provided an estimation of the elemental composition as per the composition mixed at the time of preparation.

XPS survey quantitatively explicates the compositional variance and metal ions valence state (oxidation state) in the synthesized samples. Figure 5a has shown the presence of Mg, Fe, and O elements confirming the pure $MgFe_2O_4$ phase. The appearance of C 1 s can be considered as hydrocarbon contamination with other elements [47]. In Fig. 5b, Fe 2p represents the existence of split-spin orbit components $2p_{1/2}$

Fig. 3 **a, b** SEM images and **c, d** TEM images of $MgFe_2O_4$ nanoparticles synthesized by one-pot solvothermal method

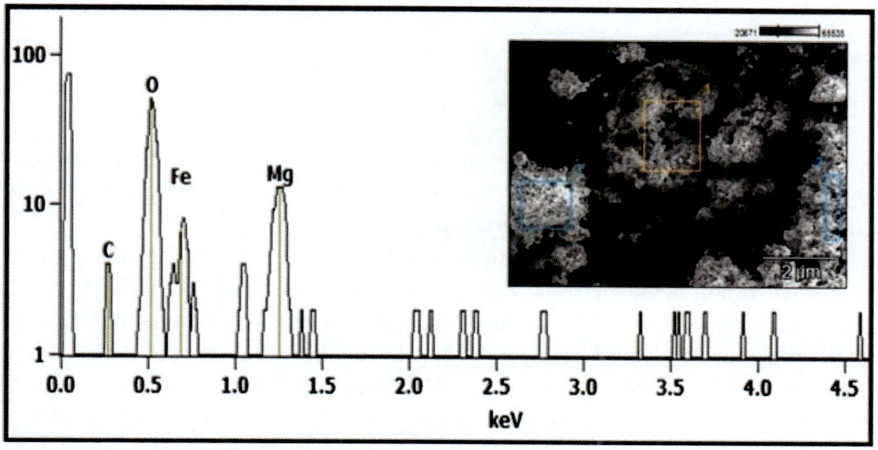

Fig. 4 EDS spectrum with SEM image on C-substrate

and $2p_{3/2}$ significantly at 709.04 and 723.09 eV, respectively. Moreover, the binding

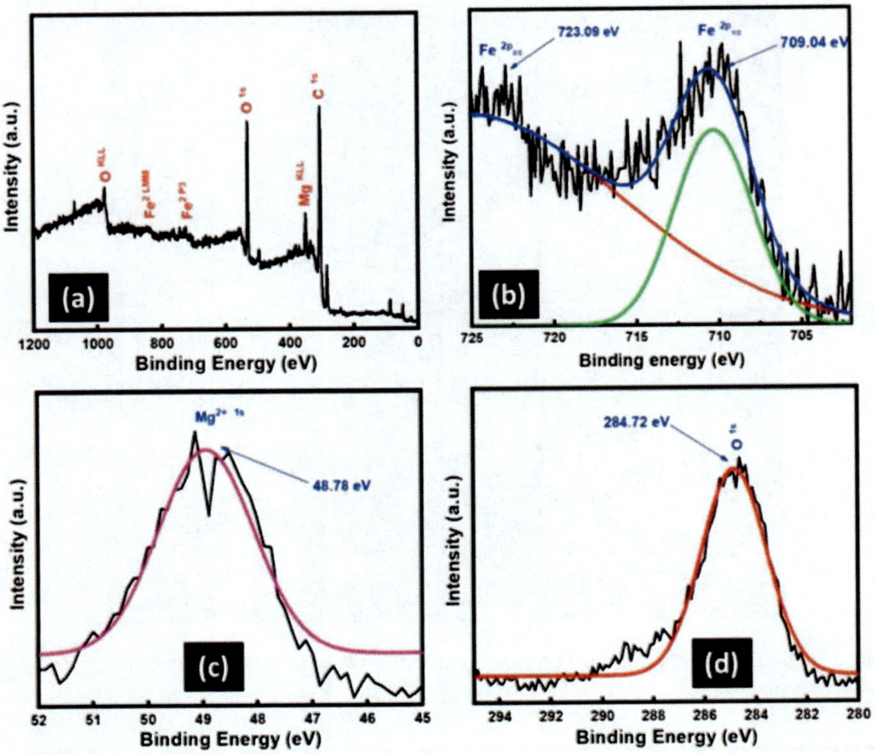

Fig. 5 **a** XPS survey scan of as prepared $MgFe_2O_4$; **b–d** High resolution XPS scan for Fe 2p, Mg 1 s, and O 1 s core levels, respectively

energy of 48.78 eV, displayed in Fig. 5c indicating the existence of Mg 1 s (comparable to literature), confirms the presence of Mg^{2+} state. Simultaneously, Fig. 5d indicates the presence of O 1 s at the binding energy of 284.72 eV.

The storage mechanism of electrode material ($MgFe_2O_4$) mainly depends on the magnetization-demagnetization process during the conversion reaction between spinel and rock-salt frame. This conversion reaction principle is influenced by the redox efficiency of electrolyte cation with electrode material in a particular potential window [48] as analyzed by cyclic voltammetry in Fig. 6a. CV measurements of $MgFe_2O_4$ against carbon are shown in Fig. 6a within the voltage window of −1.0 to 1.0 V at a different sweep rate of 5–25 mV s^{-1}. From Fig. 6a, two anodic peaks can be observed at −0.016 and 0.75 V which are attributed to the transformation of $MgFe_2O_4$ from its rock-salt ($Mg_2Fe_2O_4$) structure to spinel framework by a redox reaction between electrolyte and electrode (Eq. 4). Simultaneously, two cathodic peaks can be observed at 0.368 V and −0.462 V respectively, during the conversion reaction from spinel to rock-salt phase, which attributes to the formation of $Mg_2Fe_2O_4$ followed by the formation Fe_2O_4 and $(ClO_4)_2$ in Eqs. (2) and (3). This

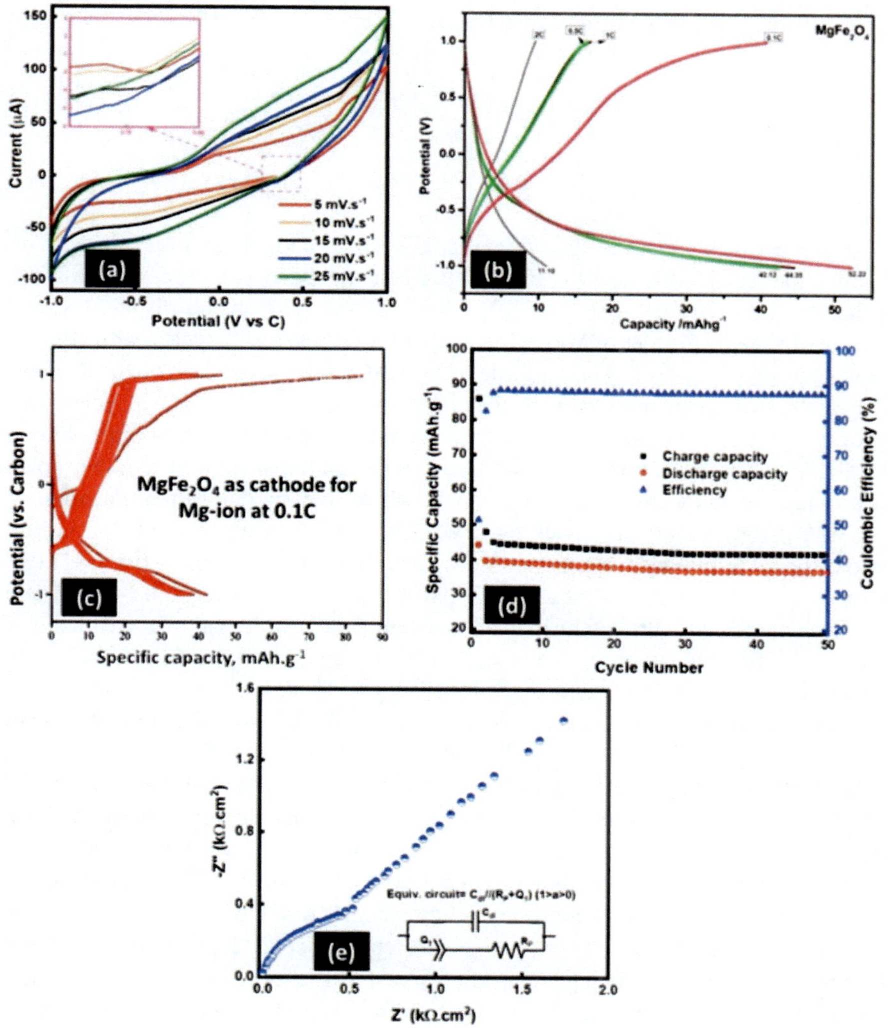

Fig. 6 **a** Cyclic voltammogram of as prepared $MgFe_2O_4$ for one cycle at different scan rates: 5, 10, 15, 20, and 25 mV s^{-1}, **b** specific capacities of $MgFe_2O_4$ at different charging rates from 0.1 to 1C for 1st cycle, **c** charge–discharge profile of $MgFe_2O_4$ at 0.1C current density, **d** cycling performance of $MgFe_2O_4$ up to 50 cycles at 0.1C current density and **e** electrochemical impedance spectra of $MgFe_2O_4$ scanned within the frequency range of 100 kHz to 1 Hz

phase conversion charging-discharging phenomena by redox transformation occur by the consumption of electrolyte cations (Mg^{2+}) to the active sites of electrode material and its reversibility [48]. Moreover, the increase in current peaks is also observed with increasing scan rate which indicates the good rate capability of the cell.

$$MgFe_2O_4 \rightleftharpoons Fe_2O_4 + Mg^{2+} + 2e^- \tag{2}$$

$$Mg(ClO_4)_2 \rightleftharpoons Mg^{2+} + (ClO_4)_2 + 2e^- \tag{3}$$

$$MgFe_2O_4 + Mg^{2+} + 2e^- \rightleftharpoons Mg_2Fe_2O_4 \tag{4}$$

The electrochemical performance of $MgFe_2O_4$ as cathode material for Mg-ion battery has been evaluated by galvanostatic charge/discharge process within the voltage window from −1.0 V to 1.0 V versus C-rate. Figure 6b displays typical charge–discharge profiles of $MgFe_2O_4$ at different C-rates for 1st cycles, while Fig. 6c represents the charge–discharge profile of $MgFe_2O_4$ at 0.1C only. Figure 6c displays the charge–discharge profiles up to 50 cycles, excluding the initial 2–3 cycles after getting stable cell voltage. The charging profiles showed an increase in potential in all the cases (Fig. 6b). But discharge profiles initially showed a fast drop in the voltage window from 1.0 V to −0.5 V, which may be due to the transition of $MgFe_2O_4$ from spinel to rock-salt framework by the Mg-insertion or magnetization into cathode. Simultaneously, the slow drop in voltage window from −0.5 to −1.0 V with discharge capacities of 52.22, 44.35, 42.12, and 11.10 mAh/g at 0.1–2.0 C-rate has been perceived. This can be indicated towards the surface reaction on cathode or electrolyte decomposition due to large volume expansion-contraction in the lattice structure of cathode with increasing C-rate [32, 49]. These high discharge capacities compared to charge capacities may be attributed to the Mg-ion insertion into spinel structure to transform into rock-salt phase [23, 50–54]. The discharge profiles at 0.1C, 0.5C, and 1C displayed a wide potential plateau near the potential window of -0.75 V with larger irreversible capacity loss due to lattice distortion in the cathode by magnesiation-demagnesiation. But the discharge plateau has been shifted to -1.0 V and got narrowed at 2C with negligible loss of irreversible capacity compared to other cases. Figure 6d illustrates the cycling stability of the $MgFe_2O_4$ as a cathode material. After initial 2-3 cycles, a sudden drop of specific capacity occurred and became stable near 40 $mAh.g^{-1}$ up to 50 cycles. This sudden reduction of specific capacity may be the cause of phase transition from spinel to rock-salt or vice-versa [32]. Consequently, the average coulombic efficiency was also stabilized after the initial 2-3 cycles, at 88%, which may be due to the formation of a solid-electrolyte interface. The common EIS (electrochemical impedance spectroscopy) equivalent circuit analysis for $MgFe_2O_4$ based Mg-ion battery in $Mg(ClO_4)_2$/AN electrolyte has been displayed in Fig. 6e where the equivalent circuit elements are R_P, C_{dl}, and Q_1 representing the polarization resistance, double-layer capacitance, and constant phase element respectively. It consists of a partial semicircle at higher frequency region and a linear line at lower frequency region which corresponds to charge transfer resistance and diffusion resistance (together polarization resistance) respectively [55]. It was speculated that due to lower conductivity of $MgFe_2O_4$ led to lower Mg^{2+} ion diffusion through SEI. Further, polarization resistance may also enhance this effect and become the cause of average specific capacity reduction of the cell after the initial few cycles.

4 Conclusions

The present work successfully explicated the one-pot facile synthesis of single-phase $MgFe_2O_4$ nanocrystals by solvothermal technique. Due to the uniform morphology and high electrolyte-ions interfacial contact, its applicability as an emerging electrode material for rechargeable Mg-ion batteries has been elucidated. Herein, $MgFe_2O_4$ was used as a cathode material, following the principle of conversion type electrode by participating in large volume expansion reaction phenomena which simultaneously allowed the Mg^{2+}-ion diffusion into the lattice as well as structural distortion. Hence, at a lower current rate (0.1C), it showed a discharge capacity of 52.22 mAh/g in 1st cycle, but after the initial few cycles, capacity decreased and became stable near 40 mAh g^{-1} up to 50 cycles with the average coulombic efficiency of 88%. These phenomena indicate that the Mg-ion battery may take initial few cycles to get stabilized because of sluggish Mg^{2+}-diffusion through formed SEI. Moreover, at a higher current rate, $MgFe_2O_4$ may start to decompose, and thus, with increasing C-rate, specific capacity decreased to 11.10 mAh g^{-1} at 2C.

Acknowledgements This work is partially supported by the Vision Group on Science and Technology, Department of Information Technology, Biotechnology and Science & Technology, Government of Karnataka under grant No. VGST/CESEM/2012-13/182. Transmission electron microscopy (TEM) and X-ray photoelectron spectroscopic (XPS) characterization of the synthesized sample were analyzed at the Institute of Multidisciplinary Research on Advanced Materials, Tohoku University, Japan.

References

1. Liu Y, Zhang N, Yu C et al (2016) $MnFe_2O4$@C Nanofibers as high-performance anode for sodium-ion batteries. Nano Lett 16:3321–3328. https://doi.org/10.1021/acs.nanolett.6b00942
2. Truong QD, Kempaiah Devaraju M, Tran PD et al (2017) Unravelling the surface structure of $MgMn_2O_4$ cathode materials for rechargeable magnesium-ion battery. Chem Mater 29:6245–6251. https://doi.org/10.1021/acs.chemmater.7b01252
3. Lin MC, Gong M, Lu B et al (2015) An ultrafast rechargeable aluminium-ion battery. Nature 520:324–328. https://doi.org/10.1038/nature14340
4. Ponrouch A, Frontera C, Bardé F, Palacín MR (2016) Towards a calcium-based rechargeable battery. Nat Mater 15:169–172. https://doi.org/10.1038/nmat4462
5. Muldoon J, Bucur CB, Gregory T (2014) Quest for nonaqueous multivalent secondary batteries: magnesium and beyond. Chem Rev 114:11683–11720. https://doi.org/10.1021/cr500049y
6. Tepavcevic S, Liu Y, Zhou D et al (2015) Nanostructured layered cathode for rechargeable Mg-ion batteries. ACS Nano 9:8194–8205. https://doi.org/10.1021/acsnano.5b02450
7. Aurbach D, Lu Z, Schechter A et al (2000) Prototype systems for rechargeable magnesium batteries. Nature 407:724–727. https://doi.org/10.1038/35037553
8. Wan LF, Prendergast D (2014) The solvation structure of mg ions in dichloro complex solutions from first-principles molecular dynamics and simulated X-ray absorption spectra. J Am Chem Soc 136:14456–14464. https://doi.org/10.1021/ja505967u
9. Carter TJ, Mohtadi R, Arthur TS et al (2014) Boron clusters as highly stable magnesium-battery electrolytes. Angew Chemie Int Ed 53:3173–3177. https://doi.org/10.1002/anie.201310317

10. Novák P, Imhof R, Haas O (1999) Magnesium insertion electrodes for rechargeable nonaqueous batteries—a competitive alternative to lithium? Electrochim Acta 45:351–367. https://doi.org/10.1016/S0013-4686(99)00216-9
11. Lu Z, Schechter A, Moshkovich M, Aurbach D (1999) On the electrochemical behavior of magnesium electrodes in polar aprotic electrolyte solutions. J Electroanal Chem 466:203–217. https://doi.org/10.1016/S0022-0728(99)00146-1
12. Levi E, Gofer Y, Aurbach D (2010) On the way to rechargeable Mg batteries: the challenge of new cathode materials. Chem Mater 22:860–868. https://doi.org/10.1021/cm9016497
13. Muldoon J, Bucur CB, Oliver AG et al (2012) Electrolyte roadblocks to a magnesium rechargeable battery. Energy Environ Sci 5:5941–5950. https://doi.org/10.1039/C2EE03029B
14. Poizot P, Laruelle S, Grugeon S et al (2000) Nano-sized transition-metal oxides as negative-electrode materials for lithium-ion batteries. Nature 407:496–499. https://doi.org/10.1038/35035045
15. Pan Y, Zhang Y, Wei X et al (2013) $MgFe_2O_4$ nanoparticles as anode materials for lithium-ion batteries. Electrochim Acta 109:89–94. https://doi.org/10.1016/j.electacta.2013.07.026
16. Sivakumar N, Gnanakan SRP, Karthikeyan K et al (2011) Nanostructured $MgFe_2O_4$ as anode materials for lithium-ion batteries. J Alloys Compd 509:7038–7041. https://doi.org/10.1016/j.jallcom.2011.03.123
17. Bock DC, Tallman KR, Guo H et al (2020) (De)lithiation of spinel ferrites Fe_3O_4, $MgFe_2O_4$, and $ZnFe_2O_4$: a combined spectroscopic, diffraction and theory study. Phys Chem Chem Phys 22:26200–26215. https://doi.org/10.1039/d0cp02322a
18. Kim C, Phillips PJ, Key B et al (2015) Direct observation of reversible magnesium ion intercalation into a spinel oxide host. Adv Mater 27:3377–3384. https://doi.org/10.1002/adma.201500083
19. Liu M, Rong Z, Malik R et al (2015) Spinel compounds as multivalent battery cathodes: a systematic evaluation based on ab initio calculations. Energy Environ Sci 8:964–974. https://doi.org/10.1039/c4ee03389b
20. Feng Z, Chen X, Qiao L et al (2015) Phase-controlled electrochemical activity of epitaxial Mg-spinel thin films. ACS Appl Mater Interfaces 7:28438–28443. https://doi.org/10.1021/acsami.5b09346
21. Yoo HD, Shterenberg I, Gofer Y et al (2013) Mg rechargeable batteries: an on-going challenge. Energy Environ Sci 6:2265–2279. https://doi.org/10.1039/C3EE40871J
22. Mao M, Gao T, Hou S, Wang C (2018) A critical review of cathodes for rechargeable Mg batteries. Chem Soc Rev 47:8804–8841. https://doi.org/10.1039/C8CS00319J
23. Okamoto S, Ichitsubo T, Kawaguchi T et al (2015) Intercalation and push-out process with spinel-to-rocksalt transition on Mg insertion into spinel oxides in magnesium batteries. Adv Sci 8:1500072. https://doi.org/10.1002/advs.201500072
24. Yokozaki R, Kobayashi H, Honma I (2020) Reductive solvothermal synthesis of $MgMn_2O_4$ spinel nanoparticles for Mg-ion battery cathodes. Ceram Int 47:10236–10241. https://doi.org/10.1016/j.ceramint.2020.10.184
25. Thackeray MM, David WIF, Bruce PG, Goodenough JB (1983) Lithium insertion into manganese spinels. Mater Res Bull 18:461–472. https://doi.org/10.1016/0025-5408(83)90138-1
26. Thackeray MM (1997) Manganese oxides for lithium batteries. Prog Solid State Chem 25:1–71. https://doi.org/10.1016/S0079-6786(97)81003-5
27. Kawai H, Nagata M, Kageyama H et al (1999) 5 V lithium cathodes based on spinel solid solutions $Li_2Co_{1+X}Mn_{3-X}O_8$: $-1 \leq X \leq 1$. Electrochim Acta 45:315–327. https://doi.org/10.1016/S0013-4686(99)00213-3
28. Liu D, Lu Y, Goodenough JB (2010) Rate properties and elevated-temperature performances of $LiNi_{0.5-x}Cr_{2x}Mn_{1.5-x}O_4$ ($0 \leq 2x \leq 0.8$) as 5 V cathode materials for lithium-ion batteries. J Electrochem Soc 157:A1269–A1273. https://doi.org/10.1149/1.3491365
29. Terada Y, Yasaka K, Nishikawa F et al (2001) In situ XAFS analysis of Li(Mn, M)2O4 (M = Cr Co, Ni) 5 V cathode materials for lithium-ion secondary batteries. J Solid State Chem 156:286–291. https://doi.org/10.1006/jssc.2000.8990

30. Amine K, Tukamoto H, Yasuda H, Fuiita Y (1996) A new three-volt spinel $Li_{1+x}Mn_{1.5}Ni_{0.5}O_4$ for secondary lithium batteries. J Electrochem Soc 143:1607. https://doi.org/10.1149/1.1836686
31. Truong QD, Kobayashi H, Nayuki K et al (2020) Atomic-scale observation of phase transition of $MgMn_2O_4$ cubic spinel upon the charging in Mg-ion battery. Solid State Ionics 344:115136. https://doi.org/10.1016/j.ssi.2019.115136
32. Truong QD, Kobayashi H, Honma I (2019) Rapid synthesis of $MgCo_2O_4$ and $Mg_{2/3}Ni_{4/3}O_2$ nanocrystals in supercritical fluid for Mg-ion batteries. RSC Adv 9:36717–36725. https://doi.org/10.1039/c9ra04936c
33. Yu S, Hong Ng VM, Wang F et al (2018) Synthesis and application of iron-based nanomaterials as anodes of lithium-ion batteries and supercapacitors. J Mater Chem A 6:9332–9367. https://doi.org/10.1039/C8TA01683F
34. Huang Y, Xu Z, Mai J et al (2017) Revisiting the origin of cycling enhanced capacity of Fe_3O_4 based nanostructured electrode for lithium ion batteries. Nano Energy 41:426–433. https://doi.org/10.1016/j.nanoen.2017.10.001
35. Bruck AM, Cama CA, Gannett CN et al (2016) Nanocrystalline iron oxide based electroactive materials in lithium ion batteries: the critical role of crystallite size, morphology, and electrode heterostructure on battery relevant electrochemistry. Inorg Chem Front 3:26–40. https://doi.org/10.1039/C5QI00247H
36. Arthur TS, Zhang R, Ling C et al (2014) Understanding the electrochemical mechanism of K-αMnO_2 for magnesium battery cathodes. ACS Appl Mater Interfaces 6:7004–7008. https://doi.org/10.1021/am5015327
37. Okubo M, Mizuno Y, Yamada H et al (2010) Fast Li-ion insertion into nanosized $LiMn_2O_4$ without domain boundaries. ACS Nano 4:741–752. https://doi.org/10.1021/nn9012065
38. Okubo M, Hosono E, Kim J et al (2007) Nanosize effect on high-rate Li-ion intercalation in $LiCoO_2$ electrode. J Am Chem Soc 129:7444–7452. https://doi.org/10.1021/ja0681927
39. Fracchia M, Manzoli M, Anselmi-Tamburini U, Ghigna P (2020) A new eight-cation inverse high entropy spinel with large configurational entropy in both tetrahedral and octahedral sites: synthesis and cation distribution by X-ray absorption spectroscopy. Scr Mater 188:26–31. https://doi.org/10.1016/j.scriptamat.2020.07.002
40. Zhandun VS, Nemtsev A V. (2021) Ab initio comparative study of the magnetic, electronic and optical properties of AB_2O_4 (A, B = Mn, Fe) spinels. Mater Chem Phys 259:124065. https://doi.org/10.1016/j.matchemphys.2020.124065
41. Sankaramahalingam A, Lawrence JB (2012) Structural, optical, and magnetic properties of $MgFe_2O_4$ synthesized with addition of copper. Synth React Inorganic Met Nano-Metal Chem 42:121–127. https://doi.org/10.1080/15533174.2011.609500
42. Cross WB, Afflcck L, Kuznetsov MV et al (1999) Self-propagating high-temperature synthesis of ferrites MFe2O4 (M = Mg, Ba Co, Ni, Cu, Zn); reactions in an external magnetic field. J Mater Chem 9:2545–2552. https://doi.org/10.1039/a904431k
43. Ahmed MA, Ateia E, Salem FM (2007) The effect of Ti^{4+} ions and gamma radiation on the structure and electrical properties of Mg ferrite. J Mater Sci 42:3651–3660. https://doi.org/10.1007/s10853-006-1349-0
44. Radwan FA, Ahmed MA, Abdelatif G (2003) Screening effect of Ti^{4+} ions on the electrical conductivity and thermoelectric power of Mg ferrite. J Phys Chem Solids 64:2465–2477. https://doi.org/10.1016/j.jpcs.2003.08.003
45. Aliyan N, Mirkazemi SM, Masoudpanah SM, Akbari S (2017) The effect of post-calcination on cation distributions and magnetic properties of the coprecipitated $MgFe_2O_4$ nanoparticles. Appl Phys A Mater Sci Process 123:446. https://doi.org/10.1007/s00339-017-1053-8
46. Bezgin F, Demirelli K (2016) Synthesis, characterization and thermal degradation kinetics of photoresponsive graft copolymers. J Thermoplast Compos Mater 29:1135–1150. https://doi.org/10.1177/0892705714563114
47. Baby JN, Sriram B, Wang SF, George M (2020) Effect of various deep eutectic Solvents on the sustainable synthesis of $MgFe_2O_4$ nanoparticles for simultaneous electrochemical determination of nitrofurantoin and 4-nitrophenol. ACS Sustain Chem Eng 8:1479–1486. https://doi.org/10.1021/acssuschemeng.9b05755

48. Jansi Rani B, Durga M, Ravi G et al (2018) Temperature-dependent physicochemical properties of magnesium ferrites ($MgFe_2O_4$). Appl Phys A Mater Sci Process 124:319. https://doi.org/10.1007/s00339-018-1749-4
49. Shimokawa K, Ichitsubo T (2020) Spinel–rocksalt transition as a key cathode reaction toward high-energy-density magnesium rechargeable batteries. Curr Opin Electrochem 21:93–99. https://doi.org/10.1016/j.coelec.2020.01.017
50. Ichitsubo T, Adachi T, Yagi S, Doi T (2011) Potential positive electrodes for high-voltage magnesium-ion batteries. J Mater Chem 21:11764–11772. https://doi.org/10.1039/c1jm11793a
51. Ichitsubo T, Okamoto S, Kawaguchi T et al (2015) Toward "rocking-chair type" Mg-Li dual-salt batteries. J Mater Chem A 3:10188–10194. https://doi.org/10.1039/c5ta01365h
52. Kotani Y, Ise R, Ishii K et al (2018) Enhanced electrochemical properties of $MgCo_2O_4$ mesocrystals as a positive electrode active material for Mg batteries. J Alloys Compd 739:793–798. https://doi.org/10.1016/j.jallcom.2017.12.315
53. Shimokawa K, Matsumoto H, Ichitsubo T (2018) Solvation-structure modification by concentrating $Mg(TFSA)_2$-$MgCl_2$-triglyme ternary electrolyte. J Phys Chem Lett 9:4732–4737. https://doi.org/10.1021/acs.jpclett.8b02209
54. Kitamura N, Tanabe Y, Ishida N, Idemoto Y (2019) The atomic structure of a $MgCo_2O_4$ nanoparticle for a positive electrode of a Mg rechargeable battery. Chem Commun 55:2517–2520. https://doi.org/10.1039/c8cc09713e
55. Harudin N, Osman Z, Majid SR et al (2020) Improved electrochemical properties of $MgMn_2O_4$ cathode materials by Sr doping for Mg ion cells. Ionics (Kiel) 26:3947–3958. https://doi.org/10.1007/s11581-020-03531-7

Methanol Electrooxidation Activity of Pt/C Catalyst Promoted by Ce-Gd-Zr-O Solid Solution

Shuchi Sharma, Malaya K. Sahoo, and G. Ranga Rao

1 Introduction

In a direct methanol fuel cell (DMFC), the chemical energy stored in liquid methanol is converted into electrical energy for a wide range of applications [1–4]. Methanol as fuel at the anode is better than its competitor H_2 as fuel in easy handling, convenient storage and distribution, and an easy method of production and purification [5]. The basic processes involved in a fuel cell are electrooxidation reactions at the anode and electro-reduction reactions at the cathode, which require electrocatalysts to speed up the reactions. However, there are chemical barriers to the commercial application of such cells which include methanol crossover and sluggish kinetic rates of methanol oxidation reaction (MOR) [6]. Platinum, which has been extensively explored as a MOR catalyst works with the initial adsorption of the methanol on its surface, followed by the addition of oxygen species leading to carbon dioxide formation as the final product. However, intermediate species like CO_{ads} are believed to poison the surface of Pt, which ultimately decreases the efficiency of the process by blocking the active sites of Pt [5, 7]. Bi-functional systems such as Pt-M, where M is another metal, metal oxides, metal sulfides, or metal carbides, can be used to overcome such problems. Metal oxides or carbides act as additional support systems, which help in scavenging the poisoning species CO_{ads} by adsorbing the oxyphilic species (-OH_{ads}). Support materials should have high surface area for dispersing stable metal nanoparticles, providing good conductivity, and retaining structural stability under electrochemical operations [8–11].

Further improvement in electrocatalytic performance can be encouraged by controlling the morphology of noble metal nanostructures and employing newer

S. Sharma · M. K. Sahoo · G. R. Rao (✉)
Department of Chemistry and DST- Solar Energy Harnessing Centre (DSEHC), Indian Institute of Technology Madras, Chennai 600036, India
e-mail: grrao@iitm.ac.in

U. K. Mudali et al. (eds.), *Recent Trends in Electrochemical Science and Technology*, Springer Proceedings in Materials 15,
https://doi.org/10.1007/978-981-16-7554-6_9

catalyst support materials. Over decades, carbon materials such as Vulcan carbon, carbon nanotubes, and reduced graphene oxide have been widely used as catalyst supports for Pt-based nano-materials in DMFCs [11–13]. The metal oxides/sulfides employed for the electrooxidation of methanol include MnO_2, CeO_2, Cu_2O, V_2O_5, TiO_2, MoO_3, MoS_2, SnO_2, and Nb_2O_5 [11, 12, 14–19].

In the present work, the preliminary results on the effect of composition-controlled solid solutions, $Ce_{0.8}Gd_{0.1}Zr_{0.1}O_{2-x}$ (CGZO-08) and $Ce_{0.3}Gd_{0.5}Zr_{0.2}O_{2-x}$ (CGZO-03), on methanol electrooxidation is reported. The methanol oxidation activity was measured in terms of onset potential, peak anodic current, electrochemical surface area, stability, and anitipoisoning activity. The experiments revealed that Pt-CGZO-03/C electrode is better than the Pt-CGZO-08/C electrode for methanol oxidation. Even a small amount of gadolinium doping in ceria lattice is expected to create intrinsic defects which can potentially increase the conductivity and electrocatalytic activity of the oxide matrix [20–22]. This concept is examined in this work.

2 Materials and Methods

2.1 Materials

Analytical grade cerium nitrate hexahydrate ($Ce(NO_3)_3{\cdot}6H_2O$ (99.9%, SD Fine, India)), gadolinium nitrate hexahydrate ($Gd(NO_3)_3.6H_2O$, 99.9%, Alfa Aesar), zirconium oxychloride hexahydrate ($ZrOCl_2.6H_2O$, 99%, Loba Chemie), cetyltrimethylammonium bromide (CTAB, $(C_{16}H_{33})N(C_5H_3)_3Br$, 99.9%, Sigma-Aldrich), urea ($CO(NH_2)_2$, 99%, Thomas-Baker), Vulcan XC-72R carbon (Cabot Corporation, USA), HCl (Rankem, India), Na_2CO_3, ethylene glycol (EG), isopropyl alcohol (SD Fine chemicals, India), chloroplatinic acid hexahydrate ($H_2PtCl_6{\cdot}6H_2O$, Aldrich, ACS reagent) with $\geq$37.50% Pt, Nafion solution (5 wt % solution in a mixture of lower aliphatic alcohols, Sigma-Aldrich) were used as received. Milli-Q water obtained from the Millipore Water Purification system was used in all the experimental procedures.

2.2 Preparation of CGZO-08 and CGZO-03/C Catalyst

The catalyst was synthesized with the method reported earlier [19]. Typically, various molar ratios of $Ce(NO_3)_3.6H_2O$, $Gd(NO_3)_3.6H_2O$, and $ZrOCl_2.6H_2O$ were dissolved in 100 mL of water. To the above solution, 10 mmol of CTAB was added as a structure-directing agent by stirring for 1 h. Then 80 mmol of urea was added as the precipitating agent and the solution was stirred for 3 h to get a homogeneous solution. The above solution was then subjected to hydrothermal treatment at 120 °C for 24 h in an electric oven. After cooling the autoclave to room temperature, the obtained precipitates were centrifuged, repeatedly washed with water, a mixture of water, and ethanol. The dried precipitates were calcined in air at 500 °C for 3 h with a heating ramp of 5 °C min^{-1} to obtain the final products $Ce_{0.8}Gd_{0.1}Zr_{0.1}O_{2\text{-}x}$ (CGZO-08) and $Ce_{0.3}Gd_{0.5}Zr_{0.2}O_{2\text{-}x}$ (CGZO-03).

2.3 *Preparation of Pt Loaded CGZO-08/C and CGZO-03 /C Electrocatalysts*

The microwave-assisted polyol method is used to prepare Pt-CGZO-08/C and Pt-CGZO-03/C electrocatalysts as reported earlier [12]. The CGZO-08 and CGZO-03 content in different electrocatalysts was fixed at 13 wt %. In brief, 20 mg of CGZO-08 and CGZO-03 and 100 mg of Vulcan XC-72 carbon black (Cabot Corporation, BET surface area of ~250 m^2 g^{-1}) in 100 mL of ethylene glycol (EG)/ isopropyl alcohol (IPA) mixture (V/V = 4:1, AR grade) was sonicated for 30 min followed by addition of 1.538 mL portion of 0.1 M H_2PtCl_6-EG solution. The mixture was then sonicated for 15 min, followed by constant stirring for 6 h. Then 1 M KOH-EG solution was used to adjust the pH of the stirring mixture ink to ~10. The mixture ink was further kept under microwave radiation for 50 s in a domestic microwave oven (Sharp NN-S327 WF, 2450 MHz, and 1100 W) to reduce H_2PtCl_6 to Pt. After cooling the solution to room temperature, the pH of the ink mixture was adjusted to ~4 by dropwise addition of 0.1 M HNO_3 solution. The ink mixture was further stirred for 12 h, which was then centrifuged, washed with Millipore water, followed by acetone (AR grade) until it became free of the Cl^- ion. The product was then dried overnight under vacuum at 60 °C. The platinum content in all the electrocatalysts was kept at 20 wt%.

2.4 *Preparation of Working Electrode and Electrochemical Measurements*

The electrocatalyst inks were prepared by ultrasonicating the mixtures of 5 mg Pt-CGZO-08/C, and Pt-CGZO-03/C composite powders each in 600 μL of distilled water, 300 μL IPA and 100 μL of Nafion solution for 2 h. A certain amount of catalyst ink was then spread on a 5 mm diameter glassy carbon electrode (electrode area of 0.196 cm^2 and polished with 0.05 μm γ-alumina micropolish, CH Instruments Inc) using the micropipette tip. The solvent on the ink was then evaporated at room temperature for 12 h which resulted in the well-dispersed layer of composite catalyst. CHI 7081C electrochemical workstation with a conventional three-electrode configuration was used for all the electrochemical measurements. A platinum foil (area = 1 × 1 cm^2) and Ag/AgCl electrode (BAS Instruments, USA) were used as the counter and reference electrodes, respectively. Before testing the electrochemical activity, the solution of 0.5 M H_2SO_4 and 1 M CH_3OH was purged with high purity N_2 for 30 min. All the electrochemical responses were recorded at a scan rate of 30 mV s^{-1}. A freshly prepared electrolyte was used for every repeat experiment.

2.5 *Physicochemical Characterizations*

All the prepared samples were first physically characterized by X-ray diffraction (XRD; Bruker D8 Advance X-ray diffractometer) using Cu Kα radiation (λ = 0.15408 nm) generated at 40 kV, 40 mA, and powder X-ray diffraction (PXRD) patterns were recorded at the scan rate of 0.01°s^{-1}. The automatic Micromeritics ASAP 2020 surface area analyzer using the Brunauer – Emmett – Teller (BET) gas adsorption method was used to carry out the multipoint nitrogen adsorption – desorption experiments at 77 K. Before physisorption measurements, the samples were degassed first at 100 °C for 2 h and then at 150 °C for 10 h in a dynamic vacuum. BET equation based on the software of the instrument was employed to calculate the specific surface area (SBET) values. Barrett-Joyner-Halenda's (BJH) method and a cylindrical pore model were used to evaluate the porosity distributions in the samples.

3 Results and Discussion

The XRD patterns of CGZO-08 and CGZO-03 samples in Fig. 1a show characteristics peaks of orthorhombic $Ce_2O(CO_3)_2{\cdot}H_2O$ phase and tetragonal phase of $Gd_2O_2CO_3$ transformed completely to their respective more stable cubic phases of oxides. The CZGO-08 catalyst shows prominent peaks at 28.6, 33.1, 47.5, 56.3,

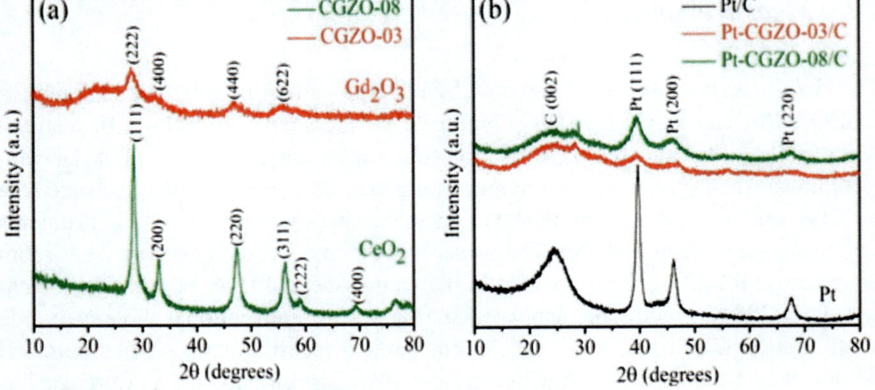

Fig. 1 XRD patterns of (a) CGZO-03 and CGZO-08, (b) Pt/C, Pt-CGZO-03/C, and Pt-CGZO-08/C

and 59.1° corresponding to, respectively, the (111), (200), (220), (311), and (222) planes. These sharp and intense XRD peaks are similar to the crystalline cubic phase of CeO_2 (PDF 81–0792). The following reactions occur during the nucleation of $Ce_2(CO_3)_2 \cdot H_2O$ under homogeneous precipitation conditions [23].

$$NH_2CONH_2 \rightarrow NH_4^+ + NCO^-$$
$$NCO^- + 3H_2O \rightarrow HCO_3^- + NH_4^+ + OH^-$$
$$HCO_3^- \rightarrow CO_3^{2-} + H^+$$
$$2Ce^{3+} + 2OH^- + 2CO_3^{2-} \rightarrow Ce_2O(CO_3)_2.H_2O$$
$$Ce_2O(CO_3)_2.H_2O \rightarrow Ce_2O(CO_3)_2 + H_2O$$
$$2Ce_2O(CO_3)_2 + O_2 \xrightarrow{\Delta} 4CeO_2 + 4CO_2$$

The CZGO-03 sample shows peaks at 29, 32.5, and 47° corresponding to (222), (400), and (440) planes, respectively. These peaks are very broad indicating the amorphous nature of the cubic Gd_2O_3 phase (PDF 65–3181) [24]. The following reactions occur during the nucleation of $Gd_2O_2CO_3 \cdot H_2O$ under homogeneous precipitation conditions [24].

$$NH_2CONH_2 \rightarrow NH_4^+ + NCO^-$$
$$NCO^- + 3H_2O \rightarrow HCO_3^- + NH_4^+ + OH^-$$
$$HCO_3^- \rightarrow CO_3^{2-} + H^+$$
$$2Gd^{3+} + 2OH^- + 2CO_3^{2-} \rightarrow Gd_2O_2CO_3$$
$$Gd_2O_2CO_3 \xrightarrow{\Delta} Gd_2O_3 + CO_2$$

Figure 1b shows the PXRD of Pt-CGZO-03/C, Pt-CGZO-08/C, and Pt/C electrocatalysts. All the electrocatalysts show a broad peak at 2θ value of 24.9° corresponding to the (002) plane of the carbon support in the catalysts while the peaks at 39.8, 46.3, and 67.6° in all the electrocatalysts confirm the presence of metallic Pt crystallites in the sample (PDF 87–0646). The Pt peaks in Pt-CGZO-03/C and Pt-CGZO-08/C samples are broad as compared to those in Pt/C samples indicating well-dispersed Pt crystallites on promoted samples [11].

The SEM images in Fig. 2 clearly show the influence of the molar ratio of metal precursors on the morphologies of the samples. The sample CGZO-08 shows dumbbell-shaped (Fig. 2a–c) particles and CGZO-03 sample shows needle-shaped morphology (Fig. 2d–f). The molar ratios of the metal precursors clearly influence the final morphologies of the oxide materials. The variation in molar ratios of precursors can affect the nucleation and growth rates of the particles under hydrothermal conditions leading to different morphologies. From the SEM images, it is seen that sharp needles of CGZO-03 have high surface porosity than that of massive dumbbell-shaped CGZO-08 which is further confirmed by BET analysis.

As shown in Fig. 3a both the catalyst samples show type IV isotherms and

Fig. 2 SEM images of **a–c** CGZO-08, and **d–f** CGZO-03

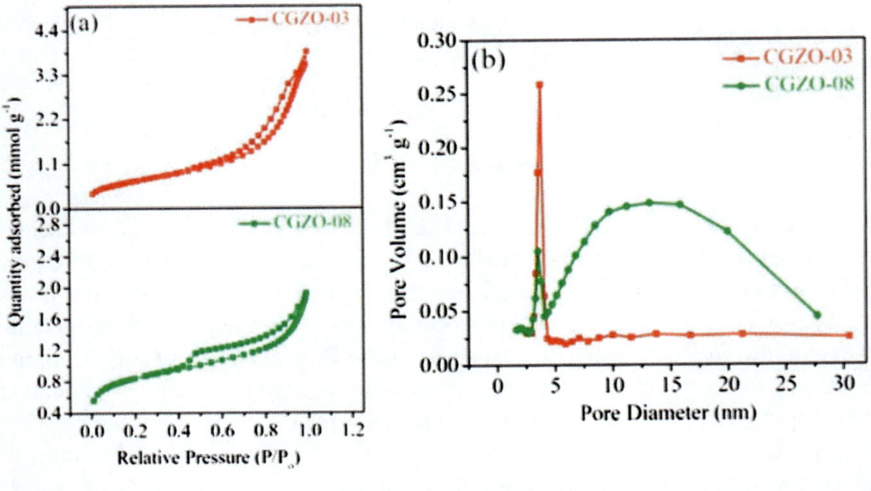

Fig. 3 **(a)** BET isotherms, **(b)** BJH pore size distribution curves of CGZO-03 and CGZO-08 samples

H4 hysteresis loops, reflecting the narrow slit-type mesopores in the samples. The BET surface area of the dumbbell-shaped CGZO-08 sample is 54 $m^2\ g^{-1}$ which is increased to 67 $m^2\ g^{-1}$ when gadolinium content is increased (CGZO-03). Figure 3b shows the pore diameter analysis of both CGZO-08 and CGZO-03. The estimated pore diameter of both CGZO-08 and CGZO-03 is nearly 3.8 nm, indicating the mesoporous nature of both the catalysts, which is beneficial for electrochemical energy conversion and storage applications.

Figure 4a and b show the electrocatalytic activity of Pt/C, Pt-CGZO-03/C, and Pt/CGZO-08/C electrodes evaluated by cyclic voltammetry (CV) technique in 0.5 M H_2SO_4 solution. The anodic current is presented in terms of both mA cm^{-2} and mA $mg_{Pt}{}^{-1}$. The forward anodic sweep shows peaks due to hydrogen desorption and Pt oxide formation while the reverse cathodic sweep shows peaks due to hydrogen adsorption and Pt oxide reduction at appropriate voltage regions [11, 14]. Under potential deposition of hydrogen process ($H^+ + e^- = H_{upd}$) appears in the region between -0.2 to 0.2 V (vs Ag/AgCl). Peaks of OH_{ad} (adsorbed hydroxyl species; $2H_2O = OH_{ad} + H_3O^+ + e^-$) on the Pt surface are observed after 0.4 V in the anodic scan, due to which the ordered layers of $PtO_x{\cdot}nH_2O/Pt(OH)_y{\cdot}nH_2O$ (x = 1, 2; y = 2,

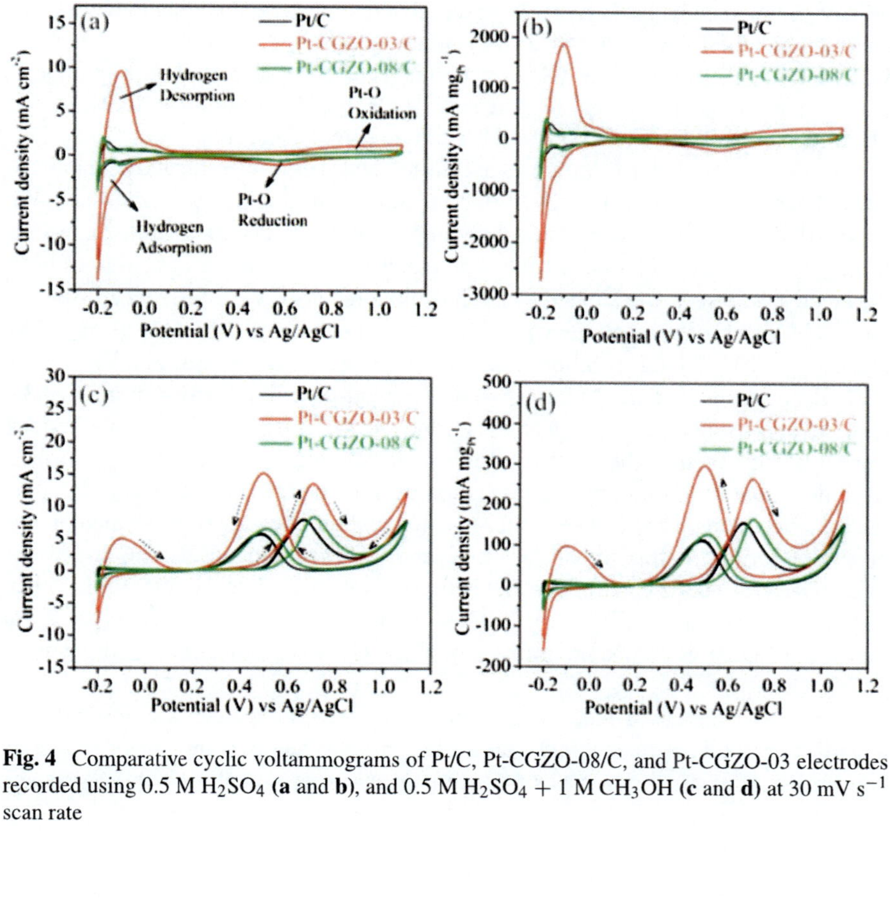

Fig. 4 Comparative cyclic voltammograms of Pt/C, Pt-CGZO-08/C, and Pt-CGZO-03 electrodes recorded using 0.5 M H_2SO_4 (**a** and **b**), and 0.5 M H_2SO_4 + 1 M CH_3OH (**c** and **d**) at 30 mV s^{-1} scan rate

4) are formed. In the cathodic scan, the reduction peak of $PtO_x \cdot nH_2O/Pt(OH)_y \cdot nH_2O$ species appears between 0.8 and 0.2 V as shown in Fig. 4a, b. All the electrocatalysts show broad hydrogen adsorption/desorption peaks. Further hydrogen adsorption/desorption on Pt is very structure sensitive. The desorption peak at -0.1 V in the anodic scan results from adsorption of hydrogen on (110) plane of Pt [11]. The Pt-CGZO-03/C shows intense hydrogen adsorption–desorption peaks as compared to Pt-CGZO-08/C and Pt/C. The electrochemical active surface areas (EAS) of all the electrode materials are calculated from cyclic voltametric analysis by using the following expression [25],

$$\text{EAS}\left(m^2 g^{-1}\right) = \frac{Q_H}{0.21 \times M_{\text{Pt}}}$$

where 0.21 mC cm^{-2} is the electrical charge associated with monolayer adsorption of hydrogen/charge required to oxidize a monolayer of H on a polycrystalline Pt electrode (assuming the surface density of 1.3×10^{15} atoms cm^{-2}) and M_{Pt} is the Pt loading on the working electrodes (5 μg cm^{-2}). The EASs of Pt nanocrystals in Pt-CGZO-08/C and Pt-CGZO-03/C electrodes are estimated to be 55 $m^2\ g^{-1}$ and 283 $m^2\ g^{-1}$, respectively. This clearly shows that irrespective of similar Pt loading on each sample, the EAS for the CGZO-03-promoted Pt/C electrode is higher than the Pt-CGZO-08/C electrode. The CV measurements in 0.5 M H_2SO_4 show higher peak current density and the higher integrated area under the H_{upd} peak for Pt-CGZO-03/C electrode compared to Pt-CGZO-08/C electrode (Fig. 4a and b).

Catalysts with higher EAS are expected to show better electrocatalytic activity [27]. Since EAS value of Pt-CGZO-03/C is more than that of Pt-CGZO-08/C, we expect Pt-CGZO-03/C electrode to show improved electrochemical performance in terms of methanol oxidation to CO_2. Methanol electrooxidation activities of both the electrocatalysts are evaluated by cyclic voltammetry (CV), linear sweep voltammetry (LSV), chronopotentiometry (CP), and chronoamperometry (CA) using 0.5 M H_2SO_4 + 1.0 M CH_3OH solution. The CV curves of Pt-CGZO-08/C and Pt-CGZO-03/C electrodes in methanol solution show two well-defined oxidation current peaks (Fig. 4c and d). During the forward scan, the oxidation peak represents the oxidation of freshly chemisorbed methanol molecules approaching the electrode surface from the bulk of the solution. The oxidation peak in the reverse scan is attributed to the removal of adsorbed CO and other residual carbonaceous species formed on the electrode during the forward sweep. It is seen that the methanol oxidation activity of the Pt-CGZO-03/C electrode is much higher compared to the activity of the other samples. The current density of the Pt-CGZO-03/C electrode (14 mA cm^{-2}, 267 mA mg_{Pt}^{-1}) is higher as compared to that of the Pt-CGZO-08/C electrode (9 mA cm^{-2}, 167 mA mg_{Pt}^{-1}).

During the methanol electrooxidation, carbonaceous species are formed and adsorbed on the Pt surface which can further prohibit the methanol adsorption and conversion. This results in a positive shift in the onset potential. The LSV taken at 10 mV s^{-1} scan rate shows that Pt-CGZO-08/C electrode has an onset potential of 0.49 V which is decreased to 0.44 V with an increase in gadolinium content (Pt-CGZO-03/C) as shown in the inset of Fig. 5.

Hence increasing gadolinium content resulted in the negative shift of the onset potential which is indicating the inhibition of the formation of carbonaceous species on the electrocatalyst. The conclusion drawn here is that Pt-CGZO-03/C electrode can easily oxidize methanol to CO_2 which is corroborated by increased electrocatalytic activity by ~2 times and decreased onset potential by 0.04 V with respect to Pt/C.

Chronopotentiometric (CP) measurements provide complementary information on the stability of electrocatalysts and antipoisoning capability for electrochemical oxidation of methanol. Figure 6a shows the change in electrode potential with time at a fixed bias current density of 5 mA cm^{-2}.

The Pt/C, Pt-CGZO-08/C, and Pt-CGZO-03 /C composite electrodes show a slow increase in electrode potentials with polarization time. A sharp rise in the potential value is noticed after some time. This sharp increase in polarization voltage is due to the poisoning effect associated with oxygen evolution or water electrolysis rather than the electrochemical oxidation of methanol. The Pt-CGZO-03/C electrode shows high catalytic activity and can withstand poisoning for a longer time. Oscillations observed in Pt-CGZO-08/C and Pt-CGZO-03/C composite electrodes are due to the formation of intermediate carbonaceous species which get adsorbed on the defect sites present at the Pt/oxide interfaces. The oscillations are more prominent in the case of Pt-$Ce_{0.3}Gd_{0.5}Zr_{0.2}O_{2-x}$ /C interface which again indicates more defect sites present in the Pt-CGZO-03 /C sample.

The stability and electrochemical activity of Pt/C, Pt-CGZO-08/C, and Pt-CGZO-03/C nanocomposite electrodes are investigated by chronoamperometry (CA) measurements. The typical chronoamperometry profiles of Pt-CGZO-08/C and Pt-CGZO-03/C composite electrodes are depicted in Fig. 6b. The CA profile of the

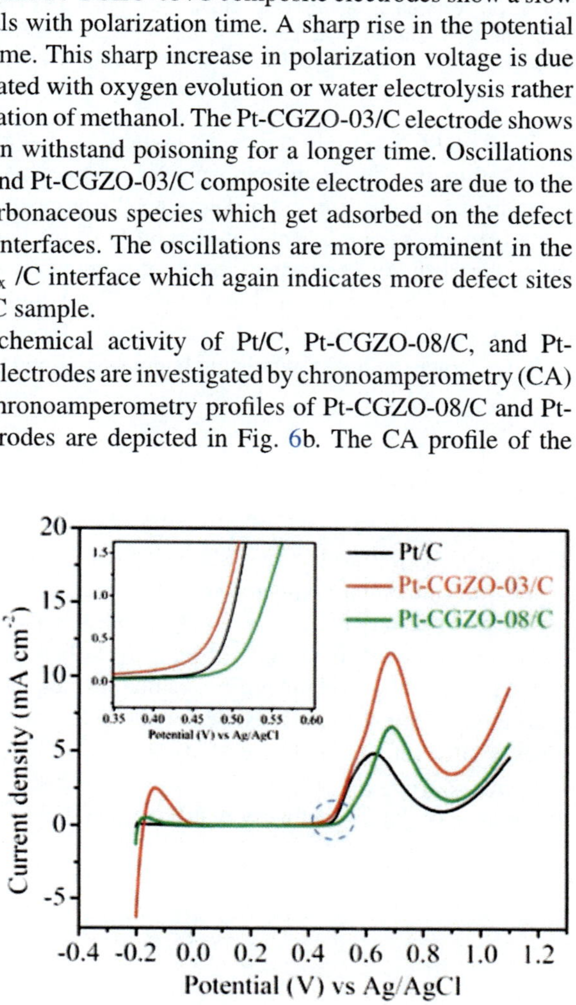

Fig. 5 Comparative LSV of Pt/C, Pt-CGZO-08/C, and Pt-CGZO-03/C sample electrodes recorded in 0.5 M H_2SO_4 + 1 M CH_3OH at 10 mV s^{-1} scan rate

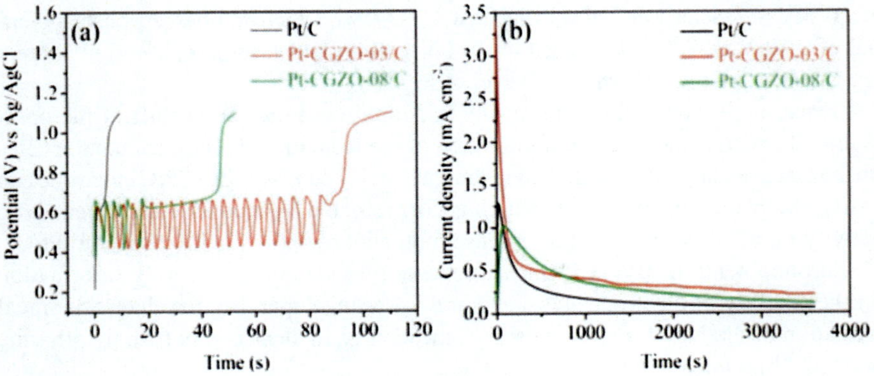

Fig. 6 **(a)** Choronopotentiometric studies of Pt/C, Pt-CGZO-08/C and Pt-CGZO-03/C at 5 mA cm^{-2}, **(b)** Chronoamperometric profiles of Pt/C, Pt-CGZO-08/C and Pt-CGZO-03/C recorded in 0.5 M H_2SO_4 + 1 M CH_3OH at − 0.2 V (vs Ag/AgCl) applied potential

Pt-CGZO-03/C electrode shows the highest current and lowest current decay with polarization time during the methanol oxidation, which is consistent with the cyclic voltammetry results in Fig. 6b. These results indicate that the Pt-CGZO-03/C electrode is much more efficient and tolerant to CO poisoning than other electrodes in the acidic medium. The present results are compared with recently reported electrocatalysts, which are presented in Table 1. These studies clearly show the profound effect of oxide promoters and the interfacial defects on the electrocatalytic activity of Pt/promoter oxide/C towards methanol oxidation [11, 22, 30].

4 Conclusions

The Pt-CGZO-03/C electrode shows much higher electrocatalytic activity and lower onset potential compared to both Pt-CGZO-08/C and Pt/C electrodes. The higher electrocatalytic activity, anti-poisoning activity, and stability of Pt-CGZO-03/C are attributed to the interfacial defects between Pt and $Ce_{0.3}Gd_{0.5}Zr_{0.2}O_{2-x}$ which promote the oxidation of poisonous species and maintain the activity. Higher gadolinium content is responsible for the defect formation and about 1.6- and 2-times higher activity of Pt-$Ce_{0.3}Gd_{0.5}Zr_{0.2}O_{2-x}$/C electrocatalyst compared to Pt-$Ce_{0.8}Gd_{0.1}Zr_{0.1}O_{2-x}$/C and Pt/C catalysts respectively. This study highlights the importance of defect chemistry in electrocatalytic activity.

Table 1 Comparison of the onset potential and peak current density of the electrocatalysts

Catalyst	Electrolyte	Scan rate (mV s^{-1})	Onset Potential (V)	Peak current density	Refs
Pt/*a*- ZrC800- XC	0.5 M H_2SO_4 + 1 M CH_3OH	20	0.33 versus Ag/AgCl	21.7 mA cm^{-2} 606.9 mA $mg_{Pt}{}^{-1}$	[9]
Pt/MnO_2 /C	0.5 M H_2SO_4 + 1 M CH_3OH	20	–	15.7 mA cm^{-2}	[12]
Pt/CeO_2-HP/C	0.5 M H_2SO_4 + 1 M CH_3OH	20	–	14.6 mA cm^{-2}	[15]
Pt–V_2O_5/C	1 M KOH + 1 M CH_3OH	50	−0.42 versus Hg/HgO	200 mA cm^{-2}	[16]
MoO_3 promoted Pt/C	0.5 M H_2SO_4 + 1 M CH_3OH	20	–	1093 mA $mg_{Pt}{}^{-1}$	[17]
Pt–Nb_2O_5 /C	0.5 M H_2SO_4 + 1 M CH_3OH	20	–	183 mA cm^{-2}	[19]
$NiCo_2O_4$-rGO,	1 M KOH + 0.5 M CH_3OH	50	0.3 versus SCE	16.6 mA cm^{-2}	[25]
Pt–V_2O_5/rGO	0.5 M H_2SO_4 + 1 M CH_3OH	20	–	26 mA cm^{-2} 928 mA $mg_{Pt}{}^{-1}$	[26]
Pt/Nb-Mo_2C-rGO	0.5 M H_2SO_4 + 1 M CH_3OH	20	0.29 versus Ag/AgCl	23.3 mA cm^{-2} 836.4 mA $mg_{Pt}{}^{-1}$	[28]
CeO_2-ZrO_2 on Pt/C	0.5 M H_2SO_4 + 1 M CH_3OH	50	–	52.17 mA $mg_{Pt}{}^{-1}$	[29]
Pt-$Ce_{0.3}Gd_{0.5}Zr_{0.2}O_{2-x}$/C	0.5 M H_2SO_4 + 1 M CH_3OH	30	0.44 versus Ag/AgCl	14 mA cm^{-2} 266 mA $mg_{Pt}{}^{-1}$	This work

Acknowledgements Shuchi Sharma received JRF and SRF fellowships from DSEHC-Solar Fuels domain, IIT Madras. The authors thank Department of Science and Technology, Government of India, for financial support through grant No.DST/TMD/SERI/HUB/1(C).

References

1. Yu EH, Scott K (2004) Development of direct methanol alkaline fuel cells using anion exchange membranes. J Power Sources 137:248–256. https://doi.org/10.1016/j.jpowsour.2004.06.004
2. Barbir F (2005) PEM fuel cells: theory and practice. Elsevier Academic Press, New York
3. Takasu Y, Fujiwara T, Murakam Y, Sasaki K, Oguri M, Asaki T, Sugimoto W (2000) Effect of structure of carbon-supported PtRu electrocatalysts on the electrochemical oxidation of methanol. J Electrochem Soc 147:4421–4427. https://doi.org/10.1149/1.1394080
4. Steele BCH, Heinzel A (2001) Materials for fuel-cell technologies. Nature 414:345–352. https://doi.org/10.1038/35104620
5. Kakati N, Maiti J, Lee SH, Jee SH, Viswanathan B, Yoon YS (2014) Anode catalysts for direct methanol fuel cells in acidic media: do we have any alternative for Pt or Pt–Ru? Chem Rev 114:12397–12429. https://doi.org/10.1021/cr400389f
6. Liu H, Zhang J (2009) Electrocatalysis of direct methanol fuel cells. Wiley-VCH Verlag GmbH and& Co. KGaA, Weinheim, pp 1–582
7. Huang H, Wang X (2014) Recent progress on carbon-based support materials for electrocatalysts of direct methanol fuel cells. J Mater Chem 2:6266–6291. https://doi.org/10.1039/C3TA14754A
8. He M, Fei G, Zheng Z, Cheng Z, Wang Z, Xia H (2019) Pt nanoparticle-loaded graphene aerogel microspheres with excellent methanol electro-oxidation performance. Langmuir 35:3694–3700. https://doi.org/10.1021/acs.langmuir.9b00021
9. Sahoo MK, Shanmugam R, Umeshbabu E, Ranga Rao G (2020) Activated ZrC promotes the methanol electro-oxidation activity and enhances poison tolerance of Pt nanoparticles in acidic medium. Chemistry Select 5:7205–7216. https://doi.org/10.1002/slct.202001581
10. Viswanath B, Patra S, Munichandraiah N, Ravishankar N (2009) Nanoporous Pt with high surface area by reaction-limited aggregation of nanoparticles. Langmuir 5:3115–3121. https://doi.org/10.1021/la802938d
11. Brouzgou A, Song SQ, Tsiakaras P (2012) Low and non-platinum electrocatalysts for PEMFCs: current status, challenges and prospects. Appl Catal B: Environ 127:371–388. https://doi.org/10.1016/j.apcatb.2012.08.031
12. Meher SK, Ranga Rao G (2013) Morphology-controlled promoting activity of nanostructured MnO_2 for methanol and ethanol electrooxidation on Pt/C. J Phys Chem C 117:4888–4900. https://doi.org/10.1021/jp3093995
13. Muhammed Musthafa OT, Sampath S (2008) High performance platinized titanium nitride catalyst for methanol oxidation. Chem Commun 67:67–69. https://doi.org/10.1039/B715859A
14. Thiagarajan V, Karthikeyan P, Manoharan R, Sampath S, Hernandez-Ramirez A, Sanchez-Castro ME, Alonso-Lemus IL, Rodriguez-Varela FJ (2018) Pt-Ru-$NiTiO_3$ nanoparticles dispersed on vulcan as high performance electrocatalysts for the methanol oxidation reaction (MOR). Electrocatalysis 9:582–592. https://doi.org/10.1007/s12678-017-0450-2
15. Meher SK, Ranga Rao G (2012) Polymer-assisted hydrothermal synthesis of highly reducible shuttle-shaped CeO_2: microstructural effect on promoting Pt/C for methanol electrooxidation. ACS Catal 2:2795–2809. https://doi.org/10.1021/cs300473e
16. Justin P, Ranga Rao G (2009) Enhanced activity of methanol electrooxidation on Pt–V_2O_5/C catalysts. Catal Today 141:138–143. https://doi.org/10.1016/j.cattod.2008.03.019
17. Justin P, Ranga Rao G (2011) Methanol oxidation on MoO_3 promoted Pt/C electrocatalyst. Int J Hydrogen Energy 36:5875–5884. https://doi.org/10.1016/j.ijhydene.2011.01.122

18. Patil SH, Anothumakkool B, Sathaye SD, Patil KR (2015) Architecturally designed Pt–MoS_2 and Pt–graphene composites for electrocatalytic methanol oxidation. Phys Chem Chem Phys 17:26101–26110. https://doi.org/10.1039/C5CP04141D
19. Justin P, Charan HK, Ranga Rao G (2010) High performance Pt–Nb_2O_5/C electrocatalysts for methanol electrooxidation in acidic media. Appl Catal B 100:510–515. https://doi.org/10.1016/j.apcatb.2010.09.001
20. Shi Y, Wu P, Du P, Cai C (2006) Solid-state electrochemistry of gadolinium hexacyanoferrate modified electrode and electrocatalytic properties of gadolinium hexacyanoferrate. Acta Physico- Chimica Sinica 22:1227–1233. https://doi.org/10.1016/S1872-1508(06)60057-0
21. Jadhav LD, Chourashiya MG, Subhedar KM, Tyagi AK, Patil JY (2009) Synthesis of nanocrystalline Gd doped ceria by combustion technique. J Alloy Compd 470:383–386. https://doi.org/10.1016/j.jallcom.2008.02.077
22. Zhang Y, Xu J, Iong Y, Ding M, Jia C (2020) Defect chemistry on electrode materials for electrochemical energy storage and conversion. Chem Nano Mat 6:1589–1600. https://doi.org/10.1002/cnma.202000437
23. Meher SK, Cargnello M, Troiani H, Montini T, Ranga Rao G, Fornasiero P (2013) Alcohol induced ultra-fine dispersion of Pt on tuned morphologies of CeO_2 for CO oxidation. Appl Catal B: Environmental 130–131:121–131. https://doi.org/10.1016/j.apcatb.2012.10.022
24. Ge W, Li Z, Lei Z, Chen T, Fu Z, Peng R, Liu M, Lu Y (2015) Synthesis of hexagonal phase $Gd_2O_2CO_3$:Yb^{3+}, Er^{3+} upconversion nanoparticles via SiO_2 coating and Nd^{3+} doping. Cryst Eng Comm 17:5702–5709. https://doi.org/10.1039/C5CE00869G
25. Umeshbabu E, Ranga Rao G (2016) $NiCo_2O_4$ hexagonal nanoplates anchored on reduced graphene oxide sheets with enhanced electrocatalytic activity and stability for methanol and water oxidation. Electrochim Acta 213:717–729. https://doi.org/10.1016/j.electacta.2016.07.161
26. Umeshbabu E, Ranga Rao G (2016) A vanadium (V) oxide nanorod promoted platinum/reduced graphene oxide electrocatalyst for alcohol oxidation under acidic conditions. Chem Phy Chem 17:3524–3534. https://doi.org/10.1002/cphc.201600667
27. Pozio A, De Francesco M, Cemmi A, Cardellini F, Giorgi L (2002) Comparison of high surface Pt/C catalysts by cyclic voltammetry. J Power Sources 105:13–19. https://doi.org/10.1016/S0378-7753(01)00921-1
28. Sahoo MK, Ranga Rao G (2020) Enhanced methanol electro-oxidation activity of Pt/rGO electrocatalyst promoted by NbC/Mo_2C Phases. Chem Select 5:3805–3814. https://doi.org/10.1002/slct.202000170
29. Wang Q, Liu Z, An S, Wang R, Wang Y, Xu T (2016) Effect of CeO_2-ZrO_2 on Pt/C electrocatalysts for alcohols oxidation. J Rare Earths 34:276–282. https://doi.org/10.1016/S1002-0721(16)60025-X
30. Ranga Rao G, Justin P, Meher SK (2011) Metal oxide promoted electrocatalysts for methanol oxidation. Catal Surv Asia 15:221–229. https://link.springer.com/article/10.1007/s10563-011-9124-x

La-and Gd-Doped CeO_2 Nanoparticles as Electrolyte Materials for Intermediate Temperature Solid Oxide Fuel Cells

Naeemakhtar Momin, J. Manjanna, S. Senthilkumar, and S. T. Aruna

1 Introduction

The solid oxide fuel cell (SOFC) is a competent contender for producing efficient, stable, low emission, and cost-effective electricity from direct oxidation of fuel with other electrochemical devices [1]. High operating temperature to achieve optimum total ionic conductivity in the order of 0.1 Scm^{-1} can bring structural and chemical instability in the commonly used yttria-stabilized zirconia (YSZ) electrolyte material to deter its mass commercialization [2]. Therefore, researchers have given much emphasis to achieve the desired optimum ionic conductivity at intermediate temperature (IT) by shifting from expensive materials to cost-effective electrolyte materials. The materials with modified space charge domain using nanotechnology can yield the desired output [3]. The different semiconductor oxides like yttria-stabilized zirconia (YSZ), Er-doped Bi_2O_3, Mg-doped $LaGaO_3$ (LSGM), alkaline, and rare-earth-doped ceria have been investigated as electrolytes for solid oxide fuel cells (SOFCs). Among them, the rare-earth-doped ceria emerges as a potential candidate for SOFC electrolyte because of its capacity to deliver high ionic conductivity at intermediate temperature with lower activation energy. The rare earth metal ions like Gd and Sm having slightly higher or same atomic radii than host ion can generate the excessive oxygen vacancies in sub-lattice. Since charge compensating defects with less distortion at the surface of the lattice provides stable phase and high ionic

N. Momin · J. Manjanna (✉)
Department of Chemistry, Rani Channamma University, Belagavi, Karnataka 591156, India

S. Senthilkumar · S. T. Aruna
Surface Engineering Division, CSIR-National Aerospace Laboratories, Bengaluru 560017, India
e-mail: ssenthil@nal.res.in

S. T. Aruna
e-mail: aruna_reddy@nal.res.in

U. K. Mudali et al. (eds.), *Recent Trends in Electrochemical Science and Technology*, Springer Proceedings in Materials 15,
https://doi.org/10.1007/978-981-16-7554-6_10

conductivity compared to pure ceria over other rare earth oxides at the intermediate temperature range [4].

The theoretical calculation performed by Butler et al. [5] suggests that the higher aliovalent cation substitution reduces the concentration of free oxygen vacancies as they bind to cation defects resulting in defect association. The predicted ionic model for rare-earth-doped cerium oxides shows a decrease in enthalpy of defects with a high ionic radius (r) of the dopant within the limit of $r_i = 0.120$ nm. Therefore, the studies on the substitution of Ce ($r_i = 0.111$ nm) by larger cations like La ($r\,i = 0.130$ nm) are limited [6]. However, Gd- and Sm-doped ceria are regarded as potential oxide ion conductors giving better performance at an intermediate temperature range and hence extensively used in SOFC applications as electrolyte materials. But the cost for 10–20 mol % of dopant amount is much expensive and is a matter of concern. Therefore, for commercial viability, the use of low-cost dopants like La-doped ceria is selected in the present study [7].

In this study, the 10 mol % of La-doped ceria composition ($Ce_{0.9}\,La_{0.1}O_{2-\delta}$) was synthesized by facile cost-effective citrate–nitrate auto-combustion method and designated as LDC10. The auto-combustion process is known to yield homogeneous, high-purity nanocrystalline ceramic oxides with a larger surface area in its most convenient and economical way [8]. The electrochemical impedance spectroscopy (EIS) was used to study the sample's electrical properties between 623 and 1023 K. The obtained results of LDC10 are compared with commercially obtained 10 mol % of Gd-doped ceria ($Ce_{0.9}Gd_{0.1}O_{2-\delta}$), which is regarded as the utmost potential electrolyte material for SOFC applications and designated as GDC10 [9]. The structural and electrical properties of LDC10 are compared with GDC10.

2 Experimental

2.1 Synthesis and Sample Preparation

The $Ce_{0.9}La_{0.1}O_{2-\delta}$ (LDC10) composition was synthesized by citrate–nitrate-based auto-combustion method. Stoichiometric amounts of precursor salts $Ce(NO_3)_3.6H_2O$ (A.R grade, HIMEDIA, Mumbai) and $La(NO_3)_3.6H_2O$ (A.R grade, LOBA Chemie, Mumbai) were dissolved at an equimolar ratio and stirred thoroughly for 1 h at room temperature. To this mixture, AR-grade citric acid ($C_6H_8O_7.H_2O$, SDFCL Limited, Mumbai) was added as the fuel to metal nitrates in the molar ratio of 1:1 and heated to 333 K. The obtained gel was further heated to 373 K, which initiated the vigorous self-propagating combustion reaction. The amorphous powder was further annealed at 973 K for 6 h to get nanocrystalline ultrafine particles [10 − 12]. The $Ce_{0.9}\,Gd_{0.1}O_{2-\delta}$ (GDC10) was procured from Fuel Cell Materials, the USA. For making the pellets, the LDC10 and GDC10 powders were uniaxially pressed to 5 MPa with 6 wt % of polyvinyl alcohol (as the binder) and sintered at 1473 K in air for 4 h with programmed, controlled heating.

2.2 *Characterization*

The Rigaku MiniFlex 600 powder X-ray diffractometer with Cu Ka radiation (λ = 0.154 nm) was used to record the XRD patterns of the powder and sintered pellets. The relative density of the pellet was calculated from the equation:

$$\text{Relativedensity} = \left(\frac{d_m}{d_{\text{th}}}\right) \times 100 \tag{1}$$

where d_m is the density of samples measured by the Archimedes method and d_{th} is the theoretical density given by

$$d_{\text{th}} = \frac{4}{N_A V}\left[(1-x)M_{\text{Ce}} + xM_{\text{La}} + \left(2 - \frac{1}{2}x\right)M_O\right] \tag{2}$$

X is the dopant ion content, V is the lattice volume of LDC10/GDC10, N_A is the Avogadro number, and $M_{\text{Ce}}/M_{\text{La}}/M_O$ is the corresponding elements' atomic weight [13]. For AC impedance measurement, the silver paste was applied on both sides of the pellet and baked at 873 K for 1 h to make electrical connections. An electrochemical system (CH Instrument, Inc CHI604D, USA) was used to record the Nyquist plots between the temperature ranges 623 and 1023 K in the frequency range of 1–10 MHz with an AC signal of 10 mA. The obtained data were fitted to the corresponding equivalent circuits [R(QR)(QR)] using the ZSimpWin software. The total conductivity (σ) was calculated using the equation:

$$\sigma = t/RA \tag{3}$$

where thickness (t), area of cross section (A), and total resistance (R) are involved [14].

3 Results and Discussion

Figure 1 shows the XRD patterns of CeO_2, GDC10, and LDC10 nanocrystalline pellets sintered at 1473 K. It has been observed that all XRD patterns exhibit the same cubic–fluorite structure. The diffraction peaks were indexed as per standard fluorite structure with space group Fm3m, and they match well with ICDD card no 34–0394. Diffraction peaks corresponding to La_2O_3 or other impurities were not observed in the XRD pattern, indicating single-phase solid solution formation. The XRD analysis of the pellets sintered at 1473 K showed a typical lattice structure aligned with the reported lattice parameter of 5.411 Å for pure CeO_2, as indicated in the JCPDS card [15] and computed from Eqs. 4 and 5:

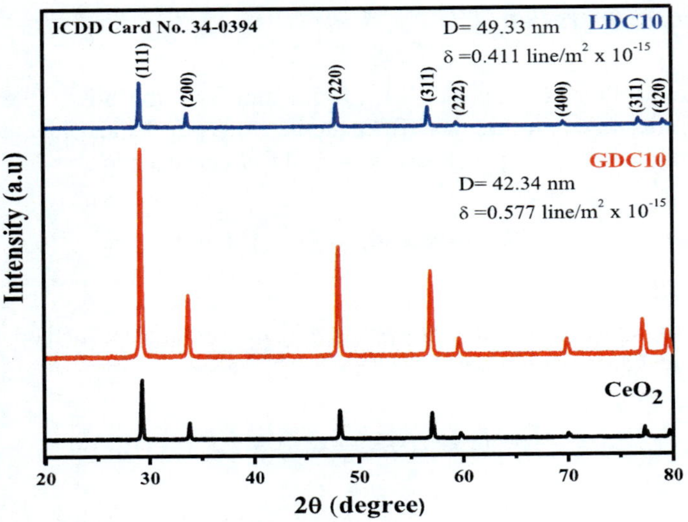

Fig. 1 XRD patterns of CeO_2, GDC10, and LDC10 pellets sintered at 1473 K

$$a = \sqrt{h^2 + k^2 + l^2} \tag{4}$$

$$d = \frac{\lambda}{2\text{Sin}\theta} \tag{5}$$

where d be the inter planar spacing and θ is the angle of diffraction.

The sharp peaks in the XRD patterns of pellets indicate higher crystallite size due to high-temperature sintering [16]. The replacement of smaller ionic radii of Ce (r_i = 0.97 Å) by a larger ionic radii Gd (r_i = 1.05 Å) and La (r_i = 1.15 Å) creates the incongruity between the ions generating the oxygen vacancies in the CeO_2 lattice. The oxygen vacancies cause the lattice strain to retain the charge equilibrium state, resulting in expanding the unit cell [17]. The crystallite sizes were computed from the preferential plane width (111) using Scherrer's formula [18],

$$D = \frac{0.9}{\beta\text{Cos}\theta} \tag{6}$$

where λ is the wavelength of the X-ray (1.5406 Å), β is the full-width at half maxima (FWHM) in radian, and θ is the diffraction angle. The obtained crystallite sizes for the GDC10 and LDC10 pellets were 75.8 nm and 51.94 nm, respectively, as obtained from Rietveld refinement analysis. A higher relative density value of more than 90% was achieved for GDC10 and LDC10 pellets. The strain generated at lattice during the doping may get localized at a sub-grain and sub-domain level near the grain boundaries, attributed to the decrease in crystallite size in doped samples [19]. The number of defects in the crystal estimated from dislocation density (δ), which is the length of dislocation per unit volume, was calculated from Eq. 7

$$\delta = \frac{1}{D^2} \tag{7}$$

where D is the average crystallite size.

The dislocation density which is a function of dopant concentration reveals the generation of a promising quality compound. Its lower values are attributed to the good crystalline nature of the sample [20]. The theoretical surface areas were calculated by using Eq. 8

$$S = \frac{6}{DX\rho} \tag{8}$$

where D is the average crystallite size and ρ is the X-ray density.

The obtained values of dislocation density and specific surface area for GDC10 and LDC10 samples are shown in Table 1.

From the obtained data, it has been observed that the density of the sample appreciably varies with the specific surface area. The increase in specific surface area with dopant content may be attributed to decrease in the crystallite size [17]. The Rietveld refinement for GDC10 and LDC10 samples was carried out using Profex 4.0.3 software. The obtained parameters from Rietveld refinement are enlisted in Table 2. The refinement results confirm negligible difference between the calculated and observed XRD lattice parameters.

Figure 2a, b shows the complex impedance spectra of GDC10 and LDC10 samples in the temperature range between 623 and 1023 K as a function of frequency.

Table 1 Structural characteristic of sintered LDC10 and GDC10 samples

Samples	Average crystallite size (D) (nm) Scherrer	Lattice constant (a = b = c) (nm)	Dislocation density (δ) (line/$m^2 \times 10^{-15}$)	Specific surface area (m^2/g)	Relative density (%)
LDC10	51.94	0.5442	0.371	16.290	92.50
GDC10	75.8	0.5418	0.174	11.016	93.00

Table 2 Rietveld refinement parameters of LDC10 and GDC10

Samples	Lattice parameters a = b = c and α = β = γ = 90°		Rietveld refinement parameters				Grain Size in nm
	a (Å)	Volume (Å)^3	R_P	R_{wp}	χ^2	GoF	
LDC10	5.44253 ± 0.000010	161.213918	14.23	14.56	1.04	1.02	51.94
GDC10	5.41873 ± 0.0000093	159.108171	6.65	12.06	3.29	1.81	75.8

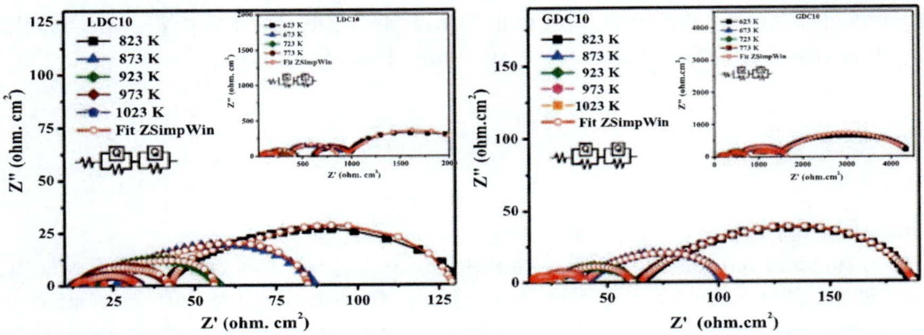

Fig. 2 Nyquist plots of LDC10 and GDC10 pellets sintered at 1473 K and measured in the temperature range of 623 K – 1023 K. (inset shows the Nyquist plots at 623–723 K)

From impedance spectra, it has been observed that the resistance decreases with rise in temperature, and the semicircles corresponding to grain and grain boundary contributions were get overlapped and depressed. Generally, the AC impedance spectrum of a solid ionic conductor shows three different arcs. The arc at high frequencies was attributed to grain (bulk) behavior, at intermediate frequencies to grain boundary (GB) behavior, and belongs to electrode behavior at lower frequencies which is more complex due to the appearance of multiple arcs. Because of different relaxation time constants for individual polarizations of conductors at a constant temperature, all arcs will not appear simultaneously. Hence, at a particular temperature, grain and GB arcs at high frequencies tend to disappear and arc corresponding to the total electrode contribution will only be observed. The grain, GB, and electrode contributions were distinguished by using an equivalent electrical circuit model [R(QR)(QR)], which is shown in Fig. 2. This equivalent circuit consists of a grain resistance (R_g) and two RC circuits in series for GB (R_{gb}) and others for electrode polarization (Re). A constant phase element (CPE = Q) is applied instead of a capacitor. This constant phase element is equivalent to the distribution of the capacitor in parallel. R_g, $R_{gb}\|Q$ represents the ionic conductivity through grain and grain boundary [21]. The total resistance of the electrolyte was calculated by using equation

$$R_t = R_g + R_{gb} \tag{9}$$

where R_g and R_{gb} be the grain and GB resistance, respectively.

The grain, grain boundary (GB), and total conductivities were obtained using Eq. 3. The capacitance values were computed using the equation $\omega RC = 1$, where ω is the angular frequency ($2\pi f$, f is the frequency in Hz) and R is the arc magnitude. Grains have a capacitance of pF, whereas GB has a capacitance of nF [22]. For all the GDC10 and LDC10 samples, the capacitance values were found to be in the range of 10^{-12}–10^{-18} F, which indicates the conduction process through grain and grain boundary. The Arrhenius Eq. (10) is used to show the influence of temperature on conductivity,

$$\sigma T = A_0 \exp\left(\frac{-E_a}{\text{kT}}\right) \tag{10}$$

where A_o, T, E_a, and K are the conductivity, pre-exponential constant, temperature, activation energy, and Boltzmann constant. Figure 3a, b shows the Arrhenius plots for grain, GB, and total conductivity for the GDC10 and LDC10 samples. The activation energy values of GDC10 and LDC10 obtained for grain, GB, and total conductivity were shown in Table 3. The GDC10 and LDC10 samples exhibit the highest total ionic conductivities of 1.96×10^{-2} Scm^{-1} and 1.01×10^{-2} Scm^{-1} at 1023 K with lower activation energy values of 0.85 and 0.70 eV, respectively. The optimum grain ionic conductivity in the order of 1.10×10^{-1} Scm^{-1} required for the operation of IT-SOFC was not achieved at 973 K for the GDC10 sample.

In contrast, the LDC10 sample did not yield optimum ionic conductivity at the intermediate temperature [773−1023 K] range. The obtained result confirms the better influence of dopant content on the grain ionic conductivity than those on GB and total ionic conductivity. The LDC10 sample shows lower values of activation energies compared to GDC10 sample. The lower activation energy of LDC10 may be attributed to close ionic radii of La and Ce elements and low binding energy [23]. The higher grain activation energy value of GDC10 than reported value may be due to SiO_2 content as impurities. Hence, the GB effect influences the total ionic conductivity in GDC10 [24]. The conductivity values of LDC10 are lower than that of GDC10 but reasonably competent in all ranges of temperatures. Dikmen et al. [6]

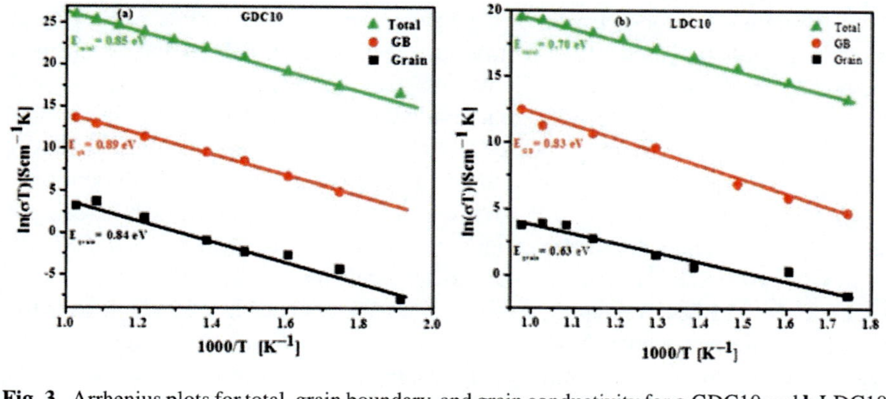

Fig. 3 Arrhenius plots for total, grain boundary, and grain conductivity for **a** GDC10 and **b** LDC10 in the temperature range 623 K–1023 K, respectively

Table 3 Activation energy of GDC10 and LDC10 samples between the temperature range of 623 K and 1023 K in eV

Sample	Activation energy, E_a (eV)		
	Grain	Grain boundary (GB)	Total
GDC10	0.89	0.84	0.85
LDC10	0.63	0.83	0.70

reported conductivity of 5.9×10^{-3} Scm^{-1} with an activation energy of 0.74 eV for LDC10 sample at 873 K prepared from the hydrothermal method. In contrast, the obtained conductivity of 3.79×10^{-3} $Scm^{-1}$with lower activation energy of 0.70 eV at 873 K for LDC10 was in good agreement with the reported. Hence, it can be concluded that the facile cost-effective auto-combustion synthesis method facilitates the synthesis of LDC with enhanced ionic conductivity at lower activation energy.

The obtained results established that the low-cost La-doped ceria nanoparticles can be the best competitor to Gd-doped ceria nanoparticles as an electrolyte for IT-SOFC applications. The comparison of conductivities of GDC10 and LDC10 in the temperature range of 623–1023 K is shown in Table 4.

Thus, the results suggest that the achievement of desired optimum total ionic conductivity in the order of 0.1 Scm^{-1}at the intermediate temperature range is still a matter of great concern, along with issues like cost of ceria-based materials. Therefore, the search for plausible electrolyte materials still finds a broad scope. Consequently, it may be concluded that auto-combustion synthesized LDC10 can be considered as a potential electrolyte material for SOFC application if desired optimum ionic conductivity is achieved at an intermediate temperature range. Nevertheless, an in-depth analysis of electrical measurements like the open-circuit voltage, ionic transference number, etc., is necessary.

4 Conclusions

Nanocrystalline La-doped ceria powder having the composition ($Ce_{0.9}La_{0.1}O_{2-\delta}$) was successfully prepared by auto-combustion method. The XRD analysis confirmed the formation of a solid solution. The AC impedance technique was used to measure the grain, grain boundary, and total ionic conductivity of GDC10 and LDC10 samples between the temperature range of 623 to 1023 K. The ionic conductivity increases with rise in temperature and activation energy decreases. The optimum grain ionic conductivity in the order of 1.10×10^{-1} Scm^{-1} was recorded for the GDC10 sample at 973 K. The highest total ionic conductivity was found to be 1.96×10^{-2} Scm^{-1} and 1.01×10^{-2} Scm^{-1} at 1023 K with a lower activation energy of 0.84 and 0.70 eV for GDC10 and LDC10 samples, respectively. The obtained ionic conductivities for the LDC10 are reasonably close with the GDC10. This work aims to approach the search and development of commercially viable new advanced electrolyte materials for IT-SOFCs.

Table 4 Total ionic conductivities of LDC10 and GDC10 samples sintered at 1473 K

Sample	Total conductivity (σ) [Scm^{-1}]								
	623 K	673 K	723 K	773 K	823 K	873 K	923 K	973 K	1023 K
GDC10	6.25×10^{-5}	1.90×10^{-4}	4.12×10^{-4}	9.03×10^{-4}	1.97×10^{-3}	3.79×10^{-3}	7.05×10^{-3}	1.21×10^{-2}	1.96×10^{-2}
LDC10	3.45×10^{-5}	1.16×10^{-4}	3.21×10^{-4}	6.82×10^{-4}	2.31×10^{-3}	3.56×10^{-3}	5.91×10^{-3}	8.89×10^{-3}	1.01×10^{-2}

Acknowledgements The authors greatly acknowledge the financial support from (i) BRNS/DAE, Government of India [37 (2)/14/20/2015/BRNS], (ii) DST-FIST, Government of India [SR/FST/CSI-273/2016], and (iii) VGST K-FIST-Level-II, Ministry of IT, BT, and Science and Technology, Government of Karnataka.

References

1. Wu W, Chen SA, Chiu YC (2016) Design and control of an SOFC/GT hybrid power generation system with low carbon emissions. Ind Eng Chem Res 55:1281–1291. https://doi.org/10.1021/acs.iecr.5b01961
2. Zhang Y, Jingjing LJ, Singh M, Hu E, Jiang Z, Raza R, Wang F, Wang J, Yang F, Zhu B (2020) Superionic conductivity in ceria-based heterostructure composites for low-temperature solid oxide fuel cells. Nano-Micro Lett 12:1–20. https://doi.org/10.1007/s40820-020-00518-x
3. Plata JJ, Márquez AM, Sanz JF (2014) Understanding the interplay of dopants, interfaces, and anionic conductivity in doped ceria/zirconia heteroepitaxial structures. Chem Mater 26:3385–3390. https://doi.org/10.1021/cm500415h
4. Jadhav LD, Chourashiya MG, Subhedar KM, Tyagi AK, Patil JY (2009) Synthesis of nanocrystalline Gd doped ceria by combustion technique. J Alloys Compd 470:383–386. https://doi.org/10.1016/j.jallcom.2008.02.077
5. Butler V, Catlow CRA, Fender BEF, Harding JH (1983) Dopant ion radius and ionic conductivity in cerium dioxide. Solid State Ionics 8:109–113. https://doi.org/10.1016/0167-2738(83)90070-X
6. Dikmen S, Shuk P, Greenblatt M (1999) Hydrothermal synthesis and properties of $Ce_{1-x}La_xO_{2-\delta}$ solid solutions. Solid State Ionics 126:89–95. https://doi.org/10.1016/S0167-2738(99)00146-0
7. Zhu B, Liu X, Zhou P (2001) Cost-effective yttrium doped ceria-based composite ceramic materials for intermediate temperature solid oxide fuel cell applications. J Mater Sci Lett 20:591–594. https://doi.org/10.1023/A:1010900829589
8. Niu B, Zhang F, Ping H, Li N, Zhou J, Lei L, Xie J, Zhang J, Wang W, Fu Z (2017) Sol-gel autocombustion synthesis of nanocrystalline high-entropy alloys. Sci Rep 7:1–7. https://doi.org/10.1038/s41598-017-03644-6
9. Zhang TS, Ma J, Cheng H, Chan SH (2006) Ionic conductivity of high-purity Gd-doped ceria solid solutions. Mater Res Bull 41:563–568. https://doi.org/10.1016/j.materresbull.2005.09.008
10. Jain SR, Adiga KC, Verneker VRP (1981) A new approach to thermochemical calculations of condensed fuel-oxidizer mixtures. Combust Flame 40:71–79. https://doi.org/10.1016/0010-2180(81)90111-5
11. Suan MSM, Johan MR, Siang CT (2012) Synthesis of $Y_3Ba_5Cu_8O_{18}$ superconductor powder by auto-combustion reaction: effects of citrate-nitrate ratio. Phys C Supercon Its Appl. 480:75–78. https://doi.org/10.1016/j.physc.2012.05.006
12. Damisih Raharjo J, Masmui Aninda RS (2017) Lestari NA (2017) Synthesis and characterization of La, Sc, Yb and Nd co-doped gadolinium doped cerium (GDC) composite electrolyte for IT-SOFC. J Phys Conf Ser 877:012077. https://doi.org/10.1088/1742-6596/877/1/012077
13. Zhang TS, Ma J, Huang HT, Hing P, Xia ZT, Chan SH, Kilner JA (2003) Effects of dopant concentration and aging on the electrical properties of Y-doped ceria electrolytes. Solid State Sci 5:1505–1511. https://doi.org/10.1016/j.solidstatesciences.2003.10.001
14. Chourashiya MG, Jadhav LD (2011) Synthesis and characterization of 10%Gd doped ceria (GDC) deposited on NiO-GDC anode-grade-ceramic substrate as half-cell for IT-SOFC. Int J Hydrogen Energy 36:14984–14995. https://doi.org/10.1016/j.ijhydene.2010.12.083

15. Feng B, Sugiyama I, Hojo H, Ohta H, Shibata N, Ikuhara Y (2016) Atomic structures and oxygen dynamics of CeO_2 grain boundaries. Sci Rep 6:1–7. https://doi.org/10.1038/srep20288
16. Kurian M, Kunjachan C (2014) Investigation of size dependency on lattice strain of nanoceria particles synthesized by wet chemical methods. Int Nano Lett 4:73–80. https://doi.org/10.1007/s40089-014-0122-7
17. Patil S, Seal S, Guo Y, Schulte A, Norwood J (2006) Role of trivalent La and Nd dopants in lattice distortion and oxygen vacancy generation in cerium oxide nanoparticles. Appl Phys Lett 88:2004–2007. https://doi.org/10.1063/1.2210795
18. Hassanzadeh-Tabrizi SA, Mazaheri M, Aminzare M, Sadrnezhaad SK (2010) Reverse precipitation synthesis and characterization of CeO_2 nanopowder. J Alloys Compd 491:499–502. https://doi.org/10.1016/j.jallcom.2009.10.243
19. Patsala P, Logothetidis S, Sygellou L, Kennou S (2003) Structure-dependent electronic properties of nanocrystalline cerium oxide films. Phys Rev B—Condens Matter Mater Phys 68:1–13. https://doi.org/10.1103/PhysRevB.68.035104
20. Tholkappiyan R, Vishista K (2015) Combustion synthesis of Mg-Er ferrite nanoparticles: cation distribution and structural, optical, and magnetic properties. Mater Sci Semicond Process 40:631–642. https://doi.org/10.1016/j.mssp.2015.06.076
21. Ali A, Raza R, Kaleemullah M, Rafique A, Wang B, Zhu B (2018) Alkaline earth metal and samarium co-doped ceria as efficient electrolytes. Appl Phys Lett 112:043902. https://doi.org/10.1063/1.5005824
22. Anjaneya KC, Nayaka GP, Manjanna J, Govindaraj G, Ganesha KN (2013) Preparation and characterization of $Ce_{1-x}Gd_xO_{2-\delta}$ (x = 0.1–0.3) as solid electrolyte for intermediate temperature SOFC. J Alloys Compd 578:53–59. https://doi.org/10.1016/j.jallcom.2013.05.010
23. Ali A, Rafique A, Kaleemullah M, Abbas G, Ajmalkhan M, Ashfaqahmad M, Raza R (2018) Effect of alkali carbonates (single, binary, and ternary) on doped ceria: a composite electrolyte for low-temperature solid oxide fuel cells. ACS Appl Mater Interfaces 10:806–818. https://doi.org/10.1021/acsami.7b17010
24. Fuentes RO, Baker RT (2008) Synthesis and properties of gadolinium-doped ceria solid solutions for IT-SOFC electrolytes. Int J Hydrogen Energy 33:3480–3484. https://doi.org/10.1016/j.ijhydene.2007.10.026

Ni–Fe Alloy Mesh as a Low Cost Oxygen Evolution Catalyst

Thimmasandra Narayan Ramesh and Chikkappa Udagani

1 Introduction

Hydrogen gas is the simplest and efficient form of energy that could be generated through photoelectrochemical reactions and photovoltaic electrolysis [1, 2]. Splitting of water requires an applied potential of 1.23 V indicating it to be a thermodynamically unfavorable process [3]. Commercial electrolyzers generally operate in the voltage range of 1.8–2 V. The kinetics of the reaction contributes to an increased overvoltage. To reduce the overpotential in the aqueous medium, water splitting is carried out either in an acidic or alkaline medium. In order to minimize the overvoltage and develop efficient catalysts, extensive research has been carried out [4–7]. Noble metals such as platinum, ruthenium oxide, and iridium oxide have been used in the past decades [8–12]. As an alternative to noble metals, molybdenum sulfide, ruthenium oxide, nickel molybdenum alloy, cobalt phosphates, metal oxides, and metal hydroxides have been explored as low cost catalysts for oxygen evolution reactions [13–18]. Nickel foam is used as a supporting material and as an electrode for commercial electrolyzers due to its porous nature [19]. Recently, nickel hydroxide and its modified forms have shown four times higher oxygen evolution reaction than the bare nickel in alkaline electrolytes. As an alternative, earth-abundant catalysts have been used in acidic/alkaline electrolyte [20, 21]. Ni–Fe layered double hydroxide (LDH) has been formed upon the incorporation of iron into the structure of nickel hydroxide in stoichiometric proportion. Ni–Fe layered double hydroxide (LDH) exhibits oxygen evolution reaction (OER) at 1.7 V in an alkaline medium on

T. N. Ramesh (✉)
Department of Studies and Research in Chemistry, University College of Science, Tumkur University, Tumakuru 572 103, India

C. Udagani
Department of PG Studies and Research in Physics, University College of Science, Tumkur University, Tumakuru 572 103, India

U. K. Mudali et al. (eds.), *Recent Trends in Electrochemical Science and Technology*, Springer Proceedings in Materials 15,
https://doi.org/10.1007/978-981-16-7554-6_11

nickel foam which is comparable to that of platinum-coated nickel foam [22]. Ni–Fe LDHs (Ni/Fe = 0.75/0.25 and Ni/Fe-0.667/0.333) are poor conductors and hence in search of alternatives, Ni–Fe alloy mesh which is being used in old computer monitors containing cathode ray tubes has been selected in this study since the Ni:Fe ratio in Ni–Fe alloy mesh is in the ratio of 0.715:0.285 falls in the range of $x = 0.25$–0.33 in Ni_{1-x}-Fe_x LDH [23–26]. To date, Ni–Fe alloy mesh e-waste material has not been explored for oxygen evolution catalyst. In this work, the proof of concept on the utilization of low-cost Ni–Fe alloy mesh as a potential catalyst for the OER in an alkaline medium has been demonstrated.

2 Experimental

Ni–Fe alloy mesh was collected from old computer monitors having cathode ray tubes (CRT). The sample has been characterized using Bruker D8 advanced X-ray powder diffractometer (CuKα source $\lambda = 1.5418$ Å) (steps: 0.05; range: 10–60°, scan rate: 4° min^{-1}). The vibrational spectrum of the sample was recorded using Bruker alpha FTIR spectrometer (ATR mode) (4000–375 cm^{-1}; 4 cm^{-1} resolution). Scanning electron microscopy (Carl Zeiss) was used to examine the morphology of the samples and energy dispersive X-ray analysis (EDAX) was used for elemental composition determination. The thickness of Ni–Fe alloy mesh was measured using Yuzuki Dtg1201 digital thickness gauge and was found to be 0.16 mm. Linear and scanning and cyclic voltammetric studies were carried out on a potentiostat/galvanostat (Technoscience Instruments, India) using two/three electrodes system in which Ni–Fe alloy mesh was used as a working electrode and Pt-Ir wire was used as the counter electrode (in two-electrode system). A saturated calomel electrode acting as the reference was used in three-electrode system in 1 M NaOH electrolyte. Linear scanning voltammetric studies were measured by scanning at different rates (10, 20, 40, 80, 100, 125, 200, and 400 mV sec^{-1}). Cyclic voltammetric studies have been carried out in 1 M NaOH solution between −2 V to + 2 V at different scan rates (10, 20, 40, 80, 100, 125, 200 and 400 mV sec^{-1}). All the measured potentials with respect to saturated calomel electrode were changed to the reversible hydrogen electrode (RHE). The DC resistivity was measured using four-probe method (Digital Micro ohmmeter"Sigma", 41/2 Digit).

3 Results and Discussion

The X-ray diffraction pattern of Ni–Fe alloy mesh is shown in Fig. 1a and very weak peaks were observed at 2θ positions that match with the cubic phase of $Ni_{1-x}Fe_y$ alloy ($x = 0.25$ to 0.75; $y = 0.75$ to 0.25) having a lattice parameter $a = 3.637$ Å [27]. Infrared (IR) spectrum was recorded for Ni–Fe alloy mesh and is given in Fig. 1b. The IR spectrum of the Ni–Fe alloy mesh exhibits a weak peak at 425 cm^{-1} corresponding

Fig. 1 **a** Powder X-ray diffraction pattern, **b** infrared spectrum, and **c** scanning electron micrograph of as received Ni–Fe alloy mesh

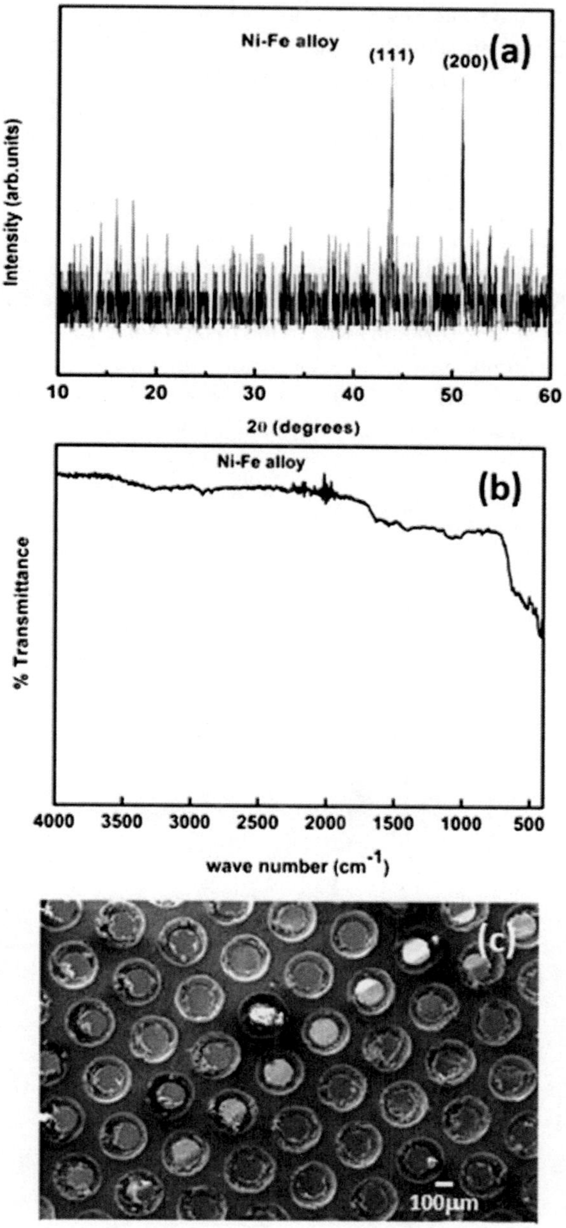

to Ni–O and several peaks in the range of 700–400 cm^{-1} corresponding to Fe–O vibrations without any OH stretching vibrations. Figure 1c shows the scanning electron micrograph of Ni–Fe alloy mesh used for the OER reaction. It shows a regular array of identical pores. The presence of these pores increases the surface area thus enhancing the electrocatalytic activity similar to nickel mesh. Figure 2a shows the EDX spectra and data of Ni–Fe alloy mesh before OER. From the EDAX, the composition of Ni:Fe was found to be in the range of from 0.70:0.30 to 0.72:0.28 which is matching well with the composition of reported Ni–Fe LDH [28–31]. One of the advantages of using Ni–Fe alloy mesh is its lower resistivity compared to Ni–Fe based layered double hydroxide. The DC resistivity of Ni–Fe alloy mesh with an area of 0.78 cm^2 was found to be 6.84 × 10^{-8} Sm^{-1}.

Figure 3a–d shows the results of linear sweep voltammetry (LSV) and cyclic voltammetric (CV) studies of Ni–Fe alloy mesh using two-electrode and three-electrode systems in 1 M NaOH. 1 M NaOH has been used in the measurements to

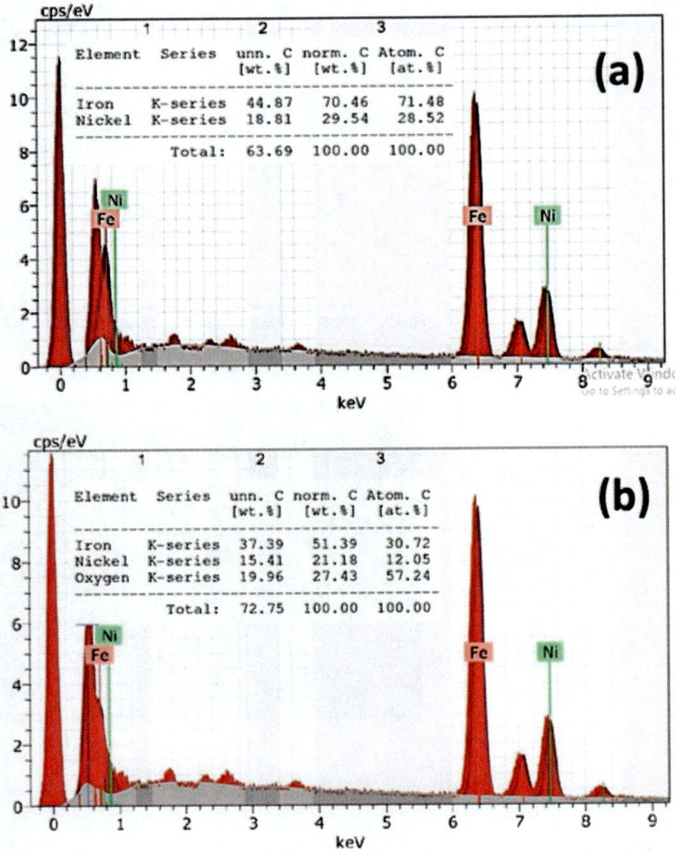

Fig. 2 EDAX spectra of Ni–Fe alloy mesh **a** before and **b** after OER

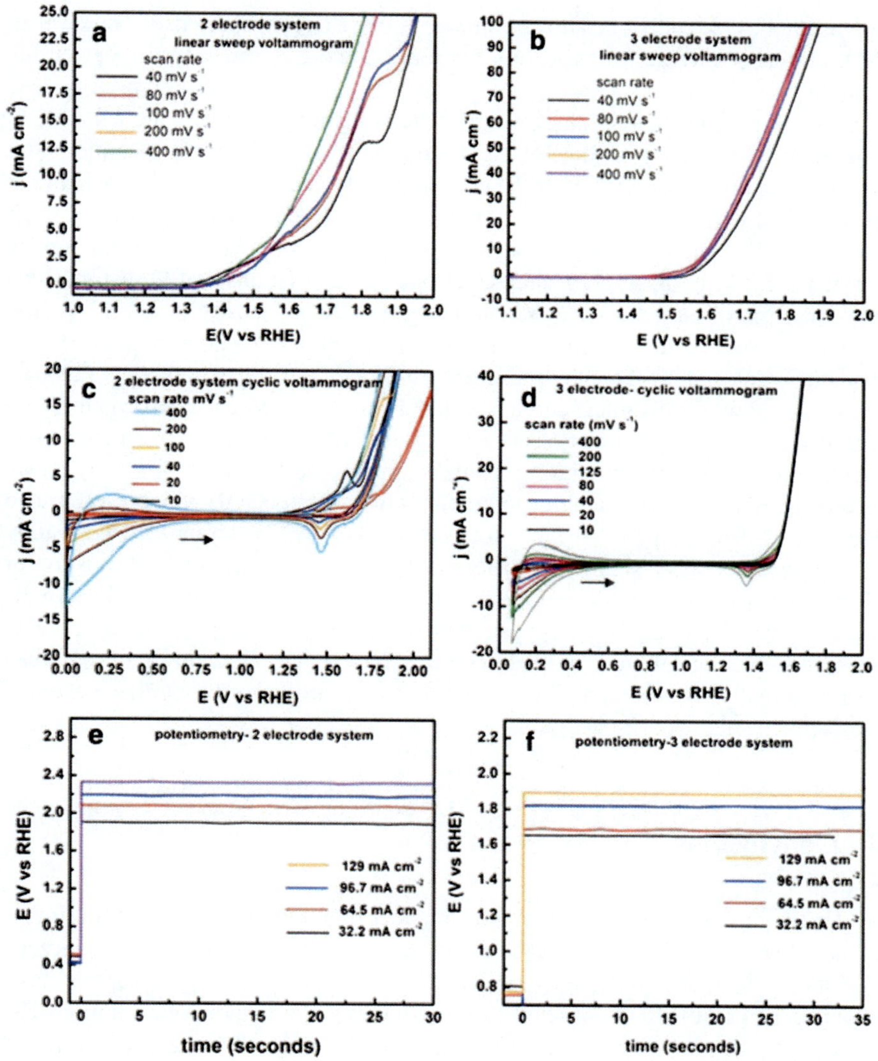

Fig. 3 Linear sweep voltammograms at different scan rates: **a** two-electrode system [Ni–Fe alloy mesh (WE) and Pt-Ir alloy (CE)], **b** three-electrode system [Ni–Fe alloy mesh (WE), Pt-Ir alloy (CE), and RHE as reference electrode]; Cyclic voltammograms at different scan rates: **c** two-electrode system [Ni–Fe alloy mesh (WE) and Pt-Ir alloy(CE)], **d** three-electrode system [Ni–Fe alloy mesh (WE), Pt-Ir alloy (CE) and RHE as reference electrode]; chronopotentiograms at different current densities: E) two-electrode system [Ni–Fe alloy mesh (WE) and Pt-Ir alloy (CE)], **e** three-electrode system[Ni–Fe alloy mesh (WE), Pt-Ir alloy(CE) and RHE as reference electrode]

compare the values of Ni–Fe alloy mesh with Ni–Fe LDH as reported by Gratzel et al. [22]. The onset of OER occurs at 1.6 V for both the two-electrode and three-electrode systems.

Since linear sweep voltammetry and cyclic voltammetric studies are dynamic in nature, and the onset potentials of OER might not be accurate, the potentiometric studies were carried out by fixing the current density and measuring the variation in the potential with time (see Figs. 3e, f). For two-electrode system, an onset of OER for Ni–Fe alloy mesh is observed at 1.7 V and with the increase in current density it reaches 2.2 V. In the case of three-electrode system, the onset of OER for Ni–Fe alloy mesh is observed at 1.63 V and at higher current density it is 1.89 V. Iron is more susceptible to oxidation in the alkaline medium compared to nickel, hence the composition of Ni–Fe alloy mesh was analyzed after the experiments [29–31]. Figure 2b shows the chemical composition indicating the presence of oxygen. There was also a change in the color of Ni–Fe alloy mesh from black to brown after OER measurements. Despite undergoing oxidation on the surface of Ni–Fe alloy mesh, its efficiency was not affected due to the formation of NiFeOOH which itself is also a good OER catalyst [32, 33]. The onset potential of OER reactions of Ni–Fe alloy mesh might not show significant differences compared to that of Pt/Pt-Ir alloy or Ni–Fe LDH [33]. One of the major limitations of Ni–Fe LDH is that it will be in the form of powder and has to be deposited on a supporting material/substrate prior to testing its electrocatalytic activity. In contrast, Ni–Fe alloy in the form of mesh can be used directly and thus can act as an alternative electrocatalyst for OER in alkaline medium.

4 Conclusions

Ni–Fe alloy mesh, e-waste from an old computer monitor having Ni:Fe ratio similar to that of Ni–Fe LDH has been used as an electrocatalyst for the OER. The oxygen evolution potential of Ni–Fe alloy mesh is comparable to that of expensive Pt/Ni/Pt-Ir alloy/Ni–Fe LDH with the onset OER at 1.63 V in sodium hydroxide solution. The electrocatalytic activity of Ni–Fe alloy mesh has been demonstrated and proposed as an alternative to the noble metal catalyst used in the practical applications of electrolysis of water in an alkaline medium.

Acknowledgements Thimmasandra Narayan Ramesh (TNR) gratefully thank Council of Scientific and Industrial Research (CSIR/01/2741/13/EMR-II/dated18-06-13;SR/FTP/PS-134/2010 dated 13-02-2014), Department of Science and Technology (DST), Government of India for providing financial support. TNR also thank Vision Group of Science and Technology (VGST), Government of Karnataka for financial support. Authors gratefully thank Tumkur University.

Conflict of Interest Authors declare no conflicts of interest.

References

1. Dias P, Mendes A (2018) Hydrogen production from photoelectrochemical water splitting. In: Meyers R (ed) Encyclopedia of sustainability science and technology. Springer, New York
2. Narayanan H, Viswanathan B, Krishnamurthy KR, Nair H (2019) Chapter 12—hydrogen from photo-electrocatalytic water splitting. Solar Hydrogen Production, Processes, Systems and Technologies, 419–486
3. Veeramani V, Yu H-C, Hu S-F, Liu R-S (2018) Highly efficient photoelectrochemical hydrogen generation reaction using tungsten phosphosulfide nanosheets. ACS Appl Mater Inter 10:17280–17286. https://doi.org/10.1021/acsami.8b03692
4. Li F, Yang H, Li W, Sun L (2018) Device fabrication for water oxidation, hydrogen generation and CO_2 reduction via molecular engineering. Joule 2:36–60. https://doi.org/10.1016/j.joule.2017.10.012
5. Yang W, Prabhakar RR, Tan J, Tilley SD, Moon J (2019) Strategies for enhancing the photocurrent, photovoltage, and stability of photoelectrodes for photoelectrochemical water splitting. Chem Soc Rev 48:4979–5015. https://doi.org/10.1039/C8CS00997J
6. Varadhan P, Fu H-C, Kao Y-C, Horng R-H, He JH (2019) An efficient and stable photoelectrochemical system with 9% solar-to-hydrogen conversion efficiency via InGaP/GaAs double junction. Nature Comm 10:1–9. https://doi.org/10.1038/s41467-019-12977-x
7. Jia J, Seitz L, Benck J, Huo Y, Chen Y, Ng J, Bilir T, Harris J, Jaramillo T (2016) Solar water splitting by photovoltaic-electrolysis with a solar-to-hydrogen efficiency over 30%. Nature Comm 7:1–6. https://doi.org/10.1038/ncomms13237
8. Kibsgaard J, Ib C (2019) Considerations for the scaling-up of water splitting catalysts. Nat Ener 4:430–433. https://doi.org/10.1038/s41560-019-0407-1
9. Tributsch H (2018) Chapter 1, the challenge of water splitting in view of photosynthetic reality and of research trends, advances in photoelectrochemical water splitting: Theory, experiment and systems analysis, energy and environment series, in David Tilley S, Lany S, van de Krol R(eds) 28:1–29
10. Nguyen PD, Duong TM, Tran PD (2017) Current progress and challenges in engineering viable artificial leaf for solar water splitting. J Sci Adv Mater Devices 2:399–417. https://doi.org/10.1016/j.jsamd.2017.08.006
11. Hossain A, Sakthipandi K, Atique Ullah AKM, Roy S (2019) Recent progress and approaches on carbon-free energy from water splitting. Nano-Micro Lett 11:1–26. https://doi.org/10.1007/s40820-019-0335-4
12. Stolarczyk JK, Bhattacharyya S, Polavarapu L, Feldmann J (2018) Challenges and prospects in solar water splitting and CO_2 reduction with inorganic and hybrid nanostructures. ACS Catal 8:3602–3635. https://doi.org/10.1021/acscatal.8b00791
13. Ghosh SK, Rahaman H (2019) Chapter 16—Noble metal–manganese oxide hybrid nanocatalysts, noble metal-metal oxide hybrid nanoparticles: fundamentals and applications. Micro Nanoo Tech 313–340
14. Li C, Baek J-B (2020) Recent advances in noble metal (Pt, Ru, and Ir)-based electrocatalysts for efficient hydrogen evolution reaction. ACS Omega 5:31–40. https://doi.org/10.1021/acsomega.9b03550
15. Mamaca N, Mayousse E, Arrii-Clacens S, Napporn TW, Servat K, Guillet N, Kokoh KB (2012) Electrochemical activity of ruthenium and iridium based catalysts for oxygen evolution reaction. Applied Cat B: Envir 111–112:376–380. https://doi.org/10.1016/j.apcatb.2011.10.020
16. Tada H, Nay S-I, Fujishima M (2018) Water splitting by plasmonic photocatalysts with a gold nanoparticle/cadmium sulfide heteroepitaxial junction: a mini review. Electrochem Comm 97:22–26. https://doi.org/10.1016/j.elecom.2018.10.005
17. Vij V, Sultan S, Harzandi AM, Meena A, Tiwari JN, Lee W-G, Yoon T, Kim KS (2017) Nickel-Based Electrocatalysts for energy-related applications: oxygen reduction, oxygen evolution and hydrogen evolution reactions. ACS Catal 7:7196–7225. https://doi.org/10.1021/acscatal.7b01800

18. Creus J, De Tovar J, Romero N, García-Antón J, Philippot K, Bofill R, Sala X (2019) Ruthenium nanoparticles for catalytic water splitting. Chemsuschem 12:2493–2514. https://doi.org/10.1002/cssc.201900393
19. Hu X, Tian X, Lin Y-W, Wang Z (2019) Nickel foam and stainless steel mesh as electrocatalysts for hydrogen evolution reaction, oxygen evolution reaction and overall water splitting in alkaline media. RSC Adv 9:31563–31571. https://doi.org/10.1039/C9RA07258F
20. Wang J, Yue X, Yang Y, Sirisomboonchai S, Wang P, Mad X, Abudula A, Guan G (2020) Earth-abundant transition-metal-based bifunctional catalysts for overall electrochemical water splitting: a review. J Alloys Comp 819:153346–153384. https://doi.org/10.1016/j.jallcom.2019.153346
21. Roger I, Shipman MA, Symes MD (2017) Earth-abundant catalysts for electrochemical and photoelectrochemical water splitting. Nat Rev Chem 1:1–6. https://doi.org/10.1038/s41570-016-0003
22. Luo J, Im J-H, Mayer MT, Schreier M, Nazeeruddin MK, Park N-G, Tilley SD, Fan HJ, Grätzel M (2014) Water photolysis at 12.3% efficiency via perovskite photovoltaics and earth-abundant catalysts. Science 345:1593–1596. https://doi.org/10.1126/science.1258307
23. Wang Q, Wang X, Tian B (2018) Catalytic performances of Ni/Fe layered double hydroxides fabricated via different methods in Fenton-like processes. Water Sci Technol 77:2772–2780. https://doi.org/10.2166/wst.2018.265
24. McCrea JL, Palumbo G, Hibbard GD, Erb U (2003) Properties and applications for electrodeposited nanocrystalline Fe-Ni alloys. Rev Adv Mater Sci 5:252–258
25. Inaba M, Teshima K, Higashinakagawa E, Ohtake Y (1988) Development of an Invar (Fe-36Ni) shadow mask for color cathode ray tubes. IEEE Trans Electron Devices 35:1721–1729. https://doi.org/10.1109/16.7378
26. Comana V, Robotinb B, Ileac P (2013) Hydrometallurgical processing and recovery of nickel from spent cathode ray tubes, e3s web of conferences 1:31004. https://doi.org/10.1051/e3sconf/20130131004
27. Kesavapillai SD, Remadevi A, Mathew P, Saju P, Kuzhichalil PS (2018) Polyol derived Ni and NiFe alloys for effective shielding of electromagnetic interference. Mater Chem Front 2:1829–1841. https://doi.org/10.1039/C8QM00264A
28. Ming G, Hongjie D (2015) A mini review of NiFe-based materials as highly active oxygen evolution reaction electrocatalysts. Nano Res 8:23–39. https://doi.org/10.1007/s12274-014-0591-z
29. Han Y, Liu Z-H, Yang Z, Wang Z, Tang X, Wang T, Fan L, Ooi K (2008) Preparation of $Ni^{2+}-Fe^{3+}$ layered double hydroxide material with high crystallinity and well-defined hexagonal shapes. Chem Mater 20:360–363. https://doi.org/10.1021/cm7023789
30. Xue BW, Zhang CH, Wang YZ, Xie WW, Li NW, Yu L (2020) Recent progress of Ni–Fe layered double hydroxide and beyond towards electrochemical water splitting. Nanoscale Adv Pub Date 06 Oct 2020. https://doi.org/10.1039/d0na00727g
31. Gong M, Li Y, Wang H, Liang Y, Wu JZ, Zhou J, Wang J, Regier T, Wei F, Da H (2013) An advanced Ni–Fe layered double hydroxide electrocatalyst for water oxidation. J Am Chem Soc 135(23):8452–8455 (2013). https://doi.org/10.1021/ja4027715
32. Song F, Busch MM, Lassalle-Kaiser B, Hsu C-S, Petkucheva E, Bensimon M, Chen HM, Corminboeuf C, Hu X (2019) An unconventional iron nickel catalyst for the oxygen evolution reaction. ACS Cent Sci 5:558–568. https://doi.org/10.1021/acscentsci.9b00053
33. Yu M, Moon G, Bill E, Tüysüz H (2019) Optimizing Ni–Fe oxide electrocatalysts for oxygen evolution reaction by using hard templating as a toolbox. ACS Appl Energy Mater 2:1199–1209. https://doi.org/10.1021/acsaem.8b01769

Design of Imprinting Matrix for Dual Template Sensing Based on Molecularly Imprinted Polymer Technology

Ritu Singh, Manjeet Harijan, and Meenakshi Singh

1 Introduction

Concerning the burden of pharmaceutical products on our society, their misuse and over usage and even of their metabolites, they should be monitored consistently and judiciously and their usage should be scaled down. Molecular imprinting (MIP) tenders such monitoring in an effortless manner [1]. MIP often named 'artificial enzymes' and/or 'artificial antibodies' is one of the most promising and flourishing techniques for sensor designing. The major advantages of MIPs compared with antibodies are their high and almost unlimited stability and the ease of preparation even at a large scale that undoubtedly supersedes antibodies in terms of costs.

Considerable interest is being devoted to the design of a tailor-made polymer matrix capable of recognizing and binding the molecular target with high selectivity [2]. Strategic integration of recognition element (MIP) with a transducer for fabrication of sensors is a requisite. Electro polymerization of monomers with analyte on electrode itself is a facile answer to such integration. A review on this electrochemical approach by Sharma et al. [3] entails such approaches used in integrating imprinting technology for the fabrication of sensing devices. The monomers generally utilized for this purpose are pyrrole, *o*-phenylene diamine, phenols, thiophene, and/or derivatives of them [3]. 3-thiophene acetic acid(3-TAA) is a suitable monomer for designing electrochemical MIP sensors as it easily generates thin films on conducting transducer surfaces. The carboxyl group is free for interaction with template molecules and does not participate in polymerization [4]. Poly(3-thiophene acetic acid) has been widely utilized for producing biomolecular hybrids [5, 6]. Imprinted polymer fabricated electrochemical sensing platforms are advantageous in terms of low cost, high sensitivity, and selectivity towards imprinted molecules [7].

R. Singh · M. Harijan · M. Singh (✉)
Department of Chemistry, MMV, Banaras Hindu University, Varanasi 221005, India
e-mail: meenakshi@bhu.ac.in

U. K. Mudali et al. (eds.), *Recent Trends in Electrochemical Science and Technology*, Springer Proceedings in Materials 15,
https://doi.org/10.1007/978-981-16-7554-6_12

Graphene comprises a single layer of carbon atoms that are sp^2 bonded, arranged two-dimensionally in a honeycomb lattice with good mechanical and thermal strength, high electrical conductivity, and large surface area making it a preferable choice for composites or nanocomposites [8–11]. Graphene/ reduced graphene oxide (rGO) being a good electrical conductor can be dispersed or mixed with polymer solutions effortlessly. MIP sensors with graphene are more sensitive and selective showing good template binding capacity [12]. rGO is well known to increase the sensitivity of sensing platforms due to the high surface area and unusual electrochemical properties [13]. During imprinting, molecular sites are introduced into the polymer matrix that can recognize the template and selectively interact with it.

To date, many research works have been performed for multiple analyte determination using the electrochemical sensing platform [14]. MIP fabricated electrochemical sensor has also been utilized by creating cavities for selective determination of two molecules simultaneously. But most of these sensors are based on the detection of either structurally or functionally similar template molecules. The present work describes the fabrication of an electrochemical sensor for two structurally and functionally different drugs in order to monitor the effect of different drugs in a patient suffering from multiple diseases. Here, antipyrine and ethionamide are used as model drugs to imprint the polymer film on the electrode for electrochemical determination. Antipyrine is used as an analgesic to reduce pain and inflammation, usually used with other antibiotics to treat ear infections [15, 16]. Ethionamide (ETA) is the most common drug that is used with other medications to treat tuberculosis. ETA functions by inhibiting the growth of Mycobacterium tuberculosis but its minimal dose requirement to treat infection is sufficiently high to avoid various side effects such as gastrointestinal disorders, hepatotoxicity, neurotoxicity, cardiovascular effects, endocrine effects, and skin reactions [17]. Both drugs are electroactive at different potential windows without any interference with each other. This study can pave the way for designing drug delivery vehicles for those patients who need multiple drugs for their treatment and also in a controlled manner.

2 Experimental

Antipyrine and 3-thiophene acetic acid (3-TAA) was purchased from sigma Aldrich, while ethionamide was purchased from MP Biomedicals. Sodium dihydrogen orthophosphate dihydrate and disodium hydrogen orthophosphate were purchased from Fisher scientific, Ethanol (AR grade) were purchased from Changshu Hongsheng Fine Chemical Co. Ltd. while methanol (HPLC grade) and acetic acid were purchased from Merck. All the reagents used were of analytical grade and double distilled water was used to prepare solutions.

Electrochemical experiments were performed with a computer-controlled electrochemical workstation (CH Instruments, CHI 210A) connected with a conventional three-electrode system containing an electrochemical cell. A glassy carbon electrode (GCE), a Pt wire, and an Ag/AgCl (3 M KCl) were used as working, counter and reference electrodes respectively. For electrochemical sensing, differential pulse voltammetry (DPV) responses for the determination of analytes were employed on imprinted polymer film coated GCE.

For cleaning, the GCE was polished with 0.05, 0.3, and 1 μm aqueous alumina slurry using smooth polishing micro-cloth pads, followed by thorough rinsing with deionized water [18, 19]. Then, the GCE was ultrasonicated in acetone and distilled water each for 10 min to remove leftover alumina or other impurities. Cleaning of GCE was assured by reproducible cyclic voltammograms obtained in 5 mM potassium ferricyanide solution (contain 0.1 mol/L KCl) solution over a potential range from − 0.2 V to + 0.6 V, with a scan rate of 50 mV/s.

Reduced graphene oxide (rGO) was synthesized by a previously reported method using graphene oxide as precursor and was used as prepared [20]. The cleaned GCE was modified with rGO. Subsequently, the template imprinted 3-TAA polymer matrix was fabricated onto rGO modified electrode by using cyclic voltammetry (CV) technique. For this purpose, the solution of 10×10^{-3} mol/L of 3-TAA (as monomer), 0.8×10^{-3} mol/L of ethionamide (template-1), 0.8×10^{-3} mol/L of antipyrine (template-2) was prepared. Next, the CV was performed to allow electro polymerization of 3-TAA monomer along with templates on rGO/GCE to form MIP adduct. After electro polymerization, the electrode was washed with double distilled water and kept to dry overnight at 4 °C (under refrigerated conditions). The MIP adduct modified electrode was washed with the methanol-acetic acid solution for 30 min by sonication, followed by subsequent washing with methanol to elute the template molecules entrapped in the polymeric matrix. Non imprinted polymer (NIP) was also prepared similarly to that of MIP without the template drug molecules.

3 Results and Discussion

3.1 Electro Polymerization of 3-Thiophene Acetic Acid

3-TAA is an advantageous monomer for imprinting as its carboxyl groups are not involved in polymer chain growing reactions and are free for interacting with complementary functional groups available on chosen analyte/template (here ethionamide and antipyrine) [4]. 3-TAA has been successfully electropolymerized to develop a polymer matrix with controlled thickness. Both MIP and NIP were synthesized using CV employing multiple cycles. The first cycle of voltammograms recorded during the electro polymerization of MIP and NIP films respectively with or without a template on rGO coated GCE are shown in Fig. 1.

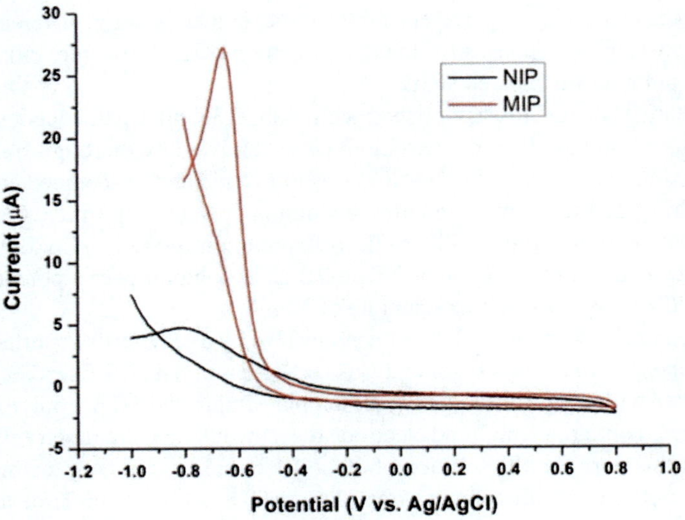

Fig. 1 Cyclic voltammogram obtained during the electro polymerization of 3-thiophene acetic acid with antipyrine and ethionamide (MIP) and without antipyrine and ethionamide (NIP)

3.2 *Extraction of Templates*

Extraction of drugs (antipyrine and ethionamide) from polymer matrix was optimized using different solvents like acetone, ethanol, methanol, and acetic acid. Acetic acid and methanol in the ratio 1:9 were found to have higher extraction efficiency. Extraction was confirmed by the appearance of a peak at about − 1.4 V for ethionamide and + 1.25 V for antipyrine in DPV of extracted solution (Fig. 2).

3.3 *Imprinting Effect of MIP*

The imprinting effect of MIP for ethionamide and antipyrine was also verified by DPV response. MIP clearly exhibited a well-defined reduction peak at about − 1.4 V for ethionamide and peak current tends to increase for the various increasing concentrations of the analyte. DPV responses for three varying concentrations of ethionamide at 0.01, 0.03 and 0.06 mM in 0.1 M PBS (pH 7) are shown in Fig. 3. On the other hand, antipyrine exhibited a distinct peak at around 1.25 V. DPV responses for various concentrations of antipyrine in 0.1 M PBS (pH 7) showed a sequential increase in increasing the concentration (Fig. 4). With the increase in concentration of analytes more imprinted sites were occupied sequentially and consequently, the corresponding increase in peak current was observed.

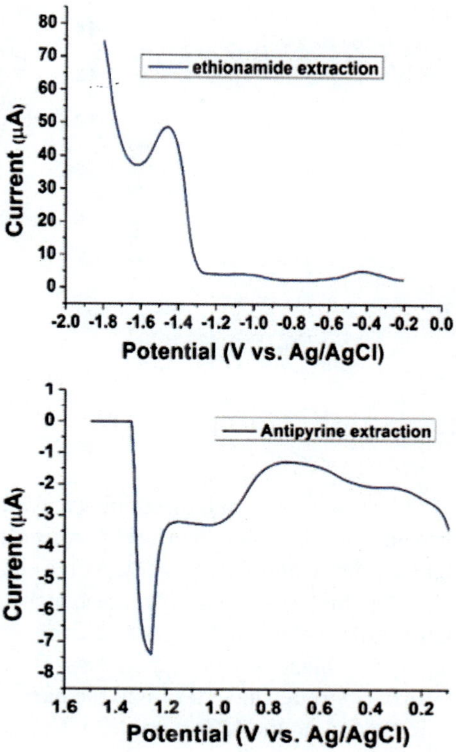

Fig. 2 DPV responses for ethionamide and antipyrine in the extracted solution of methanol and acetic acid in 0.1 M PBS buffer

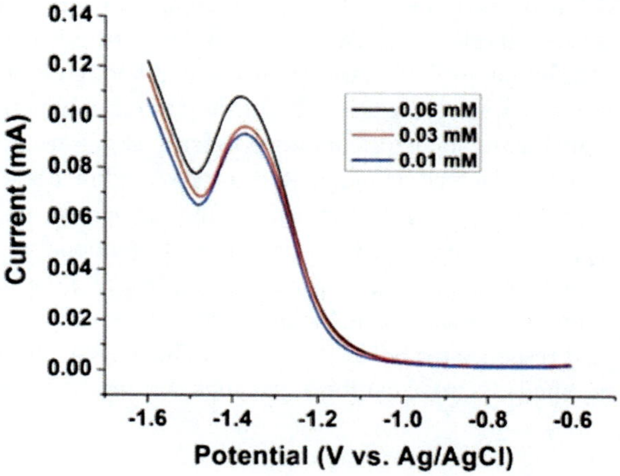

Fig. 3 DPV response for varying concentrations of ethionamide

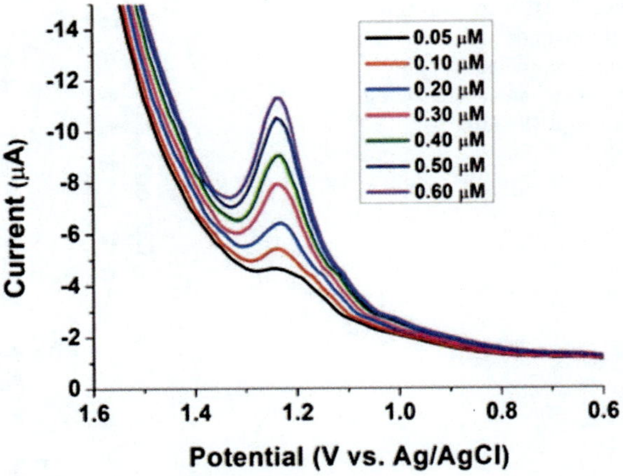

Fig. 4 DPV response in the concentration range of 0.05–0.6 μM of antipyrine

The sensor was found to exhibit excellent selectivity for imprinted analytes in the presence of analogous interfering molecules such as 4-aminoantipyrine (4 AP), 4-dimethylaminoantipyrine (4-DMAP), isoniazide, rifampicin, melphalan, phenylalanine. The highest selectivity was obtained for ethionamide and antipyrine on MIP while NIP prepared under the same experimental conditions in their absence showed negligible signals for both the drugs. rGO facilitated in improving the sensitivity towards imprinted analytes up to the nanomolar range.

Rebinding at the MIP modified GCE and NIP modified GCE was also checked. Selective recognition towards the template molecule is an overriding capability for MIP sensors. On extraction of the template from polymer, several imprinted cavities are generated in polymeric film; these specific cavities left behind contribute to selective capture of analyte molecules. Although functional groups responsible for imprinting are present in the non-imprinted polymeric matrix also, in the absence of analyte/template orientational functional groups are lacking in NIP thus not able to induce the analyte molecules towards the functional groups of the non-imprinted matrix. In the imprinted matrix, analyte molecules leave behind their 'imprint', i.e. the guided conformational orientation of functional groups which induce the analyte molecules towards them from solution. Hence, imprinted matrix developed specific 'contours' for analyte molecules during molecular imprinting. NIP showed no significant response for both the analytes. The 3-TAA imprinted polymer matrix on rGO modified electrode exhibited excellent sensitivity for both templates.

3.4 Reusability and Stability of MIP Sensor

Repeatability and reusability of the designed sensor are important for practical applications of detection platforms. The proposed sensor was found to exhibit good reproducibility even after one month, as no significant loss in peak current was observed. Repetitive use of MIP sensor with extraction (regeneration of imprinted cavity) and rebinding did not compromise the sensitivity and selectivity of the sensing matrix. The changes in current were optimized at the same concentration and an almost similar response was obtained on the electrode surface, indicating excellent reusability of the MIP sensor towards both the analytes.

4 Conclusions

This study highlights the development of a sensitive sensing platform in which rGO modified GCE was used for sensing antipyrine and ethionamide simultaneously at a single sensing platform as a model drug with exemplary selectivity and sensitivity. A dual template MIP sensing electrode where both antipyrine and ethionamide could be estimated one-after-another, without any interference was successfully fabricated. However, simultaneous analysis of these drugs at the same time appeared to be cumbersome presumably because of long-range of potential differences. The future work is focused on resolving this issue.

References

1. Whitcombe MJ, Chianella I, Larcombe L et al (2011) The rational development of molecularly imprinted polymer-based sensors for protein detection. Chem Soc Rev 40:1547–1571. https://doi.org/10.1039/c0cs00049c
2. Chen L, Wang X, Lu W et al (2016) Molecular imprinting: perspectives and applications. Chem Soc Rev 45:2137–2211. https://doi.org/10.1039/C6CS00061D
3. Sharma PS, Pietrzyk-Le A, D'Souza F, Kutner W (2012) Electrochemically synthesized polymers in molecular imprinting for chemical sensing. Anal Bioanal Chem 402:3177–3204. https://doi.org/10.1007/s00216-011-5696-6
4. Malitesta C, Guascito MR, Mazzotta E, Picca RA (2010) X-Ray photoelectron spectroscopy characterization of electrosynthesized poly(3-thiophene acetic acid) and its application in molecularly imprinted polymers for atrazine. Thin Solid Films 518:3705–3709. https://doi.org/10.1016/j.tsf.2009.10.016
5. Janáky C, Endrodi B, Kovács K et al (2010) Chemical synthesis of poly(3-thiophene-acetic-acid)/magnetite nanocomposites with tunable magnetic behaviour. Synth Met 160:65–71. https://doi.org/10.1016/j.synthmet.2009.09.034
6. Mukherjee P, Dawn A, Nandi AK (2010) Biomolecular hybrid of poly(3-thiophene acetic acid) and double stranded DNA: optical and conductivity properties. Langmuir 26:11025–11034. https://doi.org/10.1021/la101215v
7. Yang S (2019) Electrochemical sensor based on molecularly imprinted polymer-aptamer hybrid receptor for voltammetric detection of thrombin. 166(2):B23–B28. https://doi.org/10.1149/2.0131902jes
8. Kim H, Abdala AA, MacOsko CW (2010) Graphene/polymer nanocomposites. Macromolecules 43:6515–6530. https://doi.org/10.1021/ma100572e
9. Stauber T, Beltrán JI, Schliemann J (2016) Tight-binding approach to pentagraphene. Sci Rep 6:4–11. https://doi.org/10.1038/srep22672
10. Artiles MS, Rout CS, Fisher TS (2011) Graphene-based hybrid materials and devices for biosensing. Adv Drug Deliv Rev 63:1352–1360. https://doi.org/10.1016/j.addr.2011.07.005

11. Singh V, Joung D, Zhai L et al (2011) Graphene based materials: past, present and future. Prog Mater Sci 56:1178–1271. https://doi.org/10.1016/j.pmatsci.2011.03.003
12. Duan D, Yang H, Ding Y et al (2018) Three-dimensional molecularly imprinted electrochemical sensor based on Au NPs@Ti-based metal-organic frameworks for ultra-trace detection of bovine serum albumin. Electrochim Acta 261:160–166. https://doi.org/10.1016/j.electacta.2017.12.146
13. Taniselass S, Arshad MKM, Gopinath SCB (2019) Graphene-based electrochemical biosensors for monitoring noncommunicable disease biomarkers. Biosens Bioelectron 130:276–292. https://doi.org/10.1016/j.bios.2019.01.047
14. Zhang H, Gai P, Cheng R et al (2013) Self-assembly synthesis of a hierarchical structure using hollow nitrogen-doped carbon spheres as spacers to separate the reduced graphene oxide for simultaneous electrochemical determination of ascorbic acid, dopamine and uric acid. Anal Methods 5:3591–3600. https://doi.org/10.1039/c3ay40572a
15. Zou YX, Feng X, Chu ZY et al (2019) Preclinical safety assessment of antipyrine combined with lidocaine hydrochloride as ear drops. Regul Toxicol Pharmacol 103:34–40. https://doi.org/10.1016/j.yrtph.2019.01.018
16. Vesell ES (1979) The antipyrine test in clinical pharmacology: conceptions and misconceptions. Clin Pharmacol Ther 26:275–286. https://doi.org/10.1002/cpt1979263275
17. Arbex M, Castro Lima Varella M, Siquera H, Mello F (2010) Antituberculosis drugs: Drug interactions, adverse effects, and use in special situations.J Bras Pneumol 36:626–640. https://doi.org/10.1590/s1806-37132010000500016
18. Hrichi H, Louhaichi MR, Monser L, Adhoum N (2014) Gliclazide voltammetric sensor based on electropolymerized molecularly imprinted polypyrrole film onto glassy carbon electrode. Sens Actuators B Chem 204:42–49. https://doi.org/10.1016/j.snb.2014.07.056
19. da Silva H, Pacheco JG, MCS Magalhães J, Viswanathan S, Delerue-Matos C (2014) MIP-graphene-modified glassy carbon electrode for the determination of trimethoprim. Biosens Bioelectron 52:56–61. https://doi.org/10.1016/j.bios.2013.08.035
20. Srivastava J, Gupta N, Kushwaha A, Umrao S, Srivastava A, Singh M (2019) Highly sensitive and selective estimation of aspartame by chitosan nanoparticle—graphene nanocomposite tailored EQCM-MIP sensor. Polym Bull 76:4431–4449. https://doi.org/10.1007/s00289-018-2597-2

Recent Perspective and Applications of Electrode Materials for Electrochemical Sensing of Lead Ions

Monika Antil and Babankumar S. Bansod

1 Introduction

Our naturally occurring resources like water, air, and soil are badly affected because of heavy metal ions contamination due to industrial activities, fossil fuels productions, usage of fertilizers, and many other processes [1]. Many pollutants have been released regularly into the environment, but heavy metal ions are hazardous and are categorized into a different category because of their highly toxic and non-biodegradable nature [2]. Heavy metals can cause profound loss to human beings due to their non-declining nature. These heavy metal ions are difficult to degrade in the human body; they eventually accumulate within the human body [3]. Different types of heavy metals in the ecosystem, such as arsenic, cadmium, lead, mercury, etc., have threatening hazards that rely on their chemical forms and exposure levels. Lead is a valuable metal found in our environment and has wide applications in various fields. Lead is accepted as a rare case because it can quickly arise in potable water through the plumbing system and is a scary part released from natural weathering processes and industrial processes. The primary causes of lead pollution occur through automobiles, factories, lead storage battery effluents, glass manufacturing industries, lead ores mining and smelting, finishing operations, fertilizers, and additives in petroleum industries, etc. Other lead sources are occupational exposures like plumbers and fitters, chemical industry, construction workers, demolition workers, radiator repair, car technicians, battery manufacturers, recyclers, and plastic manufacturers are at risk due to lead exposure [4]. Tetraethyl and tetramethyl lead is still used to increase the octane number of gasoline in some regions like Afghanistan,

Monika Antil (✉) · B. S. Bansod
AcSIR -CSIR-Central Scientific Instrument Organisation, Chandigarh 160030, India
e-mail: monika@csio.res.in

B. S. Bansod
e-mail: scientist_babankumar@csio.res.in

U. K. Mudali et al. (eds.), *Recent Trends in Electrochemical Science and Technology*, Springer Proceedings in Materials 15,
https://doi.org/10.1007/978-981-16-7554-6_13

Table 1 Safe permissible limits of lead

EPA limit	WHO limit	EU limit	New EU limit [2020]
15 μg/L	10 μg/L [9, 10]	10 μg/L	5 μg/L [8, 11]

Algeria, Iraq, etc. Lead ions released from the automobiles exhaust are also present on the plants near the highways. The lead ions strongly affect the human body and the environment, even at sub-levels. Lead ions affect the different body parts such as the brain, nervous system, also cause cancer, skin lesions, cardiovascular diseases, neurological disorders, etc. [5, 6]. Due to its non-biodegradable nature, they increase significantly due to their tendency to accumulate in living being's vital organs [7]. Therefore, it's necessary to monitor and control lead exposures. Due to an internecine effect of lead on the environment, the World Health Organization has set its permissible safe limit as ten parts per billion (Table 1). Recently, the European Union reduced its safe permissible limit from ten to five per billion [8], affirm the demand to identify high lead exposure areas using affordable, in-situ, and accurate methods (Table 1).

2 An Electrochemical Technique for Lead Sensing

To assess the importance of this area of research, the Scopus analysis was carried out using the following search words: "Electrochemical sensing" and "Lead" (Fig. 1). From the analysis, it is evident that there is a progressive increase in the number of publications on this topic (Fig. 1a). China is the leading country researching this topic followed by India (Fig. 1b). In India, the leading institutions where the research on electrochemical sensing of lead ions is pursued are University of Madras, Birla Institute of Technology, Banaras Hindu University, Bangalore University, Indian Institute of Science, CSIR-CSIO, and others.

For monitoring lead ions in drinking water, robust and versatile electrochemical techniques have been used to provide accurate measurements with high sensitivity, high accuracy, and an extensive linear dynamic range, with inexpensive and excellent portable instrumentation [12, 13]. Among the different electrochemical techniques (Fig. 2), stripping voltammetry is a promising method for detecting trace metal ions because of its pre-concentration step [3, 14, 15].

2.1 Principle of Electrochemical Sensing of Lead

Anodic stripping voltammetry is one of the most suitable and best approaches for analyzing lead ions in water due to its high sensitivity and selectivity, ease of operation, cost-effectiveness, small equipment, easy installation, simple sample preparation, and ability to perform multi-elemental detection [15]. It is an ecofriendly and

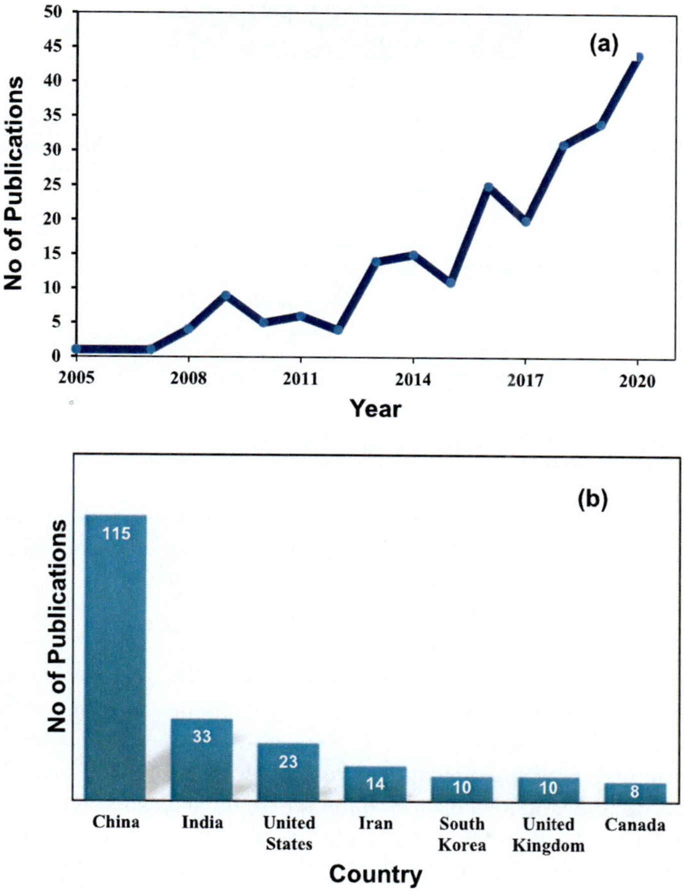

Fig. 1 Scopus analysis of the literature data performed using search words "electrochemical sensing" and "lead" as of 12th May 2021. **a** The plot of number of publications versus year of publication and **b** top countries publishing papers on the topic of electrochemical sensing of lead ions

straightforward technique based upon the redox reaction, including two steps; the pre-concentration step and the stripping action [16]. In anodic stripping voltammetry, the targeted species get reduced/deposited onto the working electrode surface by electro-deposition at a particular reduction potential known as the pre-concentration step. The reaction takes place under the pre-concentration action represented as following [17].

$$M = M^{n+} + n^{e-} \tag{1}$$

M is the targeted species, and n is the no. of ions.

After a particular period, the deposited species gets oxidized and stripped out from the electrode surface by scanning a positive potential called the stripping step.

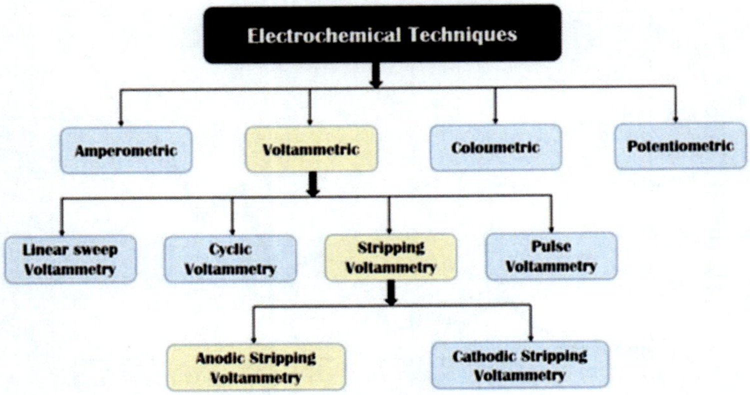

Fig. 2 Electrochemical techniques used for sensing heavy metals

In this step, during oxidation, the electrons are released that generate the current peak at the particular oxidizing potential. The reaction occurs as follows [17]:

$$M = M^{n+} + ne^- \quad (2)$$

The current peak, as shown in Fig. 3, is proportional to the sample's concentration present in the electrochemical cell solution. Based on the oxidizing potential, one can identify the sample species.

The potentiostat with three electrodes (working, counter, and a reference) is used in the anodic stripping voltammetry technique. These three electrodes play a different role in anodic stripping voltammetry, including a working electrode, which is the primary functional electrode on which targeted species are deposited and oxidized

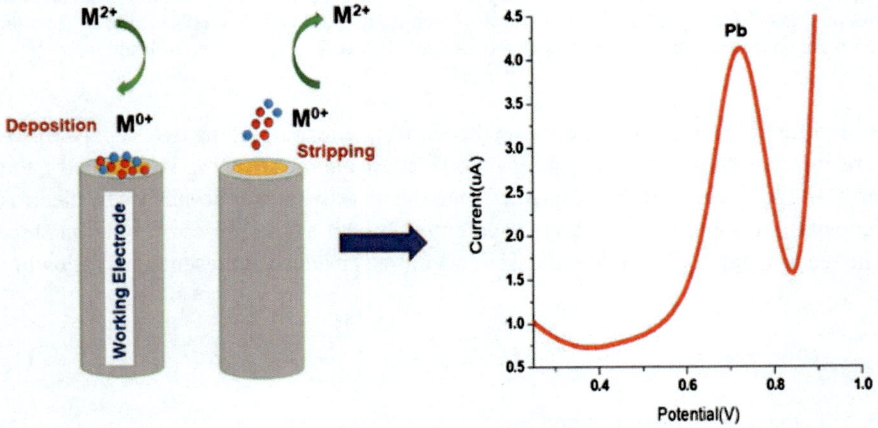

Fig. 3 Working principle of electrochemical sensing

as per the applied potential. The counter electrode is used to conduct and measure the current through the working electrode, and the reference electrode is used to keep the constant electrode potential. The working electrode's performance is critically associated with the working electrode's materials. Therefore, the selection of an appropriate working electrode for a given analyte is still a challenge. The researchers have been working on different materials to modify or enhance the working electrode's surface to get substantial results. The design and synthesis of appropriate electrode materials embellish the critical area to improve the performance and in situ heavy metal monitoring [1].

3 Perspective and Applications of Electrode Materials for Lead Sensing

The working electrode is the central part of the design sensing platform for lead ions. The electrode is modified with different materials to enhance the sensitivity. In the published reports, the electrode is modified with multiple materials that follow other properties, enhanced conductivity, enhanced porosity, surface area with more active sites, and selectivity towards a specific target species. Different conventional and modified electrodes have been used to detect lead ions in drinking water which are discussed in the next sections.

3.1 Conventional Electrodes

Different types of conventional electrodes have been used in the sensing of heavy metals. The conventional electrodes such as platinum, gold, silver, copper, nickel, and the modified electrodes such as carbon electrodes with other materials such as metallic films, nanomaterials, polymers, chemically bound molecules, etc., have been used for sensing heavy metals. In this review, the contribution of different electrode materials is mentioned for electrochemical sensing of lead ions. In reported publications, authors have used other functional groups and metallic films such as bismuth and tin films, MoS_2/rGO nano-composite, Zeolite NH_4-Y, β-cyclodextrin, etc., for the selective detection of lead ions as illustrated in Table 2.

3.2 Screen Printed Electrodes

Though conventional electrodes exhibit excellent stability, electron transfer kinetics, sensitivity, and robustness, they are expensive. In recent years, screen printed electrodes have drawn a lot of attention due to the following advantages:

Table 2 Sensing of lead ions using different conventional electrodes

Technique used	Analyte	Material used	Electrode used	Detection limit	References
Square wave anodic stripping voltammetry	Pb^{2+} & Cd^{2+}	Bismuth & Tin films	Bismuth electrode	1.5 ppb 1.3 ppb	[18]
Square wave anodic stripping voltammetry	Pb^{2+}	MoS_2/rGO nano-composite	Glassy Carbon electrode	0.005 μM	[19]
Anodic stripping voltammetry	Pb^{2+} & Cd^{2+}	Bismuth nanoparticles and Nafion	Pencil Graphite electrode	31.07 μg/L 7.31 μg/L	[20]
Anodic stripping voltammetry	Pb^{2+} & Cd^{2+}	4-azulen-1-yl-2,6-bis(2-thienyl)pyridine	Polymer-coated electrodes	0.7 nM 10 nM	[21]
Cyclic voltammetry	Pb^{2+} & Cd^{2+}	Zeolite NH_4-Y	Carbon paste electrode	3.6 μg/L 1 μg/L	[22]
Differential pulse anodic stripping voltammetry	Pb^{2+}	β-cyclodextrin	Carbon nano-tubes	2.3 ppb	[23]
Stripping voltammetry	Zn^{2+}, Cd^{2+}, Pb^{2+} & Cu^{2+}	Chromium(III) oxide	Carbon paste electrode & screen printed electrode	350 μg/L 25 μg/L 3 μg/L 3 μg/L	[24]
Anodic stripping voltammetry	Pb^{2+} & Cd^{2+}	Novel hybrid nano-composite	Multi-walled carbon nano-tubes	0.2 μg/L 0.1 μg/L	[25]
Differential pulse anodic stripping voltammetry	Pb^{2+} & Cd^{2+}	Novel bismuth oxy-carbide	Glassy carbon electrode	4.3 μg/L 3.97 μg/L	[26]

(continued)

Table 2 (continued)

Technique used	Analyte	Material used	Electrode used	Detection limit	References
Anodic stripping voltammetry	Pb^{2+}& Cd^{2+}	Magnetic nano composite ($GSHFe_3O_4$)	Glassy Carbon electrode	0.2 μg/L & 0.1 μg/L	[27]
Square wave anodic stripping voltammetry	Pb^{2+}& Cd^{2+}	Graphene oxide/κ-carrageenan/L-cysteine Nano-composite	Glassy Carbon electrode	1.08 nM 0.58 nM	[28]
Square wave anodic stripping voltammetry	Pb^{2+}& Cd^{2+}	Graphene oxide mixed ionic liquid and Nafion binder	Glassy Carbon electrode	0.42 ppb 0.03 ppb	[29]
Square wave anodic stripping voltammetry	Pb^{2+}	Nano-composite of polypyrrole and carbon nanofibers	Carbon paste electrode	0.05 μg/L	[30]

- Low-cost
- Design flexibility
- Reproducibility
- Manufacturing with different materials
- Capacity to modify the surface
- Connecting to a portable instrumentation
- Avoids tedious cleaning processes

Besides the above advantages, screen printed electrodes are compact, cost-effective, display a favorable signal-to-noise ratio, and can detect metal ions up-to-the limits set by the world health organization. Screen printed electrodes are more appealing over conventional solid electrodes because of their disposable character that avoids dreary polishing, cleaning, and activation procedures for reuse [31, 32]. Screen printed electrodes (SPEs) are the favorable and appropriate option to determine metal ions at low concentration levels [33].

Also, SPEs can be modified with new supports by depositing metallic films [34]. Researchers are focused on the design of low-cost, compact, and reliable electrochemical sensors using screen printed electrodes. Some of the important publications on sensing lead ions using the screen printed electrodes are summarized in Table 3. Foster et al. [35] deposited metal films on screen printed electrodes and determined that unmodified SPEs can detect lead at lower levels without pre-treatment of the electrode.

Wang et al. [42] developed gold-coated screen printed electrode for detecting lead ions successfully up to 0.030 μg/L detection limit. Honeychurch et al. [37] fabricated screen printed electrodes for the detection of lead ions in natural water up to 2.5 ng/mL detection limit. Masawat et al. [40] have researched the sensing of lead ions using a modified screen printed electrode with sputtered gold and detected lead ions up to 0.8 μg/L in drinking water. Martínez-Paredes et al. [41] fabricated gold nanoparticles modified screen printed electrodes for sensing lead ions up to 0.9 μg/L in water and blood samples. Mouhamed et al. [44] have researched the detection of lead ions using carbon paste screen printed electrode and found a lead detection limit of 0.138 μg/L in tap water.

3.3 Advantages of Screen–Printed Electrodes Over Conventional Electrodes

Conventional electrodes are applicable in laboratory testing or off-site work. The off-sites techniques have some disadvantages such as collecting samples, the delivery of samples to a laboratory for testing those results in changes in the conditions of samples and in turn cause accuracy and sensitivity issues (Table 4). Also, there is a cost factor in the transportation of samples. A real-time monitoring system is essential for continuous monitoring of samples to overcome these issues. For on-site monitoring, it was essential to replace the commonly used cumbersome electrodes

Table 3 Applications of screen printed electrodes for lead detection

Analyte	Technique	Sample	Electrode	Electrode material	LOD	References
Pb^{2+}	DPASV	Urine and drinking water	Screen printed electrodes	Gold-coated screen printed electrode	0.030 μg/L	[36]
Pb^{2+}	DPASV	Natural waters	Screen printed electrodes	–	2.5 ng/mL	[37]
Pb^{2+}& Cd^{2+}	LSASV	River water	Screen printed electrodes	Modified with crown-ether based membrane	1.66 μg/L 0.67 μg/L	[38]
Pb(II)	SWASV	Waste water samples	Screen printed electrodes	Gold sensor	2 μg/L	[39]
Pb(II)	SWASV	Drinking and tap water	Screen printed electrodes	Gold-sputtered SPEs	0.8 μg/L	[40]
Pb(II)	SWASV	Water and blood	Screen printed electrodes	Gold nanoparticles modified SPEs	0.09 μg/L	[41]
Pb^{2+}& Hg^{2+}	PSA	Laboratory samples	Screen printed electrodes	Gold-coated SPEs	0.6 ppb 0.5 ppb	[42]
Pb^{2+}& Cd^{2+}	SWASV	Aqueous samples	Screen printed electrodes	Filter paper strip	2.0 μg/L 2.3 μg/L	[43]
Pb^{2+}	LSV	Tab water	Screen printed electrodes	Carbon paste	0.138 μg/L	[44]

DPASV Differential pulse anodic stripping voltammetry; *LSASV* Linear sweep anodic stripping voltammetry; *SWASV* Square wave anodic stripping voltammetry; *PSA* Potentiometric stripping analysis; *LSV* Linear sweep voltammetry; *LOD* Limit of detection; *LOD* Limit of detection

with hand-held screen printed electrodes. In screen printed electrodes, all three electrodes systems, including reference, counter, and working electrodes, can be printed on a single surface that makes these electrodes very advantageous. Screen printed electrode applications support the progressive drive towards miniaturized, sensitive, compact, and portable devices and also for commercialization for real-time monitoring [48]. Screen printed electrodes provide a standard way to synthesize and fabricate innovatively designed electrochemical sensors. Due to miniaturization, cost-effectiveness, portability, and connectivity make them very pertinent for on-site determination of targeted species. The utilization of modern fabrication devices encourages the replacement of bulky electrodes with easy to use sensors.

Current smart sensors for water monitoring are developed using the screen printed electrodes to meet the compact size device and have received significant attention and motivation in the research area of heavy metal detecting sensors. For example, commercially available devices such as DropStat and 910 PSTAT mini (Fig. 4) devel-

Table 4 Significance of different electrodes in electrochemical sensing

Electrode type	Electrode material	Advantage	Disadvantage	References
Conventional electrodes	Bare electrodes	Excellent tool for the analysis of metal ions, high reproducibility, and repeatability	Safety and environmental issues	[45]
Solid bulk electrode	Gold, Silver, Platinum, etc.	Exhibit excellent stability, robust and favorable electron transfer kinetics	Not applicable for multi-element detection, meet with a higher cost	[46]
Modified electrode	Modified with different functional groups, metallic film, etc.	Sensitive, enhance a surface area, can detect multi-element simultaneously	Meet with the higher cost, cause of interference	[47]
Screen printed electrodes	Gold, silver, glassy carbon, platinum, etc.	Avoids tedious polishing, cleaning, and activation procedures commonly associated with concrete surfaces Compact size Cost-effective Best for in-situ experimentation resistant to fouling	-	[48, 49]

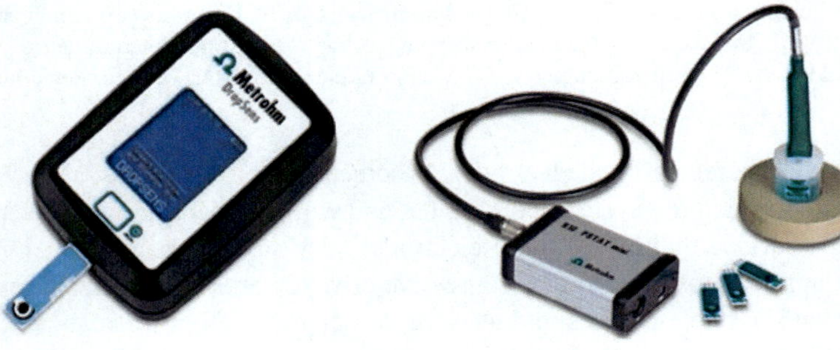

1. DropStat Sensor **2. 910 PSTAT mini**

Fig. 4 Commercially available electrochemical devices using screen printed electrodes for sensing heavy metals [50]

oped and marketed by Metro Ohm are handy devices that are potentiostat-based for sensing purposes [50].

4 Conclusions

Owing to the toxicity of heavy metal ions at trace level, a portable, reliable, sensitive, and accurate determination is imperative. Several techniques and technologies are in action for monitoring the heavy metal ions. Electrochemical techniques offer exciting opportunities due to the availability of different electrode materials. However, there are still gaps because most methods require the sample to be transported to the laboratory, making on-site analysis challenging. This review focused on the detection of lead heavy metal ions and discussed the advantages of the electrochemical technique using different types of electrodes materials. Various types of electrodes and electrode materials have materialized as motivating and encouraging in many aspects. Conventional electrodes are expensive and applicable for laboratory work. In comparison, screen printed electrodes could be inexpensive, user-friendly, and functional for on-site monitoring. Based upon the literature, it is concluded that a miniaturized, cost-effective, and portable sensor device can be fabricated using the modified or screen printed electrodes to detect trace lead ions and other heavy metal ions in water.

Acknowledgements The authors thank the Director, CSIO Director for his support and the Academy of Scientific & Innovative Research-Central Scientific Instruments Organization (AcSIR-CSIO).

References

1. Hou H, Zeinu KM, Gao S, Liu B, Yang J, Hu J (2018) Recent advances and perspective on design and synthesis of electrode materials for electrochemical sensing of heavy metals. Wiley, energy & environmental material, pp 113–131. https://doi.org/10.1002/eem2.12011
2. Buledi JA, Amin S, Haider SI, Bhanger MI, Solangi AR (2020) A review on detection of heavy metals from aqueous media using nanomaterial-based sensors. Environ Sci Pollut Res. https://doi.org/10.1007/s11356-020-07865-7
3. Guan J, Fang Y, Zhang T, Wang L, Zhu H, Du M, Zhang M (2019) Kelp-derived activated porous carbon for the detection of heavy metal ions via square wave anodic stripping voltammetry. Electro Catal 11:59–67. https://doi.org/10.1007/s12678-019-00568-9
4. Wani AL, Ara A, Usmani JA (2015) Lead toxicity: a review. Interdiscip Toxicol 8(2):55–64. https://doi.org/10.1515/intox-2015-0009
5. Dutta S, Strack G, Kurup P (2018) Gold nanostar electrodes for heavy metal detection. Sens Actuators B Chem 281:383–391. https://doi.org/10.1016/j.snb.2018.10.111
6. Flora G, Gupta D, Tiwari A (2012) Toxicity of lead: a review with recent updates. Interdiscip Toxicol 5(2):47–58. https://doi.org/10.2478/v10102-012-0009-2

7. Xie Y, Zhao S, Ye H, Yuan J, Song P, Hu S (2015) Graphene/CeO_2 hybrid materials for the simultaneous electrochemical detection of cadmium (II), lead (II), copper (II), and mercury (II). JEAC 757:235–242. https://doi.org/10.1016/j.jelechem.2015.09.043
8. Berge J, Boelens R, Vos J (2020) How the European citizens' initiative 'water and sanitation is a human right!' Changed EU discourse on water services provision. Utr Law Rev 16(2):48–59. https://doi.org/10.36633/ulr.568
9. Crawford MD, Morris JN (1967) Lead in drinking water. Lancet 290(7525):1087–1088. https://doi.org/10.1016/s0140-6736(67)90364-9
10. Sayato Y (1989) WHO guidelines for drinking-water quality. Eisei kagaku 35(5):307–312. https://doi.org/10.1248/jhs1956.35.307
11. Ferrari AGM, Carrington P, Neale SJR, C. E. Banks CE, (2020) Recent advances in portable heavy metal electrochemical sensing platforms. Environ Sci Water Res Technol 6(10):2676–2690. https://doi.org/10.1039/d0ew00407c
12. Cepriá G, Hamida S, Laborda F, Castillo JR (2009) Electroanalytical determination of arsenic (iii) and total arsenic in 1 mol L HCl using a carbonaceous electrode without a reducing agent. Anal Lett 42(13):1971–1985. https://doi.org/10.1080/00032710903082713
13. Farghaly OA, Hameed RSA, Abu-Nawwas A (2014) Analytical application using modern electrochemical techniques. Int J Electrochem Sci 9:3287–3318
14. Jaramillo O, Sukeri A, Saravia LPH (2017) Nanoporous gold microelectrode: a novel sensing platform for highly sensitive and selective determination of arsenic(III) using anodic stripping voltammetry. Electroanalysis 10:1–8. https://doi.org/10.1002/elan.201700301
15. Guo Z, Seol M, Gao C (2016) Functionalized porous Si nanowires for selective and simultaneous electrochemical detection of Cd(II) and Pb(II) ions. Electrochim Acta 211:998–1005. https://doi.org/10.1016/j.electacta.2016.06.141
16. Pizarro J, Flores E, Jimenez V (2019) Chemical Synthesis and characterization of the first cyrhetrenyl-appended calix (4) arene macrocycle and its application as an electrochemical sensor for the determination of Cu (II) in bivalve mollusks using square wave anod. Sens Actuators B Chem 281:115–122. https://doi.org/10.1016/j.snb.2018.09.099
17. Fundamentals of stripping voltammetry. Princeton applied research, application note S-6, issue 865, www.princetonappliedresearch.com
18. Hwang J, Wang X, Zhao D, Rex MM, Cho HJ, Hyoung W (2019) A novel nanoporous bismuth electrode sensor for in situ heavy metal detection. Electrochim Acta 298:440–448. https://doi.org/10.1016/j.electacta.2018.12.122
19. Sun Y, Sun J, Wang J, Pi Z, Wang L (2019) Sensitive and anti-interference stripping voltammetry analysis of Pb (II) in water using flower-like MoS_2/rGO composite with ultra-thin nanosheets. Anal Chim Acta 1063:64–74. https://doi.org/10.1016/j.aca.2019.03.008
20. Palisoc S, Jane A, Pardilla A, Racines L, Natividad M (2019) Electrochemical detection of lead and cadmium in UHT-processed milk using bismuth nanoparticles/Nafion ® -modified pencil graphite electrode. Sens Bio-Sens Res 23:100268(1–8). https://doi.org/10.1016/j.sbsr.2019.100268
21. Buica G, Ungureanu E, Birzan L, Razus AC, Popescu LM (2013) Voltammetric sensing of lead and cadmium using poly (4-azulen-1-yl-2, 6-bis (2-thienyl) pyridine) complexing films. J Electroanal Chem 693:67–72. https://doi.org/10.1016/j.jelechem.2013.01.034
22. Senthilkumar S, Saraswathi R (2009) Chemical electrochemical sensing of cadmium and lead ions at zeolite-modified electrodes: optimization and field measurements. Sens Actuators B Chem 141(1):65–75. https://doi.org/10.1016/j.snb.2009.05.029
23. Ul A, Howlader MMR, Hu N, Deen MJ (2019) Electrochemical sensing of lead in drinking water using β -cyclodextrin- modified MWCNTs. Sens Actuators B Chem 296(2):126632. https://doi.org/10.1016/j.snb.2019.126632
24. Koudelkova Z, Syrovy T, Ambrozova P, Moravec Z (2017) Determination of zinc, cadmium, lead, copper and silver using a carbon paste electrode and a screen printed electrode modified with chromium(iii) oxide. Sensor 17:1832. https://doi.org/10.3390/s17081832
25. Huang H, Chen T, Liu X, Ma H (2014) Ultrasensitive and simultaneous detection of heavy metal ions based on three-dimensional graphene-carbon nanotubes hybrid electrode materials. Anal Chim Acta 852:45–54. https://doi.org/10.1016/j.aca.2014.09.010

26. Zhang Y, Li C, Su Y, Mu W, Han X (2019) Simultaneous detection of trace Cd (II) and Pb (II) by differential pulse anodic stripping voltammetry using a bismuth oxycarbide/nafion electrode. Inorg Chem Commun 111:107672. https://doi.org/10.1016/j.inoche.2019.107672
27. Baghayeri M, Amiri A, Maleki B, Alizadeh Z, Reiser O (2018) A simple approach for simultaneous detection of cadmium (II) and lead (II) based on glutathione coated magnetic nanoparticles as a highly selective electrochemical probe. Sens Actuators B Chem 273:1442–1450. https://doi.org/10.1016/j.snb.2018.07.063
28. Priya T, Dhanalakshmi N, Thennarasu S, Thinakaran N (2018) A novel voltammetric sensor for the simultaneous detection of Cd^{2+} and Pb^{2+} using graphene oxide/κ-carrageenan/L-cysteine nanocomposite. Carbohydr Polym 182:199–206. https://doi.org/10.1016/j.carbpol.2017.11.017
29. Kumar S, Sachan S, Kumar S (2019) Ultra-trace sensing of cadmium and lead by square wave anodic stripping voltammetry using ionic liquid modified graphene oxide. Mater Sci Energy Technol 2(3):667–675. https://doi.org/10.1016/j.mset.2019.09.004
30. Oularbi L, Turmine M, Rhazi M (2017) Electrochemical determination of traces lead ions using a new nanocomposite of polypyrrole/carbon nanofibers. J Solid State Electrochem 21:3289–3300. https://doi.org/10.1007/s10008-017-3676-2
31. Biyani M, Biyani R, Takamura T (2017) DEP-On-go for simultaneous sensing of multiple heavy metals pollutants in environmental samples. Sensors (Switzerland) 17(1):1–14. https://doi.org/10.3390/s17010045
32. Hughes G, Westmacott K, Honeychurch KC, Crew A, Pemberton RM, Hart JP (2016) Recent advances in the fabrication and application of screen-printed electrochemical (bio)sensors based on carbon materials for biomedical, agri-food and environmental analyses. Biosensors 6(4):50. https://doi.org/10.3390/bios6040050
33. Sánchez-Tirado E, Salvo C, González-Cortés A, Yáñez-Sedeño P, Langa F, Pingarrón JM (2017) Electrochemical Immunosensor for simultaneous determination of interleukin-1 beta and tumor necrosis factor alpha in serum and saliva using dual screen printed electrodes modified with functionalized double–walled carbon nanotubes. Anal Chim Acta 959:66–73. https://doi.org/10.1016/j.aca.2016.12.034
34. Foster CW, De Souza AP, Metters JP, Bertotti M, Banks CE (2015) Metallic modified (bismuth, antimony, tin and combinations thereof) film carbon electrodes. Analyst 140(22):7598–7612. https://doi.org/10.1039/c5an01692d
35. Barton J, Garcia M, Santos D (2016) Screen-printed electrodes for environmental monitoring of heavy metal ions: a review. Microchim Acta 183(2):503–517. https://doi.org/10.1007/s00604-015-1651-0
36. Wang J, Tian B (1992) Screen-printed stripping voltammetric/potentiometric electrodes for decentralized testing of trace lead. Anal Chem 64(15):1706–1709. https://doi.org/10.1021/ac00039a015
37. Honeychurch KC, Hawkins DM, Hart JP, Cowell DC (2002) Voltammetric behaviour and trace determination of copper at a mercury-free screen-printed carbon electrode. Talanta 57(3):565–574. https://doi.org/10.1016/S0039-9140(02)00060-7
38. Parat C, Betelu S, Authier L, Potin-Gautier M (2006) Determination of labile trace metals with screen-printed electrode modified by a crown-ether based membrane. Anal Chim Acta 573–574:14–19. https://doi.org/10.1016/j.aca.2006.04.081
39. Noh MFM, Tothill IE (2006) Development and characterisation of disposable gold electrodes, and their use for lead (II) analysis. Anal Bioanal Chem 386(7–8):2095–2106. https://doi.org/10.1007/s00216-006-0904-5
40. Masawat P, Liawruangrath S, Slater JM (2003) Flow injection measurement of lead using mercury-free disposable gold-sputtered screen-printed carbon electrodes (SPCE). Sens Actuators B Chem 91(1–3):52–59. https://doi.org/10.1016/S0925-4005(03)00066-2
41. Martínez-Paredes G, González-Garcaiaa MB, Costa-Garcia A (2009) Lead sensor using gold nanostructured screen-printed carbon electrodes as transducers. Electroanalysis 21(8):925–930. https://doi.org/10.1002/elan.200804399

42. Wang J, Tian B (1993) Mercury-free disposable lead sensors based on potentiometric stripping analysis at gold-coated screen-printed electrodes. Anal Chem 65:1529–1532. https://doi.org/10.1021/ac00059a008
43. Shi J, Tang F, Xing H, Zheng H, Bi L, Wang W (2012) Electrochemical detection of Pb and Cd in paper-based microfluidic devices. J Braz Chem Soc 23(6):1124–1130. https://doi.org/10.1590/S0103-50532012000600018
44. Mouhamed N, Cheikhou K, Rokhy GEM, Bagha DM, Guèye MDC, Tzedakis T (2018) Determination of lead in water by linear sweep anodic stripping voltammetry (LSASV) at unmodified carbon paste electrode: optimization of operating parameters. Am J Anal Chem 09(3):171–186. https://doi.org/10.4236/ajac.2018.93015
45. Zaib M, Makshoof M, Saeed A, Farooq U (2015) Biosensors and bioelectronics electrochemical determination of inorganic mercury and arsenic- a review. Biosens Bioelectron 74:895–908. https://doi.org/10.1016/j.bios.2015.07.058
46. Ariño C, Serrano N, Díaz-Cruz JM, Esteban M (2017) Voltammetric determination of metal ions beyond mercury electrodes. a review. Anal Chim Acta 990:11–53. https://doi.org/10.1016/j.aca.2017.07.069
47. March G, Dung T, Piro B (2015) Modified electrodes used for electrochemical detection of metal ions in environmental analysis. Biosensors (Basel) 5(2):241–275. https://doi.org/10.3390/bios5020241
48. Hayat A, Marty JL (2014) Disposable screen printed electrochemical sensors: tools for environmental monitoring. Sensors (Switzerland) 14(6):10432–10453. https://doi.org/10.3390/s140610432
49. Serrano N, Castilla O, Ariño C, Diaz-Cruz MS, Díaz-Cruz JM (2019) Commercial screen-printed electrodes based on carbon nanomaterials for a fast and cost-effective voltammetric determination of paracetamol, ibuprofen and caffeine in water samples. Sensors (Switzerland), 19(18). https://doi.org/10.3390/s19184039
50. https://www.metrohm.com/nb-no/productsoverview/electrochemistry/portable%20potentiostats/29100010

Voltammetric Response of Synthesized CuO Nanoparticles Towards Dopamine

S. R. Kiran Kumar, M. S. Anantha, H. B. Muralidhara, and K. Yogesh Kumar

1 Introduction

In recent years, there is a great interest in the use of nanoparticles (NPs) and their composites for the development of modified electrodes for a variety of applications. Metal oxide nanoparticles can lower the overpotential, increase the reaction rate and sensitivity and improve the selectivity [1, 2]. Metal oxide nanoparticles (NPs) have unique features, and great attention has also been paid to the electrochemistry of metal oxide-modified electrodes [3]. Among the various metal oxide NPs, CuO has attracted particular attention because of its low cost and its properties are similar to that of other metallic NPs, and it is the simplest member of the family of copper compounds [4, 5]. It finds applications in heat transfer systems as a super-strong material [6], biosensors [7], batteries [8], solar energy [9], antimicrobial agent [10], bactericidal agents [11] and also as a catalyst [12]. CuO crystal structure possesses a narrow band gap, giving useful photocatalytic and photovoltaic properties [13].

Dopamine (DA) is one of the naturally occurring catecholamines in the mammalian central nervous system. It plays a key role in the function of the central nervous, renal, hormonal and cardiovascular systems [14]. Thus, a loss of DA-containing neurons may lead to neurological disorders, such as parkinsonism and

S. R. Kiran Kumar (✉)
Department of Chemistry, Centre for Nanosciences, K. S. Institute of Technology, Bangalore 560109, India

M. S. Anantha · H. B. Muralidhara
Centre for Incubation, Innovation, Research & Consultancy, Jyothy Institute of Technology, Bangalore 560082, India

K. Yogesh Kumar
Department of Chemistry, School of Engineering and Technology, Jain University, Jain global campus, Jakkasandra Post, Kanakpura Taluk, Ramanagra 562112, India

U. K. Mudali et al. (eds.), *Recent Trends in Electrochemical Science and Technology*, Springer Proceedings in Materials 15,
https://doi.org/10.1007/978-981-16-7554-6_16

schizophrenia [15]. Development of both sensitivity and selectivity is of equal importance in voltammetric procedures, and also this would help in the prevention and treatment of several neurodegenerative diseases.

Herein, the CuO NPs were prepared from the hydrothermal method and then used to modify the surface of the carbon paste electrode (CPE). The electrochemical behaviour of DA on this CuO-modified CPE (CuO/CPE) was studied. It was shown that the current peak for the oxidation of DA could be well resolved, and based on the different electrocatalytic activities of the modified electrode towards these species, a sensitive and selective method for the determination of DA was set up for routine analysis. The antimicrobial activity of CuO NPs was evaluated by using the agar disc diffusion method, and minimum inhibitory concentration (MIC) was determined by the dilution method.

2 Experimental

2.1 *Reagents and Stock Solution*

All the reagents including Copper (II) nitrate trihydrate ($Cu(NO)_3.3H_2O$), sodium hydroxide (NaOH), Triton X-100 ($C_{14}H_{22}O(C_2H_4O)_n$, dopamine hydrochloride (DA), disodium hydrogen phosphate (Na_2HPO_4), sodium dihydrogen orthophosphate (NaH_2PO_4), Dimethylformamide (C_3H_7NO) silicone oil and graphite powder (<20 μm particle size) were of analytical grade purchased from SD Fine Chemicals, Mumbai, India, and used as received without further purification. All the stock solutions were prepared by using double distilled water.

2.2 *Instruments*

The cyclic voltammetric (CV) measurements were performed on an electrochemical workstation (CH Instruments, Model 660D). Electrochemical experiments were carried out in a three-electrode cell system, which contained a bare carbon paste electrode (BCPE), CPE/CuO NPs, as the working electrode, an aqueous saturated calomel electrode (SCE) as the reference electrode and a Pt wire as the auxiliary electrode. X-ray diffraction (XRD) patterns were obtained on a Bruker D_2 Phaser XRD system. Surface morphology (SEM) was studied using a scanning electron microscope (JEOL JSM 840).

2.3 *Preparation of CuO Nanoparticles*

CuO NPs were synthesized via the hydrothermal method in the presence of Triton X-100. Experimental details followed are as described in our previous work [16]. In a typical procedure, 20 mL of 0.2 mol/L NaOH solution was slowly added into a 20 mL of 0.1 mol/L copper(II) nitrate trihydrate solution containing 0.004 mol/L of Triton X-100 with constant stirring. After vigorous stirring for 3 h, the mixture was autoclaved at 200 °C for 5 h. After the reaction, the system was naturally cooled to room temperature, and the precipitates were separated from the solution and thoroughly washed several times with deionized water and absolute ethanol and then dried in an oven at 50 °C for 8 h.

2.4 *Preparation of Bare Carbon Paste Electrode (BCPE) and Modified Carbon Paste Electrode (MCPE)*

The bare carbon paste electrode was prepared by hand mixing of 80% graphite powder with 20% silicone oil in an agate mortar to produce a homogenous carbon paste. The paste of around 5 mg was packed into the cavity of CPE of 3 mm in diameter and then smoothed on a weighing paper. The electrical contact was provided by a copper wire connected to the paste at the end of the tube. The modified carbon paste electrode (MCPE) was prepared by adding 2, 4, 6, 8 and 10 mg CuO NPs to the above-mentioned graphite powder and silicone oil mixture.

2.5 *Media for the Interference Studies*

The MCPE was performed using the CV technique under the optimum conditions of 40 mg/mL of dopamine concertation at 0.2 M phosphate buffer solution of pH 7.2. The interference studies were performed using $CaCl_2$, NaCl and KCl.

2.6 *Antimicrobial Activity*

2.6.1 Microbial Strains

The in vitro antimicrobial studies were carried out against six test microorganisms (two Gram-positive bacteria: *Staphylococcus aureus* (NCIM 5021) and *Bacillus subtilis* (NCIM 2999), two Gram-negative bacteria: *Escherichia coli* (NCIM 2574) and *Pseudomonas aeruginosa* (NCIM 5029), and two fungi: *Aspergillus flavus* (NCIM 524) and *Candida albicans* (NCIM 3471). Microbial strains were cultured overnight at 37 °C in nutrient and potato dextrose agar medium. All the pure microbial strains were obtained from National Chemical Laboratory (NCL), Pune, India.

The synthesized CuO was tested in vitro using the agar disc diffusion method taking Streptomycin and Fluconazole as reference drugs for bacteria and fungi, respectively. In brief, the antimicrobial potentialities of the CuO NPs were estimated by pre-sterilized filter paper discs (6 mm in diameter) impregnated with CuO NPs dissolved in 100 μg/mL was placed on the inoculated agar. The plates were incubated for 24 h at 37 °C in case of bacteria and 48 h at 28 °C in case of fungi.

2.6.2 Minimum Inhibitory Concentration (MIC)

The minimum inhibitory concentration of the CuO NPs was determined by the dilution method. The CuO NPs were dispersed in 5% DMF solution and diluted to give twofold serial concentrations of the compounds which were employed to determine the MIC with an amount of 5–75 μg/mL. The MIC value was determined as the lowest concentration of the CuO NPs inhibiting the visual growth of the microorganism on the agar plate.

3 Results and Discussion

3.1 Characterization of CuO NPs

Figure 1 shows the XRD pattern of as-prepared CuO NPs obtained by the hydrothermal process, and all diffraction peaks can be indexed as monoclinic CuO NPs in comparison with JCPDS card file No. 80-1916 ($a = 4.69$ Å, $b = 3.42$ Å and c

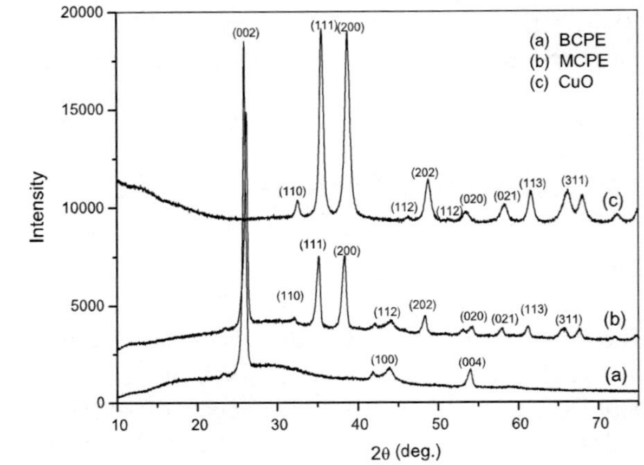

Fig. 1 Typical XRD pattern of CuO nanoparticles

Fig. 2 Typical SEM images of CuO at different magnifications

= 5.13 Å). No characteristic peaks of other impurities were detected. The diffraction peaks were very sharp, which indicates the highly crystalline nature of the NPs.

Figure 2 shows the morphologies on the CuO NPs obtained by SEM at different magnifications, both of them are composed of nanoflake structures. The size of these nanoflakes ranges from hundreds of nanometres to several micrometres. The thickness of these nanoflakes could be roughly estimated from their SEM images, which is in the range of tens of nanometres.

3.2 *Effect of CuO NPs on MCPE for Investigation of DA and AA*

In order to optimize the amount of CuO NPs on CPE, different concentrations of the CuO NPs were used to determine the response of 5×10^{-5} M DA, 1×10^{-4} M AA in a 0.2M phosphate buffer (pH 7.2) at a scan rate of 50 mV/s. The 10 mg CuO/CPE response to the maximum current as compared to the 2, 4, 6 and 8 mg of CuO NPs is shown in Fig. 3, and this optimized concentration is maintained during further investigation.

3.3 *Electrochemical Response of DA at BCPE and CuO/MCPE*

The electrochemical responses of 5×10^{-5} M DA and its voltammogram were recorded in the potential range of −0.2 to 0.6 versus SCE in the 0.2M phosphate buffer solution of pH 7.2 at the BCPE and the CuO/MCPE were measured at a scan rate of 50 mV/s by CV technique. The corresponding peak potential differences (ΔE_p = 0.0618 V) and (ΔE_p = 0.058 V) for the BCPE (dashed line) and the CuO/MCPE

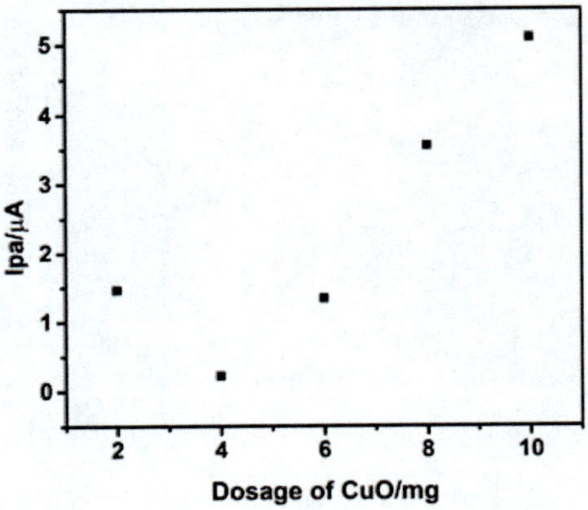

Fig. 3 CuO/CPE response to the maximum current of 5 × 10^{-5} M DA at different concentrations of CuO nanoparticles

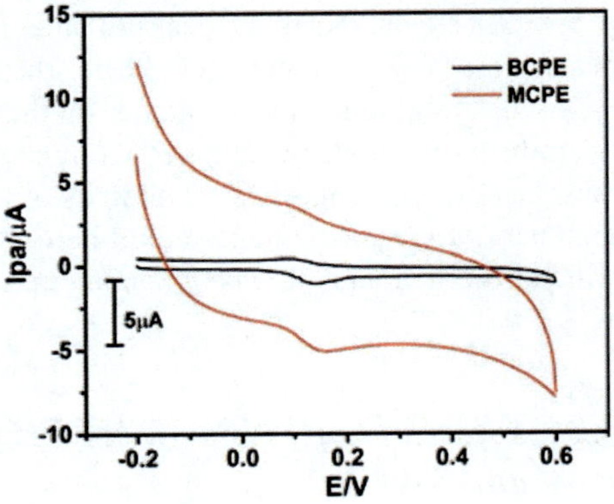

Fig. 4 Cyclic voltammograms of 5 × 10^{-5} M DA in 0.2 M phosphate buffer solution at pH 7.2 using bare CPE and CuO/MCPE at scan rate 50 mV s^{-1}

(solid line) are shown in Fig. 4. At the BCPE, the anodic peak potential (E_{pa}) and the cathodic peak potential (E_{pc}) were 0.1531 V and 0.0913 V, respectively. DA peak currents significantly increased at the CuO/MCPE with the anodic peak potential of 0.1610 V, and the corresponding cathodic peak potential is 0.1039 V. The remarkable enhancement of peak current provides clear evidence of the catalytic effect of CuO/MCPE. These results indicate that the CuO/MCPE can accelerate the rate of DA electron transfer.

3.4 *Effect of Scan Rate on the Peak Current*

The effect of scan rate for DA in phosphate buffer solution at pH 7.2 was studied by CV at CuO/MCPE. Figure 5 shows an increase in the redox peak current at a scan rate of 0.01–0.600 V/s for CuO/MCPE indicating that direct electron transfer between DA and the modified electrode surface. The graph obtained exhibited good linearity between the scan rate (v) and the redox peak current (Fig. 6) for the MCPE with

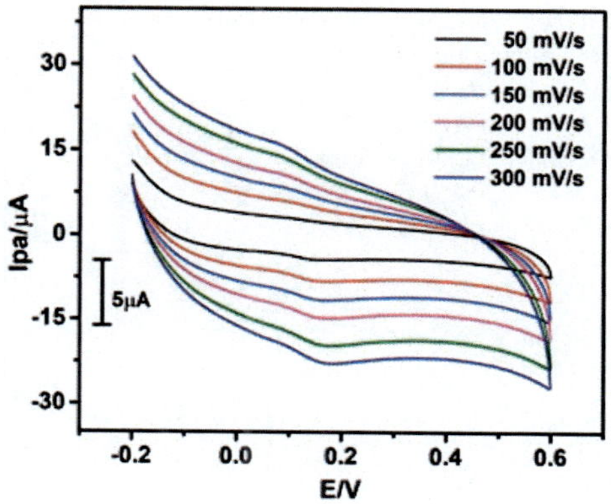

Fig. 5 Cyclic voltammograms of 5×10^{-5} M DA in 0.2 M phosphate buffer solution using CuO/MCPE at different scan rates (0.01–0.600 V s^{-1})

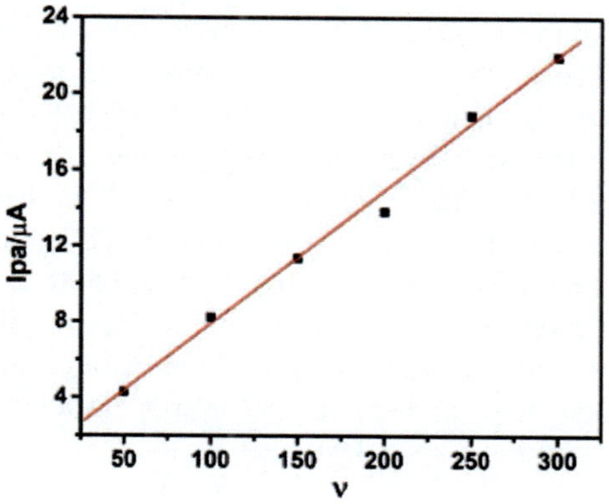

Fig. 6 Typical graph showing the linear relationship between the anodic peak current and scan rate

correlation coefficients (R^2) of 0.99664, which indicates that the electron transfer reaction was the adsorption-controlled process. The difference between the anodic peak potential and the cathodic peak potential is increasing with the scan rate.

3.5 The Effect of the Concentration Variation of Dopamine

According to electrochemical response, the redox peak current increased with increasing concentration of the analyte. By increasing the concentration of DA (1 mM to 8 mM), both the anodic peak current and cathodic peak current go on increasing (Fig. 7). At the same time, the high surface area of the CuO NPs in CuO/MCPE improved the electrode contact area of DA.

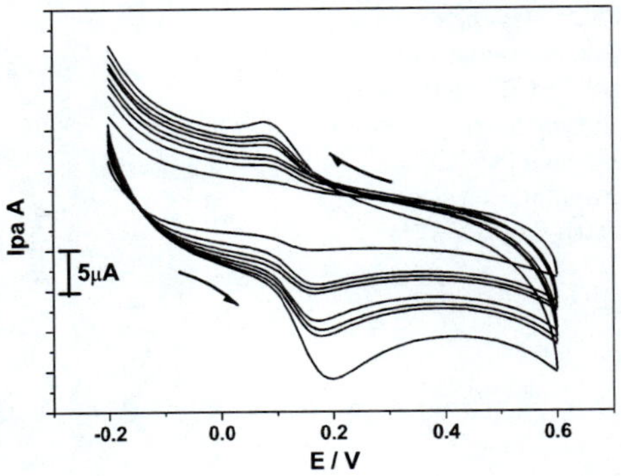

Fig. 7 Cyclic voltammograms showing different concentrations of DA (1–8 mM) in 0.2M phosphate buffer solution using CuO/MCPE

3.6 Interference Study for the Detection of DA

The influence of various foreign species as potentially interfering compounds with the determination of DA was investigated. Tolerance limit was defined as the maximum concentration of interfering species that caused an approximate relative error of ±5% for the determination of DA. After the experiments, no significant interference for the detection of DA was observed from the selected compounds such as $CaCl_2$ 4000 μM, NaCl 4000 μM and KCl 5000 μM. In the electrochemical response, the peaks remain unchanged after successive 50 cyclic voltammetric scans, confirming the good stability of CuO/MCPE. Interference studies using MCPE exhibit good selectivity and sensitivity towards the detection of dopamine even in the presence of various foreign species.

3.7 Antimicrobial Activity

The CuO NPs were evaluated for antimicrobial activity by using agar disc diffusion method [17], and minimum inhibitory concentration (MIC) was determined by dilution method [18]. CuO NPs demonstrated in vitro antimicrobial activity against the four bacterial strains belonging to the Gram-positive (*S. aureus, Bacillus subtilis,*) and Gram-negative (*Escherichia coli, Pseudomonas aeruginosa*) and two strains of fungi, namely *Aspergillus flavus, Candida albicans.* The results of the antibacterial activity of CuO NPs are presented in Table 1. The MIC is defined as the lowest concentration of nanoparticles that inhibit the growth of a microorganism. CuO NPs showed MIC at 28 and 31 μg/mL for *E. coli* and *P. aeruginosa,* respectively. According to MIC, *E. coli* and *P. aeruginosa* exhibited the highest sensitivity towards CuO nanoparticles while *B. subtilis, C. albicans* and *A. flavus* showed the least sensitivity among the tested microbes. The antimicrobial activity of the tested CuO NPs was compared to the positive control drugs, streptomycin and fluconazole.

Table 1 Antimicrobial activity of the CuO nanoparticles

Tested compounds	In vitro activity zone of inhibition in mm (MIC in μg/mL)[a]					
	Gram-positive		Gram-negative		Fungi	
	S. aureus	*B. subtilis*	*E. coli*	*P. aeruginosa*	*C. albicans*	*A. flavus*
CuO	13(50)	12(70)	13(28)	12(31)	11(75)	14(75)
Streptomycin	11.6(05)	10.4(05)	14.5(05)	13(05)	NT	NT
Fluconazole	NT	NT	NT	NT	16(05)	18(05)

[a] The values given are means of three experiments
()-MIC values
NT- denotes not tested

The antibacterial properties of CuO NPs are mainly attributed to adhesion with bacteria because of their opposite electric charges resulting in a reduction at the bacterial cell wall. It was earlier reported that the interaction between Gram-negative bacteria and CuO NPs was stronger than that of Gram-positive bacteria because of the difference in cell walls, cell structure, physiology, metabolism or degree of contact of organisms with nanoparticles. Gram-positive bacteria have thicker peptidoglycan cell membranes compared to the Gram-negative bacteria, and it is harder for CuO NPs to penetrate it, resulting in a low antibacterial response [19].

4 Conclusions

The synthesized CuO/MCPE improved the sensitivity of the electrode and is acting as a good electrochemical sensor for the detection of DA. Cyclic voltammetry measurements revealed a reasonably fast electron transfer and good stability of the electrode in phosphate buffer solution. The developed CuO-based electrode is cheaper, easy to prepare and possesses high selectivity. It has good reproducibility of the voltammetric response, and thus, the developed electrode is very useful in the construction of simple medical devices for the diagnosis of dopamine deficiency. It is expected that because of its good electrocatalytic behaviour, the developed CuO/MCPE may be used for the development of biosensors and other electroanalytical devices.

Conflict of Interest The authors declare that they have no known competing financial interests or personal relationships that could have appeared to influence the work reported in this paper.

References

1. Zare HR, Memarzadeh F, Ardakani MM, Namazian M, Golabi SM (2005) Norepinephrine-modified glassy carbon electrode for the simultaneous determination of ascorbic acid and uric

acid. Electrochim Acta 50:495–3502. https://doi.org/10.1016/j.electacta.2004.12.026
2. Ren W, Luo HQ, Li NB (2006) Simultaneous voltammetric measurement of ascorbic acid, epinephrine and uric acid at a glassy carbon electrode modified with caffeic acid. Biosens Bioelectron 21(7):1086–1092. https://doi.org/10.1016/j.bios.2005.04.002
3. Kim HJ, Sohn HJ, Kim S, Yi SN, Ha DH (2011) Molecular detection based on the electrical conductance of gold nanoparticle arrays. Sens Actuators B Chem 156(2):990–993. https://doi.org/10.1016/j.snb.2011.03.018
4. Ramgir N, Datta N, Kaur M, Kailasaganapathi S, Debnath AK, Aswal DK, Gupta SK (2013) Metal oxide nanowires for chemiresistive gas sensors: issues, challenges and prospects. Colloids Surf A Physicochem Eng Asp 439:101–116. https://doi.org/10.1016/j.colsurfa.2013.02.029
5. Ruparelia JP, Chatterjee AK, Duttagupta SP, Mukherji S (2008) Strain specificity in antimicrobial activity of silver and copper nanoparticles. Acta Biomater 4(3):707–716. https://doi.org/10.1016/j.actbio.2007.11.006
6. Nasir FM, Mohamad AY (2006) Heat transfer of CuO-water based nanofluids in a compact heat exchanger. ARPN J Eng Appl Sci 11(4):2517–2523
7. Liu R, Kulp EA, Oba F, Bohannan EW, Ernst F, Switzer JA (2005) Epitaxial electrodeposition of high-aspect-ratio Cu_2O (110) nanostructures on InP (111). Chem Mater 17(4):725–729. https://doi.org/10.1021/cm048296l
8. Chandrasekaran S (2013) A novel single step synthesis, high efficiency and cost effective photovoltaic applications of oxidized copper nano particles. Sol Energy Mater Sol Cells 109:220–226. https://doi.org/10.1016/j.solmat.2012.11.003
9. Snoke D (1996) Coherent exciton waves. Science 273:1351–1352. https://doi.org/10.1126/science.273.5280.1351
10. Amiri M, Etemadifar Z, Daneshkazemi A, Nateghi M (2017) Antimicrobial effect of copper oxide nanoparticles on some oral bacteria and candida species. J Dent Biomater 4(1):347
11. Azam A, Ahmed AS, Oves M, Khan MS, Memic A (2012) Size-dependent antimicrobial properties of CuO nanoparticles against Gram-positive and-negative bacterial strains. Int J Nanomed 7:3527. https://doi.org/10.2147/IJN.S29020
12. Zhou K, Wang R, Xu B, Li Y (2006) Synthesis, characterization and catalytic properties of CuO nanocrystals with various shapes. Nanotechnology 17(15):3939. https://doi.org/10.1088/0957-4484/17/15/055
13. Li J, Sun F, Gu K, Wu T, Zhai W, Li W, Huang S (2011) Preparation of spindly CuO microparticles for photodegradation of dye pollutants under a halogen tungsten lamp. Appl Catal A Gener 406(1–2):51–58. https://doi.org/10.1016/j.apcata.2011.08.007
14. Kumar SK, Prashanth MK, Muralidhara HB, Kumar KY, Mamatha GP (2016) Synthesis and characterization of copper oxide nanoparticles: In the study of voltammetric response of biomolecules. Surf Eng Appl Electrochem 52(5):469–474. https://doi.org/10.3103/S1068375516050082
15. Reyes S, Fu Y, Double KL, Cottam V, Thompson LH, Kirik D, Paxinos C, Watson D, Cooper HM, Halliday GM (2013) Trophic factors differentiate dopamine neurons vulnerable to Parkinson's disease. Neurobiol Aging 34(3):873–886. https://doi.org/10.1016/j.neurobiolaging.2012.07.019
16. Kumar KY, Muralidhara HB, Nayaka YA, Hanumanthappa H, Veena MS, Kumar SK (2014) Hydrothermal synthesis of hierarchical copper oxide nanoparticles and its potential application as adsorbent for Pb (II) with high removal capacity. Sep Sci Technol 49(15):2389–2399. https://doi.org/10.1080/01496395.2014.922101
17. Ramesh kumar N, Ashok kumar M, Subramanian EH, Ilavarasan R, Sridhar SK (20030 Synthesis of 6-fluoro-1, 4-dihydro-4-oxo-quinoline-3-carboxylic acid derivatives as potential antimicrobial agents. Eur J Med Chem 38(11–12):1001–1004. https://doi.org/10.1016/S0223-5234(03)00151-X
18. Lennette E, Spaulding E, Truant JP (1978) Manual of clinical microbiology, 2nd edn. American Society for Microbiology, Washington, DC. USA, p 970

19. Tawale JS, Dey KK, Pasricha R, Sood KN, Srivastava AK (2010) Synthesis and characterization of ZnO tetrapods for optical and antibacterial applications. Thin Solid Films 519(3):1244–1247. https://doi.org/10.1016/j.tsf.2010.08.077

Electrochemical and Photocatalytic Applications of ZnO Nanoparticles Synthesized Using the Leaf Extract of *Ricinus Communis*

Mamata C. Naik, Jyothi H. Kini, and B. E. Kumaraswamy

1 Introduction

Nanoparticles are the primary source of various nanostructured devices or materials; hence, they are considered to be the basic elements of nanotechnology [1]. Nanoparticles have gained worldwide importance due to their diverse optical, electronic, thermal, and catalytic properties [2]. Different methods such as precipitation method [3], sol–gel method [4], hydrothermal method [5], solvothermal method [6], and chemical vapor deposition method [7] are available for the synthesis of different sized ZnO nanoparticles. In particular, the plant-mediated synthesis of nanoparticles has represented a very promising route which could be an alternative to chemical and physical methods for the synthesis of nanoparticles.

Synthesis of metal oxide nanoparticles from plant extract is one of the popular approaches for synthesizing ZnO nanoparticles using different parts of plants such as leaves, seeds, stem, fruits, and flowers because of the phytochemicals produced by them. This approach has attracted a special attention these days because of their simplicity, cost-effectiveness, biocompatibility, non-toxicity, safety, and environment-friendly nature. Biosynthesis routes help us to produce nanoparticles with better-defined size and also morphology as compared to some of the physicochemical methods [8].

M. C. Naik
Post Graduate Department of Chemistry, St. Xavier's College, Mapusa, Goa, India

J. H. Kini (✉)
Department of PG Studies and Research in Industrial Chemistry, Sir M.V. Govt. Science College, Bommanakatte, Bhadravathi, Karnataka 577302, India
e-mail: jyothikini.867@ka.gov.in

B. E. Kumaraswamy
Department of PG Studies and Research in Industrial Chemistry, Jnana Sahyadri, Kuvempu University, Shankaraghatta, Shivamogga, Karnataka, India

U. K. Mudali et al. (eds.), *Recent Trends in Electrochemical Science and Technology*, Springer Proceedings in Materials 15,
https://doi.org/10.1007/978-981-16-7554-6_17

The *Ricinus communis* is commonly known as castor oil plant. Castor plant shows antioxidant, anti-inflammatory, antimicrobial, and antihistamine properties. Leaves are used for treating painful urination, lumps, heart disease, and colic pain in the bladder. It is best known for the treatment of jaundice and liver-related problems [9]. The *Ricinus communis* plant extract especially leaf extract contains essential phytochemicals such as terpenoids, flavonoids, ketones, aldehydes, amides, and carboxylic acids that can be easily extracted. These phytochemicals facilitate the production of ZnO nanoparticles and enhance its bioactivity. The plant extract reduces metal ions to metal nanoparticles in much lesser time. Therefore, it serves as an excellent source for metal as well as metal oxide nanoparticle synthesis. Leaf extracts act as both reducing and stabilizing agents in nanoparticle synthesis [10–13].

Quantitative determination of dopamine in human physiological fluids is gaining tremendous importance in both clinical and biochemical diagnosis. Dopamine is one of the most important neurotransmitters widely existing in the brain for the message transfer in the mammalian central nervous system. Dopamine detection has become important and has been given tremendous attention because a change in the dopamine levels in the body helps in understanding the brain functions such as learning and memory formation. Abnormal concentration of dopamine in the body may lead to various diseases like Parkinson's disease, Schizophrenia, and HIV infection. Therefore, investigation of simple and rapid determination methods of dopamine is currently the subject of great interest by chemists and neuroscientists in bioanalytical and biomedical areas of research. The development of voltammetric sensors for determining dopamine and other catecholamines present in the mammalian central nervous system has received tremendous attention. The determination of concentration of dopamine in different solutions is usually studied by electrochemical methods because dopamine can be easily oxidized by electrochemical methods [14–17]. Nano-sized ZnO can be synthesized in different morphologies such as nanorods, nanotubes, nanobelts, nanoflakes, and nanoflowers [18]. The applications of biosynthesized ZnO nanoparticles include antibacterial agents, cancer treatment, targeted drug delivery, biosensors, gene therapy, DNA analysis, etc. [19].

The main objectives of this study were the synthesis of ZnO nanoparticles, use them for modifying a glassy carbon electrode, and study the electrochemical behavior of the modified electrode in the presence of dopamine by using cyclic voltammetric technique. The photocatalytic activity of the synthesized ZnO nanoparticles has also been studied through methylene blue photodegradation.

2 Experimental

2.1 *Materials and Methods*

2.1.1 Reagents and Chemicals

Zinc acetate dihydrate (Loba Chemie), sodium hydroxide pellets (Loba Chemie), dopamine, phosphate buffer pH 6.6, distilled water were used along with the extracts and air-dried *Ricinus communis* leaves.

2.1.2 Preparation of Leaf Extract

25 g of leaves were weighed and washed with running tap water and dried under shade. The dried leaves were placed in a 250 mL beaker containing 100 mL distilled water and heated for 1 h at 60 °C. The extract was filtered through a sterile filter paper. The filtered extract was used as a source for the synthesis of ZnO nanoparticles.

2.1.3 Synthesis of ZnO Nanoparticles

About 0.25 g of zinc acetate was dissolved in 50 mL distilled water. 1 mL of the extract was added dropwise, and the resulting mixture was stirred for 20 min using a magnetic stirrer. To adjust the pH of the solution to pH 7, sodium hydroxide (2 M) was added dropwise. A white crystalline precipitate of zinc hydroxide was obtained. The precipitate was then filtered, washed with distilled water, and dried in a hot air oven at 60 °C for 1 h to obtain the zinc oxide nanoparticles [20].

2.1.4 Electrochemical Behavior of Dopamine

The electrochemical experiments were carried out using an electrochemical analyzer (CH-instrument, model no.CH1200B). The standard three-electrode cell was used to perform all the electrochemical experiments. ZnO modified glassy carbon electrode was used as the working electrode. The reference electrode was aq. Ag/AgCl, and the counterelectrode was platinum wire.

2.1.5 Preparation of ZnO Modified Glassy Carbon Electrode

50 mg of ZnO was accurately weighed and mixed in 0.1 mL distilled water with constant stirring for about one and half minute. One drop of this mixture was mounted over a previously cleaned bare glassy carbon electrode followed by 3–5 min of drying

without any disturbance. This coated electrode was used as a ZnO modified glassy carbon electrode to study the electrochemical behavior of dopamine [21, 22].

2.1.6 Photocatalytic Activity

The photocatalytic activity was carried out using a UV–Vis spectrophotometer (Thermo Scientific Evolution 201) at 400–800 nm range. Methylene blue solution was prepared by taking concentration of 10 mg/L (10 ppm). From this, 100 mL solution was taken in the beaker and 10 mg of ZnO nanoparticles were added to it. The mixture was stirred for 30 min. The solution was exposed to sunlight. The absorbance was measured after every 30 min of sunlight exposure for 3 h using UV–Vis spectroscopy [23].

3 Results and Discussion

3.1 Powder XRD

Figure 1 represents the X-ray diffraction pattern of synthesized ZnO nanoparticles. The diffraction peaks observed at 31.57, 34.21, 36.32, 47.36, 56.32, 62.63, 66.05, 67.47, 68.68, 72.11, and 76.32° that can be indexed to (100), (002), (101), (102), (110), (103), (200), (112), (201), (004), and (002) crystal planes, respectively which can correspond to the hexagonal wurtzite crystalline phase of ZnO (JCPDS card No. 36-1451) [24]. Further, it also confirms the absence of impurities as it does not contain any characteristic XRD peaks other than ZnO peaks. The average size of the crystallites was calculated using Debye–Scherrer formula.

$$D = \frac{k\lambda}{\beta \cos\theta}$$

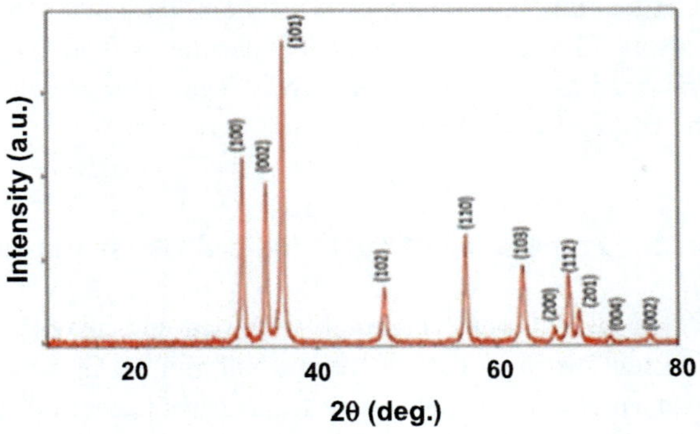

Fig. 1 X-ray diffraction pattern of ZnO nanoparticles synthesized from leaf extract of *Ricinus communis*

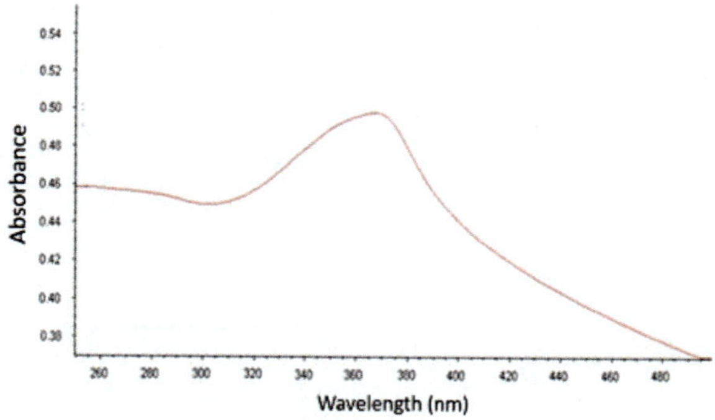

Fig. 2 UV–Visible spectra of ZnO nanoparticles synthesized from leaf extract of *Ricinus communis*

where D is the average crystallite size, λ is the wavelength of X-rays, θ is the diffraction angle, and β is the full width at half-maxima (FWHM) [25]. The crystallite size calculated using XRD was found to be 19.04 nm.

3.2 UV–Visible Spectra

Figure 2 represents the absorption spectrum of ZnO nanopowder. The UV–Vis absorption spectrum of ZnO nanoparticles synthesized from *Ricinus communis* exhibits a strong absorption band at about 367 nm corresponding to the intrinsic band-gap absorption of ZnO that happens because of electron transitions taking place from the valence band to the conduction band (O2p–Zn3d). An excitonic absorption peak is seen at about 254 nm which corresponds to the extract of Ricinus communis leaf. The energy band gap of ZnO nanoparticles was estimated by using Tauc formula:

$$E = hc/\lambda \tag{1}$$

where c is the velocity of light (3×10^8 m/s), h is Planck's constant (6.626×10^{-34} Js), and λ is the wavelength (367.06 nm). The band gap energy of synthesized ZnO nanoparticles was found to be 3.37 eV. Since the ZnO NPs shows strong absorption in the UV region, it can be used for various medical applications [26, 27].

3.3 FTIR Spectroscopy

Figure 3 represents the FTIR spectrum of ZnO nanoparticles synthesized from Ricinus communis leaf extract. The broad peak at 3412 cm^{-1} corresponds to O–H stretching and hydrogen-bonded groups in alcohol or phenolic or water molecules in the leaf extract. The peak at 2071 cm^{-1} indicates the –C=C stretching vibration of alkynes. The strong absorption peaks at 1636 cm^{-1} correspond to the stretching

Fig. 3 FTIR spectrum of *Ricinus communis* derived ZnO nanoparticles

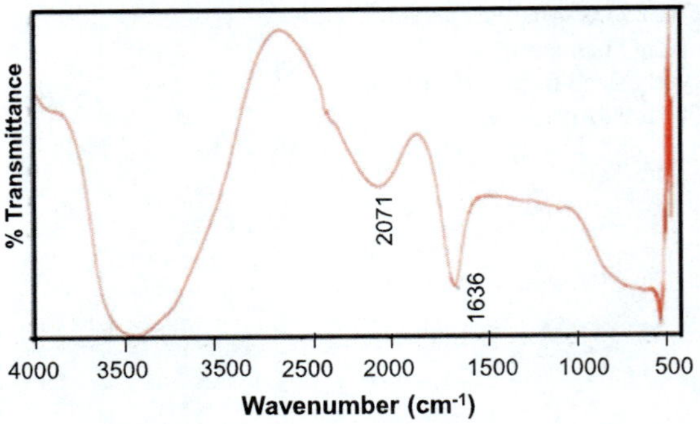

vibration of C=O hydroxyl (or) carboxyl groups on the surface of the sample. The vibrational peaks observed between 400 cm^{-1} and 600 cm^{-1} correspond to metal oxides [M–O] [28, 29].

3.4 Scanning Electron Microscopy (SEM)

The morphology of the synthesized ZnO nanoparticles was determined by SEM. Figure 4a, b represents the typical images of synthesized ZnO nanoparticles. SEM image demonstrates uniform structure and size of ZnO nanoparticles. However, some agglomerates are also seen. The ZnO nanoparticle size is in the range from 60 to 100 nm [30].

Fig. 4 SEM images of synthesized ZnO nanoparticles of *Ricinus communis* at **a** 30 kX and **b** 10 kX magnifications

3.5 *Electrochemical Behavior of Dopamine Using ZnO Modified Glassy Carbon Electrode*

Cyclic voltammogram of ZnO modified glassy carbon electrode was recorded in the presence of 2 mM dopamine and 0.2 M phosphate buffer solution in pH 6.6 at different scan rates (20–100 mV/s) (Figs. 5 and 6). With an increase in scan rate, the redox peak current increased simultaneously along with enlarged peak separation. Furthermore, the square root of scan rate was plotted against cathodic as well as anodic peak currents (μA) which showed linear increase with the increase in scan rate. The graph obtained was nearly straight lines as shown in Fig. 7. From the two linear plots, the linear regression equation can be described as $I_{pa} = 2.2919x + 5.5615$ ($R^2 = 0.9276$); $I_{pc} = -1.5764x + 1.2202$ ($R^2 = 0.9884$) and $I_{pa} = 0.356x + 7.7734$ ($R^2 = 0.8917$); $I_{pc} = -0.6759x + 1.2975$ ($R^2 = 0.9379$), respectively. From this result, it could be interpreted that for all scan rates considered, the electrochemical oxidation of dopamine at ZnO modified glassy carbon electrode is adsorption-controlled [31, 32].

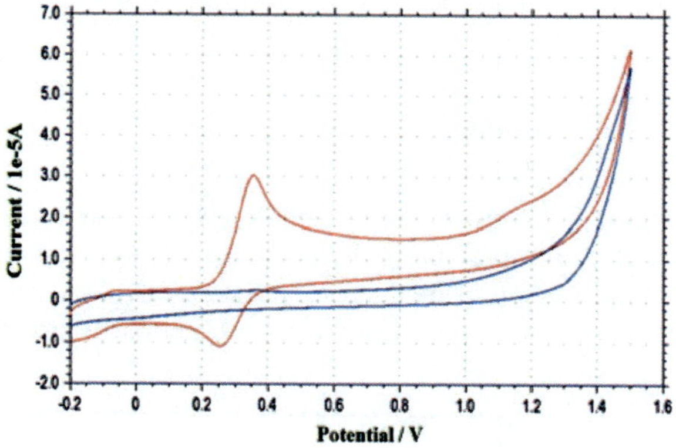

Fig. 5 Cyclic voltammogram of glassy carbon electrode in presence of 2 mM dopamine and 0.2 M phosphate buffer solution in pH 6.6 at scan rate of 100 mV/s

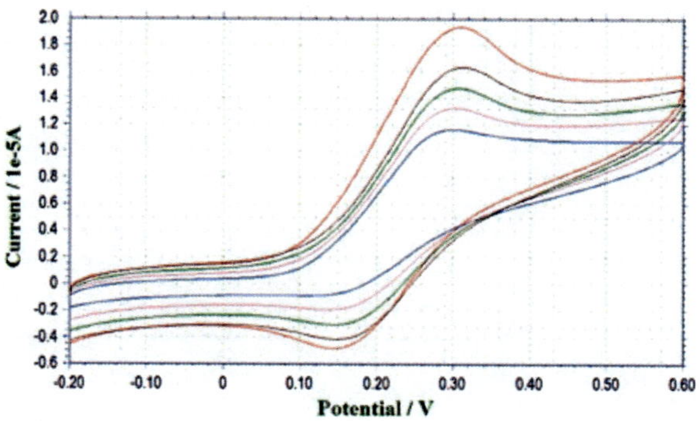

Fig. 6 Cyclic voltammogram overlay of ZnO modified glassy carbon electrode in the presence of 2 mM dopamine and 0.2 M phosphate buffer solution in pH 6.6 at various scan rates

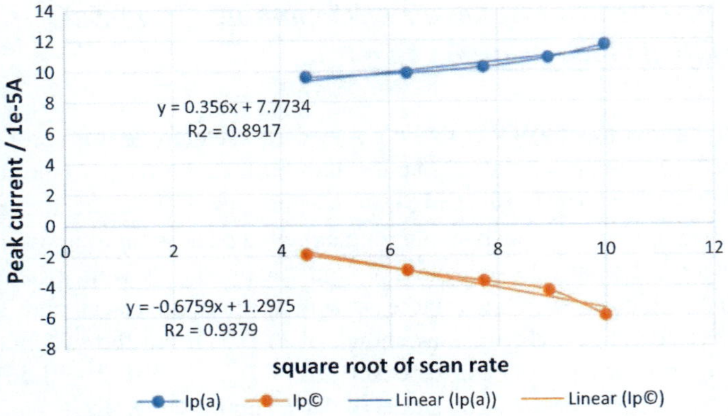

Fig. 7 Linear relationship between anodic and cathodic peak current vs. square root of scan rate

The linear increase in the redox peak currents of dopamine with the effect of increasing scan rates from 20 to 100 mV/s using ZnO modified glassy carbon electrode is a proof for the excellent electrocatalytic activity.

3.6 Photodegradation

Figure 8 represents the UV–Vis spectra of methylene blue with the time interval of 30 min representing the photodegradation of methylene blue. Figure 8 shows % degradation of methylene blue with ZnO nanoparticles exposed to sunlight by varying time. The % degradation of methylene blue was found to be 90% within

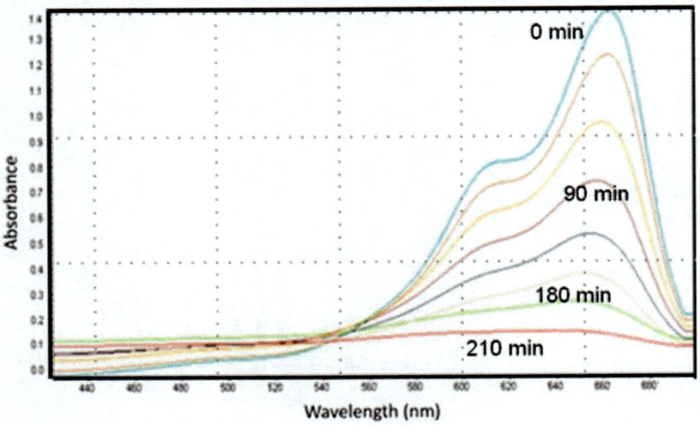

Fig. 8 UV–Vis absorption spectra of methylene blue dye solution degradation in the presence of synthesized ZnO nanoparticles

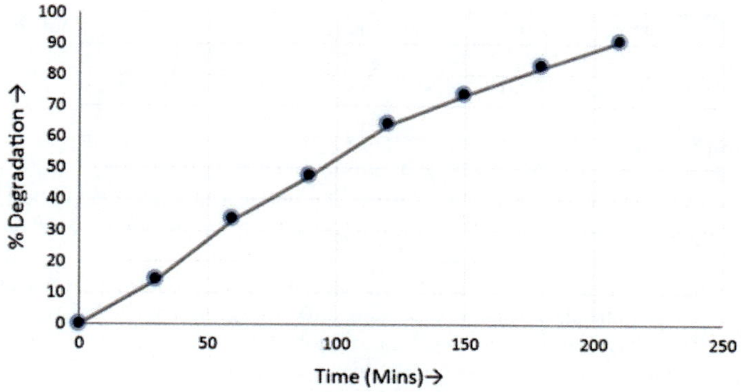

Fig. 9 Plot of % methylene blue dye degradation using 10 mg of synthesized ZnO at different time intervals

a span of 210 min. The enhanced photodegradation of ZnO is attributed to better surface area to mass ratios and light dispersion on the surface of synthesized ZnO [33–35] (Fig. 9).

4 Conclusions

ZnO nanoparticles were synthesized successfully from the aqueous extract of leaves of *Ricinus communis*. The ZnO nanoparticles were analyzed using FTIR, XRD, UV–Vis spectroscopy, and SEM. The prepared ZnO nanoparticle-modified glassy carbon electrode was more effective in investigating the electrochemical behavior of dopamine. ZnO modified glassy carbon electrode exhibits an adsorption-controlled process and enhanced electrochemical sensing property for the detection of dopamine. The photocatalytic activity of the synthesized ZnO showed methylene blue degradation of over 90% within a span of 210 min.

References

1. Annu AA, Ahmed S (2018) Green synthesis of metal, metal oxide nanoparticles, and their various applications. In: Handbook of ecomaterials. Springer International Publishing, Cham, 1–45. https://doi.org/10.1007/978-3-319-48281-1_115-1
2. Maruthupandy M, Zuo Y, Chen JS, Song JM, Niu HL, Mao CJ, Shen YH (2017) Synthesis of metal oxide nanoparticles (CuO and ZnO NPs) via biological template and their optical sensor applications. Appl Surf Sci 397:67–174. https://doi.org/10.1016/j.apsusc.2016.11.118
3. Ghorbani HR, Mehr FP, Pazoki H, Rahmani BM (2015) Synthesis of ZnO nanoparticles by precipitation method. Orient J Chem 31:1219–1221. https://doi.org/10.13005/ojc/310281
4. Hasnidawani JN, Azlina HN, Norita H, Bonnia NN, Ratim S, Ali ES (2016) Synthesis of ZnO nanostructures using sol-gel method. Procedia Chem 19:211–216. https://doi.org/10.1016/j.proche.2016.03.095

5. Li D, Qin Q, Duan X, Yang J, Guo W, Zheng W (2013) General one-pot template-free hydrothermal method to metal oxide hollow spheres and their photocatalytic activities and lithium storage properties. ACS Appl Mater Interf 5(18):9095–9100. https://doi.org/10.1021/am4024059
6. Hosokawa S (2016) Synthesis of metal oxides with improved performance using a solvothermal method. J Ceram Soc Jpn 124(9):870–874. https://doi.org/10.2109/jcersj2.16109
7. Raoufi D (2013) Synthesis and microstructural properties of ZnO nanoparticles prepared by precipitation method. Renew Energy 50:932–937. https://doi.org/10.1016/j.renene.2012.08.076
8. Palomo JM, Filice M (2016) Biosynthesis of metal nanoparticles: novel efficient heterogeneous nanocatalysts. Nanomater 6(5):84. https://doi.org/10.3390/nano6050084
9. Kuete V (2014) Physical, hematological, and histopathological signs of toxicity induced by African medicinal plants. In: Toxicological survey of African medicinal plants. Elsevier, pp 635–657. https://doi.org/10.1016/B978-0-12-800018-2.00022-4
10. Mittal AK, Chisti Y, Banerjee UC (2013) Synthesis of metallic nanoparticles using plant extracts. Biotechnol Advan 31:346–356. https://doi.org/10.1016/j.biotechadv.2013.01.003
11. Fakhari S, Jamzad M, Kabiri Fard H (2019) Green synthesis of zinc oxide nanoparticles: a comparison. Green Chem Lett Rev 12(1):19–24. https://doi.org/10.1080/17518253.2018.1547925
12. Jayappa MD, Ramaiah CK, Kumar MAP, Suresh D, Prabhu A, Devasya RP, Sheikh S (2020) Green synthesis of zinc oxide nanoparticles from the leaf, stem and in vitro grown callus of Mussaenda frondosa L: characterization and their applications. Appl Nanosci 10(8):3057–3074. https://doi.org/10.1007/s13204-020-01382-2
13. Hashemi S, Asrar Z, Pourseyedi S, Nadernejad N (2016) Green synthesis of ZnO nanoparticles by Olive (Olea europaea). IET Nanobiotechnol 10(6):400–404. https://doi.org/10.1049/iet-nbt.2015.0117
14. Jeevanandam J, Chan YS, Danquah MK (2016) Biosynthesis of metal and metal oxide nanoparticles. Chem Bio Eng Rev 3(2):55–67. https://doi.org/10.1002/cben.201500018
15. Reddy S, Kumara Swamy BE, Vasan HN, Jayadevappa H (2012) ZnO and ZnO/polyglycine modified carbon paste electrode for electrochemical investigation of dopamine. Anal Methods 4(9):2778–2783. https://doi.org/10.1039/C2AY25203A
16. Shashanka R, Kumara Swamy BE (2020) Biosynthesis of silver nanoparticles using leaves of Acacia melanoxylon and their application as dopamine and hydrogen peroxide sensors. Phy Chem Res 8(1):1–18. https://doi.org/10.22036/PCR.2019.205211.1688
17. Kuskur CM, Kumara Swamy BE, Shivakumar K, Jayadevappa H, Sharma SC (2019) Poly (sunset yellow) sensor for dopamine: a voltammetric study. J Electroanal Chem 840:52–59. https://doi.org/10.1016/j.jelechem.2019.03.031
18. Huang DQ, Chen C, Wu YM, Zhang H, Sheng LQ, Xu HJ, Liu ZD (2012) The determination of dopamine using glassy carbon electrode pretreated by a simple electrochemical method. Int J Electrochem Sci 7:5510–5520
19. Agarwal H, Kumar SV, Rajeshkumar S (2017) A review on green synthesis of zinc oxide nanoparticles–An eco-friendly approach. Resour -Eff Technol 3(4):406–413. https://doi.org/10.1016/j.reffit.2017.03.002
20. Gowsalya V, Santhiya E, Chandramohan K (2017) Synthesis, characterization of ZnO nanoparticles from Thespesia populnea. Indian J Appl Res 7(10):542–543
21. Aquisman AE, Wee BS, Chin SF, Kwabena DE, Michael KO, Bakeh T, Sylvester DS (2020) Synthesis, characterization, and antibacterial activity of ZnO nanoparticles from organic extract of cola nitida and cola acuminata leaf. Int J Nanosci Nanotechnol 16(2):73–89
22. Xia C, Wang N, Wang L, Guo L (2010) Synthesis of nanochain-assembled ZnO flowers and their application to dopamine sensing. Sens Actuators B: Chem 147(2):629–634. https://doi.org/10.1016/j.snb.2010.04.005
23. Balram D, Kuang-Yow L, Neethu S (2018) A novel electrochemical sensor based on flower shaped zinc oxide nanoparticles for the efficient detection of dopamine. Int J Electrochem Sci 13(2018):1542–1555. https://doi.org/10.20964/2018.02.06

24. Nava OJ, Luque PA, Gómez-Gutiérrez CM, Vilchis-Nestor AR, Castro-Beltrán A, Mota-González ML, Olivas A (2017) Influence of Camellia sinensis extract on Zinc Oxide nanoparticle green synthesis. J Mol Struct 1134:121–125. https://doi.org/10.1016/j.molstruc.2016.12.069
25. Karaköse E, Çolak H, Duman F (2017) Green synthesis and antimicrobial activity of ZnO nanostructures punica granatum shell extract. Green Process Synth 6(3):317–323. https://doi.org/10.1515/gps-2016-0190
26. Chennimalai M, Do JY, Kang M, Senthil TS (2019) A facile green approach of ZnO NRs synthesized via ricinus communis L. leaf extract for Biological activities. Mater Sci Eng C 103:109844. https://doi.org/10.1016/j.msec.2019.109844
27. Elumalai K, Velmurugan S (2015) Green synthesis, characterization and antimicrobial activities of zinc oxide nanoparticles from the leaf extract of Azadirachta indica (L.). Appl Surf Sci 345:329–336. https://doi.org/10.1016/j.apsusc.2015.03.176
28. Senthilkumar SR, Sivakumar T (2014) Green tea (Camellia sinensis) mediated synthesis of zinc oxide (ZnO) nanoparticles and studies on their antimicrobial activities. Int J Pharm Pharm Sci 6(6):461–465
29. Alamdari S, Sasani GM, Lee C, Han W, Park HH, Tafreshi M, Ara MHM (2020) Preparation and characterization of zinc oxide nanoparticles using leaf extract of sambucus ebulus. Appl Sci 10(10):3620. https://doi.org/10.3390/app10103620
30. Senthilkumar N, Nandhakumar E, Priya P, Soni D, Vimalan M, Potheher IV (2017) Synthesis of ZnO nanoparticles using leaf extract of Tectona grandis (L.) and their anti-bacterial, anti-arthritic, anti-oxidant and in vitro cytotoxicity activities. New J Chem 41(18):10347–10356. https://doi.org/10.1039/C7NJ02664A
31. Fooladsaz K, Negahdary M, Rahimi G, Habibi-Tamijani A, Parsania S, Akbari-dastjerdi H, Asadi A (2012) Dopamine determination with a biosensor based on catalase and modified carbon paste electrode with zinc oxide nanoparticles. Int J Electrochem Sci 7:9892–9908
32. Manjunatha JG, Kumara Swamy BE, Mamatha GP, Chandra U, Niranjana E, Sherigara BS (2009) Cyclic voltammetric studies of dopamine at lamotrigine and TX-100 modified carbon paste electrode. Int J Electrochem Sci 4:187–196
33. Chan SHS, Yeong WuT, Juan JC, Teh CY (2011) Recent developments of metal oxide semiconductors as photocatalysts in advanced oxidation processes (AOPs) for treatment of dye waste-water. J Chem Technol Biotechnol 86(9):1130–1158. https://doi.org/10.1002/jctb.2636
34. Aminuzzaman M, Ying LP, Goh WS, Watanabe A (2018) Green synthesis of zinc oxide nanoparticles using aqueous extract of Garcinia mangostana fruit pericarp and their photocatalytic activity. Bull Mater Sci 41(2):1–10. https://doi.org/10.1007/s12034-018-1568-4
35. Kočí K, Obalová L, Matějová L, Plachá D, Lacný Z, Jirkovský J, Šolcová O (2009) Effect of TiO_2 particle size on the photocatalytic reduction of CO_2. Appl Catal B: Environ 89(3–4):494–502. https://doi.org/10.1016/j.apcatb.2009.01.010

Trisodium 2-Hydroxypropane-1,2,3-Tricarboxylate Encapsulated Nanocontainer-Based Template-Free Electrochemical Synthesis of Multidimensional Copper/Copper Oxide Nanoparticles

Mona Saini, Nutan Rani, Asifa Mushtaq, Rini Singh, Seema Rawat, Manoj Kumar, and Kalawati saini

1 Introduction

Copper nanoparticles (NPs) are of great importance due to their wide application in anti-viral/antibacterial coatings [1–3], CO_2 adsorption [4–6], methanol oxidation [7], electronics [8], and sensor [9]. Copper is a good alternative to more costly metallic silver and gold nanoparticles and can be used for antibacterial, photocatalytic, and

M. Saini · N. Rani · K. saini (✉)
Department of Chemistry, Miranda House, University of Delhi, Patel Chest Marg, New Delhi 110007, India
e-mail: kalawati.saini@mirandahouse.ac.in

N. Rani
e-mail: nutan.rani@mirandahouse.ac.in

A. Mushtaq
Department of Botany and Microbiology, Garhwal University, H.N.B, Srinagar, Uttarakhand 241674, India

R. Singh
Graduate School of Engineering, Hiroshima University, 1-4-1 Kagamiyama, Higashi-Hiroshima 739-8527, Japan
e-mail: rini@hiroshima-u.ac.jp

S. Rawat
School of Life Sciences, Central University of Gujarat, Sector-30, Gandhinagar, Gujarat 382030, India
e-mail: seema.rawat@cug.ac.in

M. Kumar
Department of Physics, Malaviya National Institute of Technology, Jaipur, Rajasthan 302017, India

U. K. Mudali et al. (eds.), *Recent Trends in Electrochemical Science and Technology*, Springer Proceedings in Materials 15,
https://doi.org/10.1007/978-981-16-7554-6_18

biomedical applications. Also, copper nanoparticles have higher catalytic efficiency than other nanoparticles [10, 11]. The properties of copper nanoparticles can be controlled easily by changing the synthesis pathway. Copper nanoparticles have been synthesized using various routes like chemical reduction, biological and non-biological methods, chemical wet method (solvents thermal decomposition), green synthesis (surfactant-less), and other modified methods [12–16]. They can easily oxidize to form copper oxides. To avoid oxidation, these syntheses are performed in non-aqueous media at a low concentration of precursor and under an inert atmosphere either in nitrogen or in argon atmosphere.

Template-assisted electrochemical synthesis of semiconductor nanowires has been reported by Sisman et al. [17]. Lithographically patterned metallic nanowire electrodeposition (LPNE) has been reported by Menke et al. [18], which is an example of bottom-up electrochemical synthesis. Template-based electrochemical synthesis of copper (Cu) nanowires has been reported by Gupta et al. [19] and could sense CH_2Cl_2. Synthesis of copper oxide nanoparticles using carbon nanotubes as templates has been reported by Wu et al. [20]. Template-free synthesis of copper oxide has been carried out by using the precipitation method [22]. Simple template-free synthesis of $Cu(OH)_2$ and CuO nanostructures in a solution phase has been demonstrated by Mehdizadeh et al. [22]. Nanostructure and microstructures of polypyrrole have been synthesized by an electrochemical template-free process on a nickel (Ni) electrode [23].

In recent years, copper oxide nanostructured materials like nanorods [24], nanowires [25, 26], nanoribbons [27], nanobelts [28], and micro-sheets [29] are attracting the attention of researchers due to their potential biomedical applications. The clusters of metallic Cu have been prepared using reverse micelles as microreactors [30]. Heterostructures of CuO/polypyrrole (CuO/PPy) and CeO_2 polypyrrole (CeO_2/PPy) have been reported using the metal–organic decomposition (MOD) technique [31]. A variety of CuO nanostructures have been synthesized using high-temperature (at 393 K and 423 K) approaches like the hydrothermal method. The flower-like nanostructures of CuO have been synthesized using a hydrothermal method in a domestic microwave oven [32]. Electrochemical methods of preparation of Cu NPs have a low environmental impact and are cost-effective as reported by Saini et al. [33, 34]. Cu_2S nanoparticles have been electrochemically synthesized using cyclic voltammetry [35]. Potentiostatic electrochemical deposition (ECD) of copper sulfate solution within the nanochannels of porous anodic alumina templates has been used to fabricate copper nanowires [36]. Nanocomposites of CuO with multi-walled carbon nanotubes (MWCNT) and metallic Cu have been synthesized via electrochemical route [37, 38]. The synthesis of Cu_8O has been reported by Guan et al. [39]. The single-crystal X-ray structure of CuO has been determined at 196 K and room temperature by Asbrink et al. [40]. Electrochemical synthesis of copper oxide nanoparticles and nanorods has been reported in the literature [41, 42]. The synthesized nanorods have been studied for photocatalytic activity. Copper oxide (CuO) has been used for antibacterial application by Ren et al. [43]. Cu NPs have been studied for antibacterial activities against E. Coli by Raffi and coworkers [44]. Copper nanoparticles have also been reported as antibacterial agents by Mahmoodi et al. [45]. The antibacterial activity of copper/C nanocomposites synthesized via the green route has been reported by Bhavyasree et al. [46].

In the electrochemical synthesis route, copper salt is not used; hence, pure materials can be synthesized without using any reducing agent. Also, one can synthesize crystalline materials using the direct electrochemical method. The advantages of this method are: (i) no need of washing synthesized material with alcohol and (ii) no need of annealing of sample for a long time. Thus, electrochemical synthesis follows the principle of green chemistry and is very promising for the synthesis of materials. The template-free electrochemical method is cost-effective, industrially feasible, and eco-sustainable process. In this paper, the synthesis of Cu/copper oxide nanoparticles (NPs) by electrochemical reduction using trisodium citrate as a capping agent is reported. The process is devoid of chemical reducing agents such as hydrazine hydrochloride, sodium borohydride, amino acids, cetyl trimethyl ammonium bromide, and N-benzyl-N-dodecyl-N-bis(2-hydroxyethyl) ammonium chloride. The NPs have been characterized using powder X-ray diffractometry (PXRD), scanning electron microscopy (SEM), energy dispersive spectroscopy (EDS), transmission electron microscopy (TEM), and X-ray photoelectron spectroscopy (XPS). To the best of the authors' knowledge, the developed protocol (electrochemical method at constant applied potential) for the synthesis of copper, cuprous oxide, and cupric oxide NPs is not reported in the literature. These nanoparticles can be used for photocatalytic and antibacterial activities. Herein the antimicrobial activities of the synthesized copper/copper oxide NPs against gram positive as well as gram negative bacteria are discussed.

2 Experimental

2.1 Chemicals and Reagents

All chemicals used for the synthesis were of analytical reagent grade. Trisodium citrate (TSC, 99%) was purchased from Merck Limited, India. The copper metal strip (99.9% pure, metal basis) was procured from Alfa Aesar, and a Pt strip (99.99%) used as a reference electrode was purchased from Sinsil International. All other chemicals were used as such without any further purification. AR grade NaOH was used to maintain the basic pH of the solution. A DC power supply (Keithley 2231A-30–3 triple channel) was used to synthesize the NPs.

2.2 Preparation of Copper/Copper Oxide Nanostructures

Nanoparticles of copper/copper oxide nanomaterials were prepared with various concentrations of TSC (50,100, 150, 200, 250 mM) and at various pH values of reaction solutions 2.11, 4.22, 6.5, 7.8, 8.5, 13.11 with 2.55 mM of TSC to study the effect of pH and concentration of capping agent on the morphology of nanoparticles. The samples prepared at various pH values 2.1, 4.22, 7.8, and 8.5 with 2.55 mM of TSC and also with the above-mentioned other concentrations of TSC have been already reported in the literature [33]. So, herein the synthesis of copper/copper oxide nanostructures using 2.55 mM of TSC and at pH 13.11 is reported. The trisodium citrate (TSC) capped copper nanostructures (Cu and Cu_2O NPs) were synthesized via electrochemical route at pH 13.11, temperature of 373 K, and applied potential of 6.7 V. The concentration of TSC used was 2.55 mM. The pH 13.11 was maintained by the addition of 0.1 M of NaOH solution.

The electrochemical cell was formed by immersing the copper strip (anode, working electrode) and platinum strip (cathode, reference electrode) in 100 mL solutions of TSC. The TSC was used as the capping agent. Both electrodes were connected with a DC power supply, and the desired potential (6.7 V) was applied for a fixed time interval of two hours. The electrolysis of copper was carried out in the air for copper and copper oxide NPs using an electrochemical cell equipped with a magnetic stirrer at 450 rpm. The solution turned brownish red in 30 min after applying the set potential in the reaction setup which indicates the formation of Cu/Cu_2O NPs in the electrolyte solution. Then it turned blackish-red indicating the formation of all three types of NPs (Cu, Cu_2O, and CuO) in the electrolytic solution. The particles started accumulating or depositing at the Pt cathode when the potential was being applied. As soon as the power supply was stopped, the deposited $Cu/Cu_2O/CuO$ NPs got stripped off from the cathode in the form of precipitate and settled down at the bottom of the electrochemical cell. The precipitate was then filtered with Whatman filter paper number 42 and was washed several times with deionized water and then dried under vacuum. The powder was collected and used for further characterization.

2.3 *Materials Characterization*

Powder X-ray diffraction (PXRD) pattern of the dried powder was recorded using Bruker D8 Advance diffractometer equipped with Ni-filter and Cu Kα radiation. The data was collected in the 2θ range of 10–70° with a step size of 0.02° and a step time of 1 s. The surface morphology was studied by transmission electron microscopy (TEM) by using Tecnai G^2 F 20 TWIN TMPSeries microscope, Model FEG 200 kV. A carbon-coated copper grid was used for getting the TEM images of NPs. A freshly sonicated solution of 5 μL was spread with a 10 μL pipette on the carbon-coated side of the copper grid and dried under a bulb. Scanning electron microscopy (SEM) with energy dispersive spectroscopy (EDS) (JEOL JSM 6610 model no. at 20 kV) was used to obtain the morphology and composition of the synthesized powder. X-ray photoelectron spectroscopy (XPS, Omicron Nanotechnology, monochromatized Al, 1486.6 eV) was performed with XPS instrument having several features (ESCA + Omicron Nano Technology: Small spot XPS for high-speed depth profiling, Dual

Beam Charge Neutralization + Fully integrated Software Control and Automation, Rapid Quantification within Multipak or CASAXPS).

2.4 Antibacterial Activities

The antibacterial potential of Cu/Cu_2O/CuO nanoparticles against the pathogens, viz. *Escherichia coli, Pseudomonas aeruginosa, Staphylococcus aureus, Streptococcus pneumoniae,* was evaluated by broth dilution method [47]. A stock solution of 2 mg/mL of nanoparticles was prepared by dispersing them in pre-sterilized deionized water by ultra-sonication. The varying concentrations ranging from 0.2 μg/mL to 200 μg/mL were prepared from stock solution and were added to the test tubes containing varying amounts of sterile Luria Bertani broth [48]. 50 μL of overnight culture (0.5 McFarland turbidity standards) of test pathogens was added to these tubes aseptically. The tubes were incubated at 37 °C for 24 h. The bacterial cultures without any test solution and the tubes with only sterile media were kept as positive and negative controls, respectively. The results were recorded by measuring the optical density of the inoculated broth at 600 nm. Minimum inhibitory concentration (MIC) was recorded as the lowest concentration of the test sample inhibiting the growth of the inoculated test pathogen.

3 Results and Discussion

During the electrochemical reduction, Cu/Cu_2O/CuO is formed following the nucleation of the NPs by electrochemical reaction followed by the growth process. When the potential is applied, copper anode first gets oxidized into + 2 oxidation state.

$$Cu \rightarrow Cu^{2+} + 2e^{-} \tag{1}$$

This Cu^{2+} species in the presence of citrate capping agent and applied potential reduces back to Cu^{o} oxidation state (i.e., nucleation). The formation of copper oxides depends upon the applied potential, temperature, pH, and type of atmosphere (inert N_2/Ar/air). The formed nuclei are of different shapes of the NPs. The Cu^{o} nuclei then undergo a growth process to generate NPs. The growth process is controlled by the diffusion of growth species which in turn depends upon the concentration of trisodium citrate.

3.1 Powder X-ray Diffraction (PXRD) Analysis

The XRD pattern of the synthesized $Cu/Cu_2O/CuO$ nanoparticles is shown in Fig. 1. The 2θ values at 43.65° and 50.43° correspond to (111) and (200) planes of cubic (face-centered) Cu (JCPDS, PDF, File No. 04–0836) phase. The other peaks at 2θ values of 32.49, 38.97, 47.29, 49.55, 51.40, 53.81, 58.39, 63.41, and 66.53° correspond to planes (110), (111), (-112), (-202), (112), (020), (202), (-113), and (022), respectively, match the monoclinic CuO (JCPDS, PDF, File No. 41–0254, 45–0937 and 80–1917) as reported in the literature [34, 35]. Herein, all three JCPDS file numbers are given for CuO nanoparticles. The other peaks at 2θ values of 33.08, 38.52,41.45, 44.74, 49.31, 51.67, 53.78, and 58.63° corresponding to planes (103), (004), (014), (220), (024), (105), (214), and (303) belong to copper oxide as reported by R. Guan [39] [(JCPDS, PDF, File No.78–1588)]. No impurity diffraction peaks have been detected which confirms the high purity of the product obtained by this method. The observation of diffraction peaks intensity for all the CuO nanoparticles indicates their high crystallinity. The intensity of peaks for primitive Cu_2O NPs at 2θ values 29.7°(110), 36.6°(111), 42.4°(200), and 61.4°(220) is found to be extremely low which shows that these NPs are less crystalline than the CuO NPs. It means PXRD analysis of the reaction products obtained at basic pH conditions does not contain pure CuO phase as reported by Nikam [49]. Based on XRD, it is clear that the peak intensity of CuO is found to be more as compared to Cu_2O and Cu NPs, respectively.

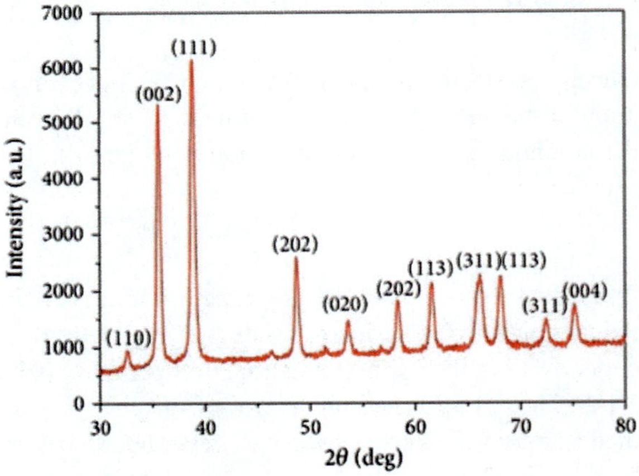

Fig. 1 Powder XRD pattern of $Cu/Cu_2O/CuO$ nanoparticles with 2.55 mM trisodium citrate at 6.7 V, and at pH 13.11

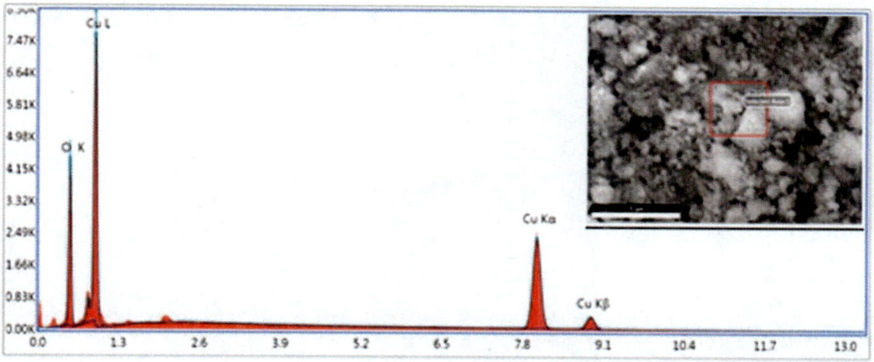

Fig. 2 EDS of the synthesized $Cu/Cu_2O/CuO$ nanoparticles

3.2 EDS and TEM Analyses

The EDS of the synthesized nanoparticles is shown in Fig. 2. The EDS data confirmed the formation of copper/copper oxides nanoparticles. The atomic ratios of oxygen and copper were found to be 59.52 and 40.48%, respectively.

The sonicated aqueous solution of NPs was used for analyzing the morphology of the synthesized sample. Figure 3a–f shows the TEM images of $Cu/Cu_2O/CuO$ NPs obtained at pH 13.11. These micrographs show different shapes of the nanoparticles such as leaf or feather, spherical, block-type, and rod-shaped. The leaf-shaped or feather-shaped particles have a length in the range of 5.24 μm and width in the range of 323 nm as shown in Fig. 3a. The block-type particles having one dimension as 135 nm as shown in Fig. 3b, c demonstrate the presence of all types of NPs with bigger sizes of rods having the dimension of 3.41 μm (length) and 125 nm (width).

The rod-shaped particles have a length of 205 nm and a width of 27.9 nm as shown in Fig. 3d. The spherical particles are also deposited on the surface of rods (Fig. 3d). The shape of spherical particles varies from 11.5 nm to 43.5 nm as shown in Fig. 3d–f and shows that some particles are agglomerated in a specific manner.

3.3 X-ray Photoelectron Spectroscopy (XPS) Analysis

XPS was used to determine the oxidation states and the surface chemical composition of the synthesized nanoparticles. Figure 4 shows the XPS spectra of the electrochemically synthesized $Cu/Cu_2O/CuO$ NPs at applied potential of 6.7 V and pH of 13.11. The analysis of XPS peaks confirms the formation of a mixture of cuprous oxides (Cu_2O) and cupric oxides (CuO), respectively. Figure 4 demonstrates the deconvoluted XPS spectra of the Cu 2p core level. Doublet peaks positioned at binding energy of 932.43 eV and 952.72 eV in Cu_2O corresponding to Cu $2p_{3/2}$ and Cu $2p_{1/2}$, respectively. While the other doublet peaks are assigned corresponding to Cu $2p_{3/2}$

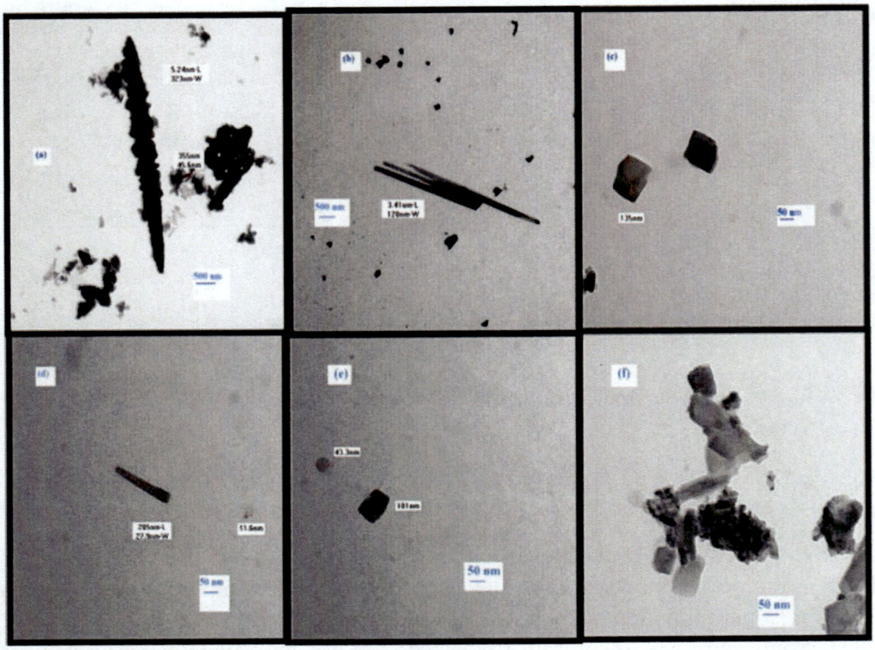

Fig. 3 TEM of $Cu/Cu_2O/CuO$ nanoparticles exhibiting various morphologies of the nanostructures

and Cu $2p_{1/2}$ in CuO at binding energy of 934.83 eV and 954.73 eV, respectively. The satellite peaks at 940.01 and 944.25 eV correspond to Cu $2p_{3/2}$ and Cu $2p_{1/2}$ in CuO and indicate the existence of an unfilled Cu 3d shell. The peak is assigned at a binding energy of 531.66 eV corresponding to O 1 s, and the calibration peak is obtained corresponding to C1s at binding energy of 288.50 eV and 284.97 eV, respectively. The obtained XPS data match well with the reported data in the literature [50–52]. Copper in CuO exists in the + 2 state with $3d^9$ configuration, while Cu_2O exists in + 1 state with 3d ^{10} configuration as reported in the literature [53].

From the XPS results, we can estimate the ratio of Cu_2O and CuO on the basis of peak area. The fitting of high-resolution spectra of elements helps to quantify the ratio of oxidation states. The area ratio of CuO and Cu_2O is 7.3: 1. It is quite difficult to distinguish only Cu metal because of the similar binding energy of Cu and Cu(I) oxide. The peak position is in agreement with the reported literature [54, 55]. The % ratio of $Cu/Cu_2O/CuO$ in the mixture of synthesized material with XPS study was not possible.

Based on XPS peak-fit values of the Cu 2p core level measured on the surface of the synthesized NPs, the predominant phase was found to be CuO NPs and having adsorbed oxygen on the surface which can enhance its catalytic activity. This adsorbed oxygen can be used to enhance the photocatalytic and antibacterial activities.

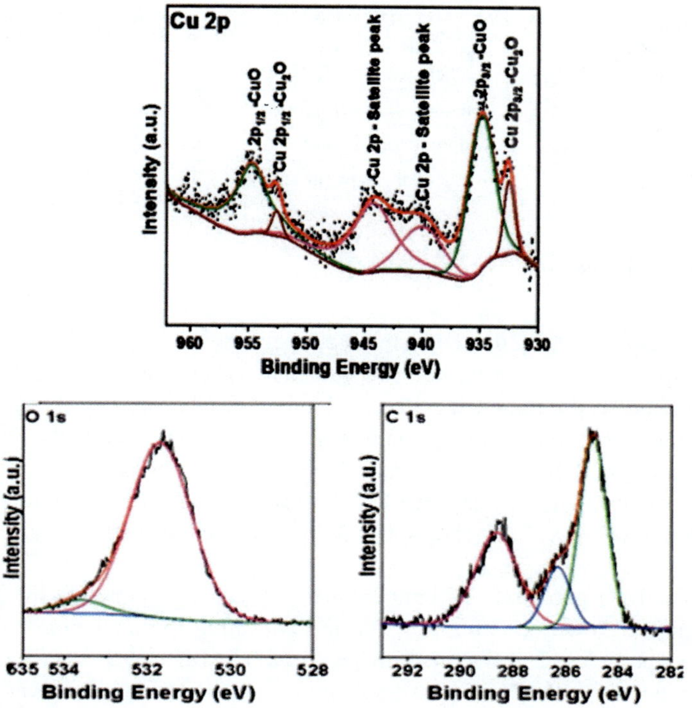

Fig. 4 Cu, O, and C XPS peaks of $Cu/Cu_2O/CuO$ nanoparticles synthesized from electrochemical route using 2.55 mM trisodium citrate at pH 13.11, 6.7 V at 373 K

3.4 Antibacterial Activity

It is well known that the CuO NPs show significant antibacterial activity against gram negative and gram positive bacteria. Herein, the antimicrobial activities of *Escherichia coli* (*E.Coli*) and *Pseudomonas Aeruginosa* have been studied. These are gram negative bacteria. The antimicrobial activities on *Staphylococcus Aureus* and *Streptococcus Pneumoniae* which are gram positive bacteria have also been studied. This study has been carried out using the broth dilution method. A higher value of minimum inhibitory concentration (MIC) indicates that a higher concentration of NPs is required to inhibit bacterial growth. The synthesized $Cu/Cu_2O/CuO$ nanoparticles do not affect the growth of *Pseudomonas Aeruginosa*. It means no inhibition (NI) of bacterial growth, as evident from Table 1. The MIC values were 155 μg/mL for *E.Coli* and 150 μg/mL for *Staphylococcus Aureus* as well as for *Streptococcus Pneumoniae*. These synthesized NPs show no inhibition on the growth of *Pseudomonas Aeruginosa*. These results are in good agreement with the reported results in the literature [56].

Table 1 Antibacterial activities of $Cu/Cu_2O/CuO$ nanoparticles with minimum inhibitory concentration (MIC, in μg/mL)

Isolates	$Cu/Cu_2O/CuO$ (MIC in μg/mL)
Escherichia coli (E. Coli)	155
Pseudomonas aeruginosa	No inhibition
Staphylococcus aureus	150
Streptococcus pneumoniae	150

4 Conclusions

The electrochemical reduction behavior of copper ions, as well as its nucleation and growth on a platinum electrode, has been studied in the aqueous solution of TSC. The copper electrode is the source of copper + 2 ions, and subsequently, it reduces to Cu at the platinum electrode and in a solution of 2.55 mM TSC having pH 13.11 in the electrochemical cell. In a basic medium, these Cu NPs again oxidize into Cu_2O and CuO, respectively. The PXRD and EDS analysis confirmed the formation of both types of oxides. The XPS analysis showed the peak area ratio of CuO and Cu_2O as 7.3: 1

Both electrode potential and temperature play an important role in tuning the nucleation and growth kinetics and also in controlling the final morphologies of copper nanoparticles. The obtained images from TEM reveal the exact shape and size of synthesized NPs. This is a special and unique case of one-step template-free synthesis, in which we have obtained four different morphologies of NPs. These synthesized NPs show very good antimicrobial activities against *E.Coli*, *Staphylococcus Aureus,* and *Streptococcus Pneumoniae.* The MIC values of $Cu/Cu_2O/CuO$ nanoparticles have been found from 150 μg/mL to 155 μg/mL for inhibiting the growth of the above-mentioned bacteria.

The electrochemical synthesis is a novel technique that is simple and environment-friendly than the conventional chemical reduction methods. Thus, the electrochemical synthesis is an ideal process as it consumes less energy and has a high output (yield 80–90%) and easy to control the process parameters.

Acknowledgements The authors thank Department of Chemistry, Indian Institute of Technology, for providing the facility for PXRD. The authors also wish to thank the Department of Physics, MNIT, Jaipur, India, for XPS analysis. The authors also wish to thank Kusuma School of Biological Science, the Indian Institute of Technology for providing the facility for TEM and the University of Delhi (USIC), for providing the facility for SEM with EDS analysis.

References

1. Lakshmanan M, Anumakonda VR, Gobi SK, Suchart S, Kumar J, Parameswaranpillai RS (2019) Preparation of cellulose/copper nanoparticles bionanocomposite films using a bioflocculant polymer as reducing agent for antibacterial and anticorrosion applications. Compos B

Eng 175:107177. https://doi.org/10.1016/j.compositesb.2019.107177

2. Jagaran K, Singh M (2021) Nanomedicine for COVID-19: Potential of copper nanoparticles. Biointerface Res Appl Chem 11(3):10716–10728. https://doi.org/10.33263/BRIAC113.1071610728
3. Raha S, Mallick R, Basak S, Duttaroy A (2020) Is copper beneficial for COVID- 19 patients? Med. Hypotheses 142. https://doi.org/10.1016/j.mehy.2020.109814
4. Smyrnioti M, Tampaxis C, Steriotis T, Ioannides T (2020) Study of CO_2 adsorption on a commercial $CuO/ZnO/Al_2O_3$ catalyst. Catal Today 357:495–502. https://doi.org/10.1016/j.cattod.2019.07.024
5. Larmier K, Liao WC, Tada S, Lam E, Verel R, Bansode A, Urakawa A, Vives C, Copéret C (2017) CO_2-to-methanol hydrogenation on zirconia-supported copper nanoparticles: Reaction intermediates and the role of the metal–support interface. Angew Chem 56:2318. https://doi.org/10.1002/anie.201610166
6. Bersani M, Gupta K, Mishra AK, Lanza R, Taylor SFR, Islam H, Hollingsworth N, Hardacre C, de Leeuw NH, Darr JA (2016) Combined EXAFS, XRD, DRIFTS, and DFT study of nano copper-based catalysts for CO_2 hydrogenation. ACS Catal 6(9):5823. https://doi.org/10.1021/acscatal.6b01529
7. Studt F, Behrens M, Kunkes EL, Thomas N, Zander S, Tarasov A, Schumann J, Frei E, Varley JB, Abild-Pedersen F, Nørskov JK, Schlögl R (2015) The mechanism of CO and CO_2 hydrogenation to methanol over Cu-based catalysts. Chem Cat Chem 7(7):1232. https://doi.org/10.1002/cctc.201590041
8. Sagadevan S, Murugasen P (2015) Electrical properties of copper oxide nanoparticles. Nano Res 30:1. https://doi.org/10.4028/www.scientific.net/JNanoR.30.1
9. Kang X, Mai Z, Zou X, Cai P (2007) A sensitive nonenzymatic glucose sensor in alkaline media with a copper nanocluster/multiwall carbon nanotube-modified glassy carbon electrode. Anal Biochem 363(1):143. https://doi.org/10.1016/j.ab.2007.01.003
10. Roudriguze JA, Lui P, Hrbek J, Evans J, Parez M (2007) Water gas shift reaction on Cu and Au nanoparticles supported on $CeO_2(111)$ and ZnO(0001): intrinsic activity and importance of support interactions. Angew Chem Int Ed Engl 46(8):1329. https://doi.org/10.1002/anie.200603931
11. Ali ZI, Ghazy OA, Meligi G, Saleh HH, Bekhit M (2018) Radiation-induced synthesis of copper/poly(vinyl alcohol) nanocomposites and their catalytic activity. Adv Polym Technol 201:21675. https://doi.org/10.1002/adv.21675
12. Dang TMD, Le TTT, Fribourg-Blanc E, Dang MC (2011) The influence of solvents and surfactants on the preparation of copper nanoparticles by a chemical reduction method. Adv Nat Sci Nanosci Nanotechnol 2:025004. https://doi.org/10.1088/2043-6262/2/2/025004
13. Ghorbani HR (2014) Biological and non-biological methods for fabrication of copper nanoparticles. Chem Eng Commun 202(11):1463–1467. https://doi.org/10.1080/00986445.2014.950732
14. Zhong CJ, Mott D, Galkwoski J, Wang L, Luo J (2007) Synthesis of size-controlled and shaped copper nanoparticles. Langmuir 23(10):5740. https://doi.org/10.1021/la0635092
15. Khodashenas B, Ghorbani HR (2014) Synthesis of copper nanoparticles: an overview of the various methods. Korean J Chem Eng 31:1105–1109. https://doi.org/10.1007/s11814-014-0127-y
16. Wu S-H, Chen DH, Synthesis of high-concentration Cu nanoparticles in aqueous CTAB solutions. J Colloid Interface Sci 273:165. https://doi.org/10.1016/j.jcis.2004.01.071
17. Sisman I (2011) Template-assisted electrochemical synthesis of semiconductor nanowires. Chapter in book: Nanowires - Implementations and Applications, Source-InTech. https://doi.org/10.5772/20551
18. Menke EJ, Xiang C, Thompson MA, Yang LC, Penner RM (2006) Lithographically patterned nanowire electrodeposition. Nat Mater 5(11):914. https://doi.org/10.1038/nmat1759
19. Gupta J, Arya S, Singh A, Verma S, Sharma A, Singh B, Tomar A (2020) Template based electrochemical synthesis of copper (Cu) nanowires as CH_2Cl_2. Sensor Integr Ferroelectr 204:63. https://doi.org/10.1080/10584587.2019.1674990

20. Wu HQ, Wei XW, Shao MW, Gu JS, Qu MZ (2002) Synthesis of copper oxide nanoparticles using carbon nanotube as templates. Chem Phys Lett 364(1):152. https://doi.org/10.1016/S0009-2614(02)01301-5
21. Wahab R, Ahmad N, Alam M, Aldahmash AB, Abdulaziz A Al- Khedhairy (2016) Template free synthesis of copper oxide nanoparticles prepared via precipitation process. Asian J Chem 28(12):2622. doi:https://doi.org/10.14233/ajchem.2016.20029
22. Mehdizadeh R, Hasanzadeh M, Sanati S, Saghatforoush LA (2012) Simple template-free solution route for the synthesis of $Cu(OH)_2$ and CuO nanostructures and application for electrochemical determination three ß-blockers. J Exp Nanosci 9(8):763. https://doi.org/10.1080/17458080.2012.714479
23. Chebil S, Monod MO, Fisicaro P (2014) Direct electrochemical synthesis and characterization of polypyrrole nano- and micro-snails. ElectrochimActa 123:527. https://doi.org/10.1016/j.electacta.2014.01.058
24. Panigrahi S, Kundu S, Ghosh SK, Nath S, Praharaj S, Basu S, Pal T (2006) Selective one-pot synthesis of copper nanorods under surfactantless condition. Polyhedron 25(5):1263–1269. https://doi.org/10.1016/j.poly.2005.09.006
25. Wang X, Xiaofeng W, Yuan L, Huang K, Feng S (2017) Ultra-low reflection CuO nanowire array in-situ grown on copper sheet. Mater Design 113:297–304. https://doi.org/10.1016/j.matdes.2016.10.029
26. Kaur M, Muthe KP, Despande SK, Choudhury S, Sing JB, Verma N, Gupta SK, Yakhmi JV (2006) Growth and branching of CuO nanowires by thermal oxidation of copper. J Cryst Growth 289(2):670–675. https://doi.org/10.1016/j.jcrysgro.2005.11.111
27. Zhu CL, Chen CN, Hao LY, Hu Y, Chen ZY (2004) Template-free synthesis of $Cu_2Cl(OH)_3$ nanoribbons and use as sacrificial template for CuO nanoribbon. J Cryst Growth 263(1–4):473–479. https://doi.org/10.1016/j.jcrysgro.2003.11.003
28. Song X, Yu H, Sun S (2005) Single-crystalline CuO nanobelts fabricated by a convenient route. J Colloid Interf Sci 289:588. https://doi.org/10.1016/j.jcis.2005.03.074
29. Yang Q, Yan PX, Chang JB, Feng JJ, Yue GH (2007) Growth of bicrystal CuO microsheets from aqueous solution. Phys Lett A 361:493. https://doi.org/10.1016/j.physleta.2006.07.056
30. Lisiecki I, Pileni MP (1993) Synthesis of copper metallic clusters using reverse micelles as microreactors. J Am Chem Soc 115(10):3887–3896. https://doi.org/10.1021/ja00063a006
31. Galembeck A, Alves OL (1999) Planar heterostructures oxide/conducting polymer (CuO/polypyrrole and CeO/polypyrrole. Synth Met 102:1238. https://doi.org/10.1016/S0379-6779(98)01439-8
32. Volanti DP, Keyson D, Cavalcante LS, Simoes AZ, Joya MR, Longo E, Varela JA, Pizani PS, Souza AG, Volanti P, Keyson D, Cavalcante LS, Souza AG (2008) Synthesis and characterization of CuO flower-nanostructure processing by a domestic hydrothermal microwave. J Alloys Compd 459:537–542. https://doi.org/10.1016/j.jallcom.2007.05.023
33. Saini K, Devnani H, Bhat M, Ingole PP (2017) Anisotropic plasmonic copper/copper oxide nanostructures by DC electrophoretic dissolution of copper in water for plasmonic sensing of glucose. J Electrochem Soc 164(13):B674–B680. https://doi.org/10.1149/2.1391713jes
34. Saini K, Ingole PP, Bhatia SS, Rani N (2018) Rod-shaped copper (Cu, C_u2O) nano catalyst for the facile oxidation of methanol. Adv Mater Lett 9:36–41. https://doi.org/10.5185/amlett.2018.1714
35. Fotouhi L, Rezaei M (2009) Electrochemical synthesis of copper sulfide nanoparticles. Microchim Acta 167:247. https://doi.org/10.1007/s00604-009-0234-3
36. Gao T, Meng G, Wang Y, Sun S, Zhang L (2002) Electrochemical synthesis of copper nanowires. J Phys Condens Matter 14:355–363. https://doi.org/10.1088/0953-8984/14/3/306
37. Pandey P, Merwyn S, Agarwal GS, Pant SC (2012) Electrochemical synthesis of multi-armed CuO nanoparticles and their remarkable bactericidal potential against waterborne bacteria. J Nanopart Res 14(1):1–13. https://doi.org/10.1007/s11051-011-0709-0
38. Zhang QB, Hua YX (2014) Electrochemical synthesis of copper nanoparticles using cuprous oxide as a precursor in choline chloride–urea deep eutectic solvent: nucleation and growth mechanism. Phys Chem Chem Phys 16:27088. https://doi.org/10.1007/s11051-011-0709-0

39. Guan R, Hashimoto H, Kuo KH (1984) Electron-microscopic study of the structure of metastable oxides formed in the initial stage of copper oxidation. II Cu8O Acta Cryst B40:560–566. https://doi.org/10.1107/S010876818400269X
40. Asbrink S, Waskowska A (1991) CuO: X-ray single-crystal structure determination at 196 K and room temperature. J Phys Condens Matter 3:8173. https://doi.org/10.1088/0953-8984/3/42/012
41. Maria Starowicz M (2019) Electrochemical synthesis of copper oxide particles with controlled oxidation state, shape and size. Mater Res Express 6: 0850a3. 1088/2053–1591/ab239d
42. Mousali E, Zaanjanchi MA (2019) Electrochemical synthesis of copper (II) Oxides nanorods and their application in photocatalytic reactions. J Solid State Electrochem 23:925. https://doi.org/10.1007/s10008-019-04194-9
43. Ren G, Hu D, Cheng EW, Vargas-Reus MA, Reip P, Allaker RP (2009) Characterisation of copper oxide nanoparticles for antimicrobial applications. Int J Antimicrob Agents 33:587–590. https://doi.org/10.1016/j.ijantimicag.2008.12.004
44. Raffi M, Mehrwan S, Bhatti TM, Akhter JI, Hameed A, Yawar W, Masood ul Hasan M (2010) Investigations into the antibacterial behavior of copper nanoparticles against Escherichia coli. Ann Microbiol 60:75–80. https://doi.org/10.1007/s13213-010-0015-6
45. Mahmoodi S, Elmi A, Nezhadi HS (2018) Copper nanoparticles as antibacterial agents. Mol Pharm Org Process Res 6:1. https://doi.org/10.4172/2329-9053.1000140
46. Bhavyasree PG, Xavier TS (2020) Green synthesis of copper oxide/carbon nanocomposites using the leaf extract of Adhatoda vasica Nees, their characterization and antimicrobial activity. Heliyon 6:e03323. https://doi.org/10.1016/j.heliyon.2020.e03323
47. Garcia L (2010) Broth microdilution MIC Test, p 25–41. In Clinical Microbiology Procedures Handbook, 3rd edn. ASM Press, Washington, DC. https://doi.org/10.1128/9781555817435.ch5.2
48. Muniyan A, Ravi K, Mohan U, Panchamoorthy R (2017) Characterization and in vitro antibacterial activity of saponin-conjugated silver nanoparticles against bacteria that cause burn wound infections. World J Microbiol Biotechnol 33:147. https://doi.org/10.1007/s11274-017-2309-3
49. Nikam AV, Kashmir AA, Krishnamoorthy K, Kulkarni AA, Prasad BLV (2014) pH-dependent single-step rapid synthesis of CuO and Cu_2O nanoparticles from the same precursor. Cryst Growth Des 14:4329. https://doi.org/10.1021/cg500394p
50. Espinos JP, Morales J, Barranco A, Caballero A, Holgado JP, Gonzalez-Elipe AR (2002) Interface effects for Cu, CuO, and Cu_2O deposited on SiO_2 and ZrO_2. XPS determination of the valence state of copper in Cu/SiO_2 and Cu/ZrO_2 catalysts. J Phys Chem B 106:6921. https://doi.org/10.1021/jp014618m
51. Wang P, Ng YH, Amal R (2013) Embedment of anodized p-type Cu_2O thin films with CuO nanowires for improvement in photoelectrochemical stability. Nanoscale 5:2952. https://doi.org/10.1039/c3nr34012k
52. Ji JY, Shih PH, Yang CC, Chan TS, Ma YR, Yun Wu SY (2009) Spontaneous self-organization of Cu_2O/CuO core–shell nanowires from copper nanoparticles. Nanotechnology 21:045603. https://doi.org/10.1088/0957-4484/21/4/045603
53. Ghijsen J, Tjeng LH, Van Elp J, Eskes H, Westerink J, Sawatzky GA, Czyzyk MT (1988) Electronic structure of Cu_2O and CuO. Phys Rev B: Condens Matter 38:11322. https://doi.org/10.1103/PhysRevB.38.11322
54. Gupta A, Ramen Jamatia R, Patil RA, Ma Y-R, Pal AK (2018) Copper oxide/reduced graphene oxide nanocomposite-catalyzed synthesis of flavanones and flavanones with triazole hybrid molecules in one pot: a green and sustainable approach. ACS Omega 3:7288–7299. https://doi.org/10.1021/acsomega.8b00334
55. Gao Y, Yang F, Yu Q, Fan R, Yang M, Rao S, Lan Q, Yang Z, Zhenquan Y (2019) Three-dimensional porous Cu@Cu_2O aerogels for direct voltammetric sensing of glucose. Microchim Acta 186:192. https://doi.org/10.1007/s00604-019-3263-6
56. Padil VVT, Cernik M (2013) Green synthesis of copper oxide nanoparticles using gum karaya as a biotemplate and their antibacterial application. Int J Nanomedicine 8:889–898. https://doi.org/10.2147/IJN.S40599

Facile Fabrication of Stable Superhydrophobic and Conductive Carbon Black Coating

Edna Richard and S. T. Aruna

1 Introduction

Carbon-based superhydrophobic materials perform as excellent candidates for water filtration, oil–water separation, oil-spill cleanup, gas separation, etc. [1]. Some research groups even extended the application of carbon-based superhydrophobic surfaces for preparing conductive transparent films and also for electromagnetic interference shielding [2–4]. Sansotera et al. [5] used linear perfluoropolyether (PFPE) peroxide for the chemical linkage of fluorinated PFPE chains on the surface of carbon cloth and commercial conductive carbon black to make it superhydrophobic. Conductivity measurements showed that the covalent linkage of fluorinated chains weakly modified the electrical properties of the conductive carbon black, even when the surface properties changed. Shen et al. [6] fabricated superhydrophobic conductive Ketjen black-high density polyethylene (HDPE) surface by a single-step pressing method where the fillers were pressed into the polymer surface leaving it partially exposed to form a high static water contact angle (WCA) of 160°. Das et al. [3] presented a method to prepare a superhydrophobic polymeric surface using carbon nanofibers as fillers. The effectiveness of the coatings as electromagnetic interference (EMI) shielding materials was explored for the first time. A static water contact angle of 158° with high electrical conductivity of 309 S/m was obtained for the substrate with an EMI shielding effectiveness up to 25 dB in the X-band (8.2–12.4 GHz).

E. Richard
Department of Chemistry, St. Xavier's College, Thumba 695586, Thiruvananthapuram, Kerala, India

S. T. Aruna (✉)
Surface Engineering Division, Council of Scientific and Industrial Research-National Aerospace Laboratories, Post Bag No. 1779, Bangalore 560 017, India
e-mail: aruna_reddy@nal.res.in

U. K. Mudali et al. (eds.), *Recent Trends in Electrochemical Science and Technology*, Springer Proceedings in Materials 15,
https://doi.org/10.1007/978-981-16-7554-6_19

Bao et al. [2] tailored the morphology of raspberry-like superhydrophobic carbon black-polystyrene colloidal suspension by adjusting the polarity and the concentration ratio of the filler polymer suspension. Mittal et al. [7] fabricated water-repelling mesoporous carbon nanocapsule (MCC)/polyvinylidene fluoride (PVDF) film by a facile solution approach. The electrically conducting film exhibited a water contact angle of 160° and a sliding angle of 5°. The films also exhibited high thermal stability up to 350 °C in the oxidative atmosphere and also retained superhydrophobicity at highly corrosive acidic and basic conditions. Caffrey et al. [8] produced an electrically conducting superhydrophobic poly(dimethylsiloxane) (PDMS) surface by adding multiwall carbon nanotubes. The microstructure/nanostructure was replicated onto the polymer surface using an ultrafast laser micro-texturing process to obtain a WCA of 161° without any additional coating. An improvement in the conductivity was observed by controlling the amount of MWCNT added to PDMS, and at a bulk loading of 5 wt %, there was an improvement in conductivity over pure PDMS. Asthana et al. [9] prepared superhydrophobic and superoleophobic coatings by drop-casting suspensions with different kinds of carbon nanoparticles in combination with a fluoropolymer suspension on a glass substrate as well as on Whatman filter paper. The ability of the coatings to resist dynamic water impalement and electrical conductivity was also evaluated. The coatings containing a 1:1:2 ratio of carbon black: graphene nanoparticles: polymer showed both excellent impalement resistance and electrical conductivity [9]. Tian et al. [10] developed superhydrophobic silicone rubber using a variety of particles with dissimilar features, such as SiO_2, copper, $BaTiO_3$, carbon black for imparting roughness to the silicon rubber matrix.

The objective of the present study is to fabricate superhydrophobic conductive carbon black surfaces on both glass substrate and cotton fabric. A very few studies are available on carbon black superhydrophobic surfaces. Carbon black being a promising candidate due to its physicochemical properties and conductive nature was used in the present study to impart roughness and to produce a conductive superhydrophobic surface. The fluorosurfactant was used to impart low surface energy and to improve the adhesion of the coating. The stability of the superhydrophobic carbon black coating with respect to pH, temperature, and water immersion was also studied. Further, the possibility of using carbon black superhydrophobic cotton fabric to separate petroleum ether water mixture is also investigated.

2 Experimental

The carbon black (CB) used in this work was a commercially available conductive CB (Raven 430 ultra from Columbian carbon Black, USA) with a particle size of 82 nm and a surface area of 31 m^2/g. Zonyl, the fluorosurfactant used in the present study, was purchased from Rishichem Private Ltd. and absolute ethanol was purchased from Merck.

The superhydrophobic conductive carbon black coatings were prepared as follows: 0.1 g of CB was dispersed in 5 mL ethanol by ultrasonicating for 30 min.

Then 5 mL of 1% zonyl was added to it and carefully stirred. The coatings on glass slides were prepared by spraying the solution on clean dry glass slides using a spray gun with a compressed air pressure of 20 ± 5 psi. The superhydrophobic cotton fabric was prepared by immersing the cleaned fabric (2″ × 2″) in the above fresh solution and padded it with a glass rod on either side for 10 min. Both the fabric and the coating on glass slides were allowed to dry at room temperature and then heated at 100 °C for 1 h to expel the trapped solvent. The unmodified coatings were prepared without the zonyl surfactant for comparison. The schematic depicting the preparation of superhydrophobic conductive carbon black coating on the glass slide is shown in Fig. 1.

The surface morphology of the superhydrophobic coatings and the fabrics was examined under secondary electron mode using a field emission scanning electron microscope (FESEM, Carl Zeiss, Supra 40 VP) equipped with an energy dispersive X-ray analysis (EDX) system. Water contact angles (WCA) of all the coatings and the fabrics were measured using a contact angle analyzer (Phoenix 300 Plus from M/s Surface Electro Optics). The tangent fitting mode was used in this instrument for the determination of WCA. The volume of Milli-Q water drop was 8 μL in the study. Five measurements of WCA on the coatings were taken, and the mean value was reported. Water sliding angle (SA) measurements were made using a simple in-house fabricated setup consisting of a smooth and planar platform on which the sample can be fixed. The angle at which the water drop slides was measured by using a protractor attached to the instrument. The adhesion of the coatings was measured

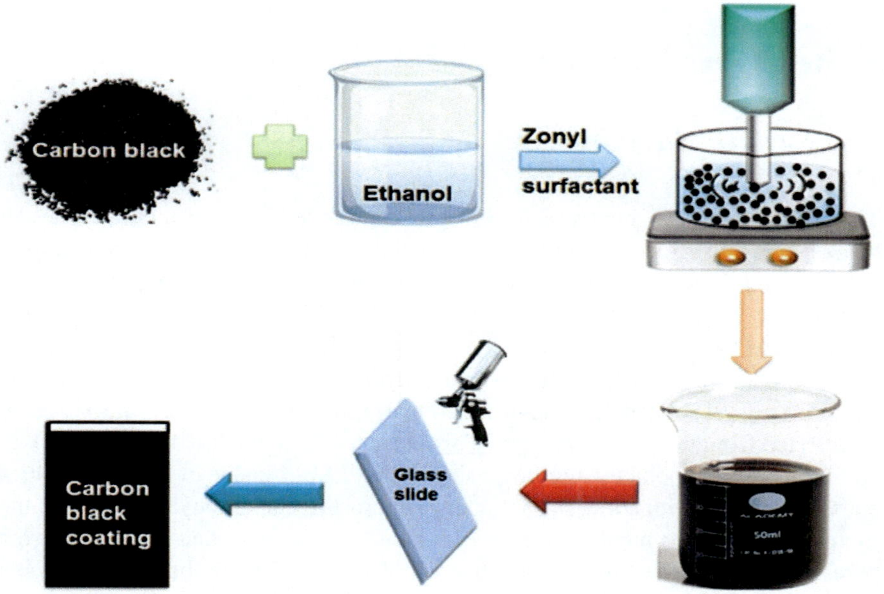

Fig. 1 Schematic depicting the steps involved in the preparation of superhydrophobic conductive carbon black coating

using a cross-hatch cutter, Model Elcometer 107 according to the ASTM D3359-02 standard test method. Fourier transform infrared (FTIR) spectra were recorded on Bruker Alpha-P spectrometer. The surface roughness of the fabrics was measured by using a 3D profilometer (Nano Map 500LS from AEP Technology). To check the stability of the coatings in water, the coatings were immersed in double distilled water, taken out periodically at regular intervals (1 day), dried at room temperature for 30 min, and WCA and SA were measured. The conductivity of the coatings was measured by DC four-probe conductivity using Keithley multimeters.

A mixture of petroleum ether (PE) and water (dyed with methyl blue for easy observation) in different ratios ranging from 1:4 to 1:3 (water/petroleum ether volume ratios) was poured onto the superhydrophobic carbon black fabric that was fixed on a funnel placed over a conical flask. Only the PE could pass through the fabric and water stayed on the surface. The gravity-driven separation was over within 10 min. The water was then collected and weighed to calculate the separation efficiency (η) which is defined as the ratio of the weight of water collected to that initially added as given by the equation shown in [11, 12]:

$$\eta = \left(\frac{m_1}{m_0}\right) \times 100$$

where m_0 and m_1 are the masses of the water before and after the separation process, respectively.

3 Results and Discussion

3.1 Surface Microstructure

The FESEM surface microstructure of the unmodified and the modified carbon black coatings is shown in Fig. 2. A large number of particles aggregate with non-homogeneous pores are observed for the carbon black coating on the glass substrate without any modification (Fig. 2a), and the WCA of this smooth unmodified coating was around 129°. Moreover, these pores were capable of producing roughness while entrapping the air into its cavities to make it hydrophobic without any modification. Asthana et al. [9] noticed hydrophilic functional groups on the surface of carbon black as detected from the FTIR spectrum. This prevents the unmodified carbon black to attain superhydrophobicity. Figure 2b shows the FESEM image of the carbon black coating modified with a low surface energy fluorosurfactant (zonyl 8740). With the use of surfactant, the number of pores will reduce as the surfactant partially fills the pores and anchors the particles well to each other as well as to the glass substrate. Thus, apart from rendering low surface energy to the surface, the surfactant acts as a suitable matrix for holding the carbon black nanoparticles firmly. A smooth coating of zonyl was hydrophobic with a WCA of 102°. But after introducing the carbon black

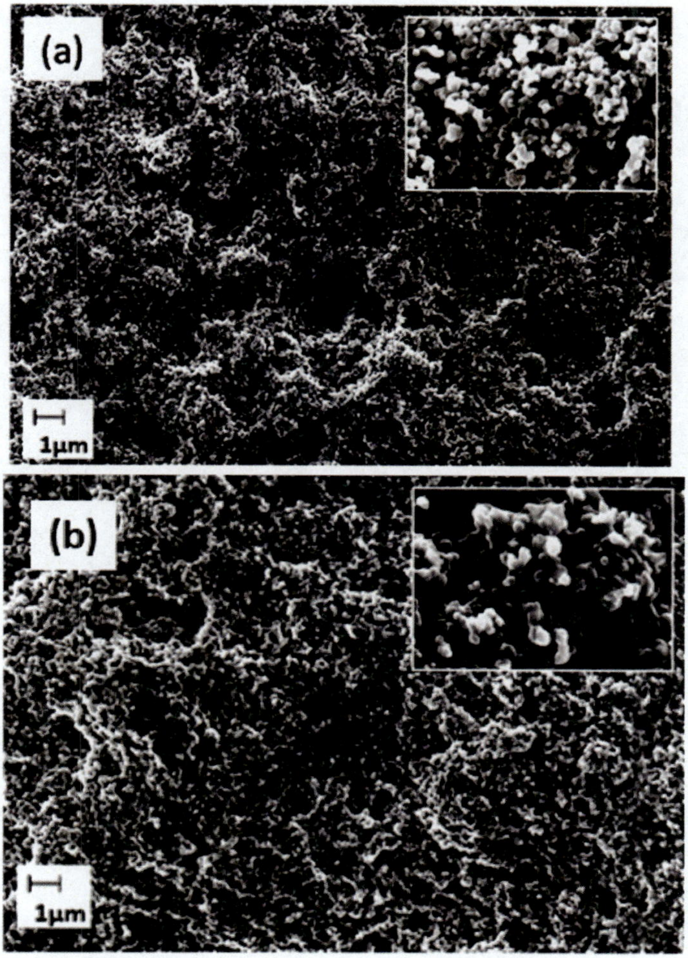

Fig. 2 FESEM images of the surfaces of **a** unmodified CB and **b** CB/Zonyl coating (6 kX (inset 25 kX))

fillers, it turned superhydrophobic with a WCA of 157°. Hence, superhydrophobicity is attributed to the combined effect of the roughness produced by the agglomerated fillers as well as the low surface energy rendered by the fluorosurfactant.

Figure 3a reveals the particle size of the carbon black nanoparticles, and it is seen that the particle size ranges from 50–100 nm. Figure 3b–d shows the FESEM surface morphologies of the cotton fabric immersed in the carbon black–zonyl mixture for 10 min. The figure shows the surface FESEM image of the modified cotton fabric. Figure 3b shows an even surface coverage with a uniform density of carbon black nanoparticles coating the yarns of the fabric. This shows that immersing the fabric in the solution for 10 min is sufficient to uniformly coat the fabric with the modified carbon black nanoparticles. Figure 3c shows a magnified image displaying the

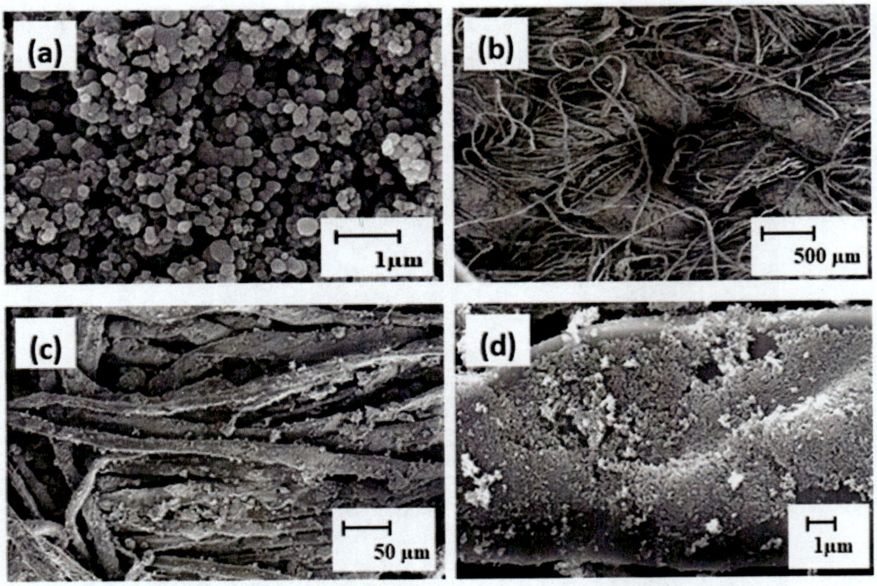

Fig. 3 FESEM images of **a** carbon black particles (**100,000** X) and **b**, **c**, **d** modified superhydrophobic CB fabric at different magnifications **b** 120 X **c** 1000 X and **d** 10,000 X

uniform distribution of particles across every yarn. Even though the macro-yarns induce a regular surface roughness, the agglomerated particles carrying the pores on each yarn still enhance the overall surface roughness adding to its ability to trap more air. The yarns carrying the uniform conductive particles (Fig. 3d) overlap with each other that will add to the overall conductivity of the fabric. Similar morphologies have been reported for fluorinated carbon black nanocomposite coating [5].

3.2 *The Effect of Fluorosurfactant on WCA of the Modified CB*

The WCA of the coatings with different concentrations of the fluorosurfactant was studied as shown in Fig. 4. The WCA of smooth unmodified carbon black on a glass substrate with a thickness of 12 μm was around 129° which increased to 153 ± 2° after modification with 1% ethanolic solution of fluorosurfactant. On this surface, the measurement of sliding angle showed a value less than 2°. The WCA of the coatings with different volume % of the fluorosurfactant remained greater than 150° for all the concentrations up to 10% fluorosurfactant. The highest WCA of 155° was observed for the coating modified with 5% zonyl solution. The water sliding angle was also less than 2° for all the measurements. All the coatings were oleophilic. Previously, Lakshmi et al. [13] had reported a significant increase in oleophobicity

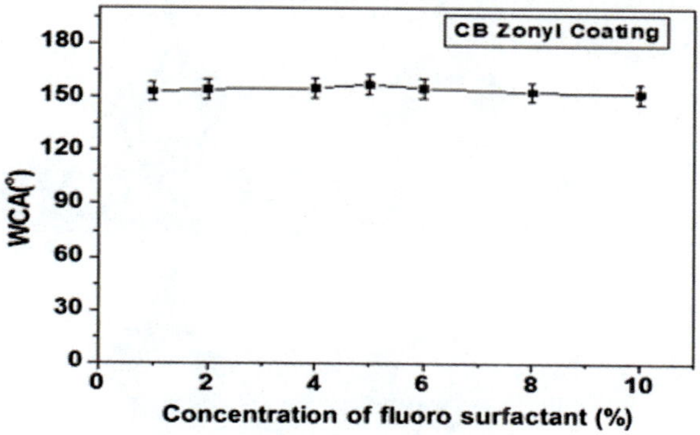

Fig. 4 Effect of fluorosurfactant concentration on WCA of the CB/zonyl coating

with an increase in the concentration of zonyl using silica as fillers. In the present work, the oleophilicity can be attributed to the presence of hydrophilic moieties on the surface of the carbon black as well as its porous nature.

The adhesion of unmodified CB coating to glass substrate was 0B which improved to 2B after modification with the 1% zonyl solution. The adhesion to glass substrate was further improved and remained constant at a value of 4B with an increase in the volume concentration of zonyl (5%) (Fig. 5). The non-adhesive behavior of the surface is evident in Fig. 5a. The water drop placed on the CB coating modified with 5% zonyl solution showed no signs of adhesion of water droplets on its surface (Fig. 6).

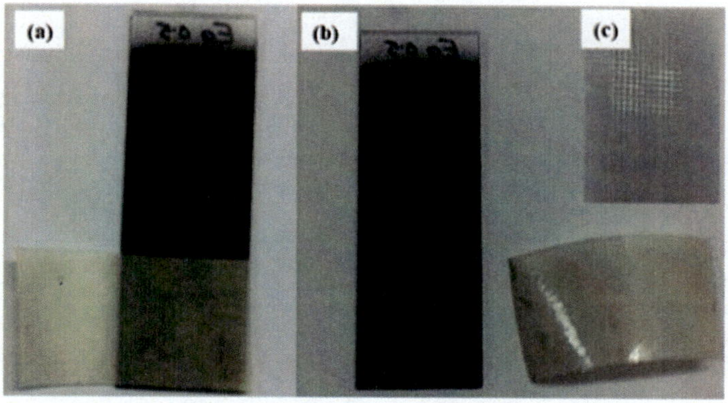

Fig. 5 Photographs of coatings **a** before and **b** after tape peel test **c** inset shows lattice (4B) obtained after the tape peel off test

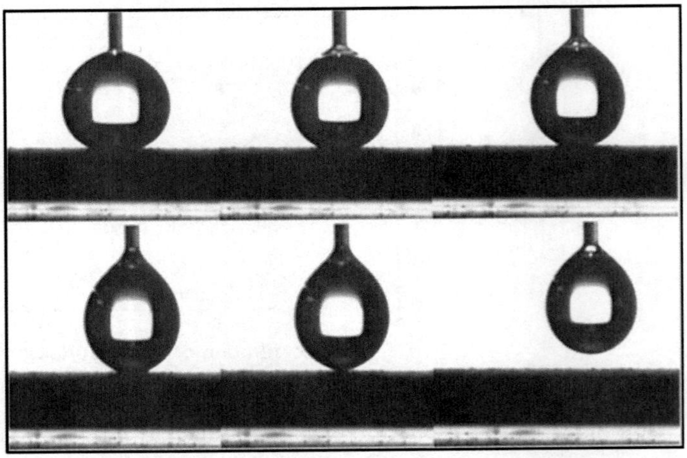

Fig. 6 Images depicting the non-adhesive behavior of a water drop on the superhydrophobic carbon black coating modified with 5% zonyl solution

3.3 Effect of Temperature on Superhydrophobic CB Coating

In order to study the thermal stability of the coating, the modified CB coatings with 12 μm thickness were subjected to heat treatment from room temperature to 250°C and the WCA of the coatings were measured and are shown in Fig. 7. It is seen that the surface is close to superhydrophobic with a WCA of 142° before the heat treatment. The heat treatment at 100°C expelled the solvents so as to achieve superhydrophobicity with a WCA of 157°. Further, the WCA remained almost constant till 200°C after which it showed a decrease and reached to 112° at 250°C. This is attributed to the stability of fluorosurfactant which is only stable up to 200°C, and above this temperature, the surfactant starts to decompose.

Fig. 7 Effect of heat treatment on WCA of the modified carbon black coating with zonyl

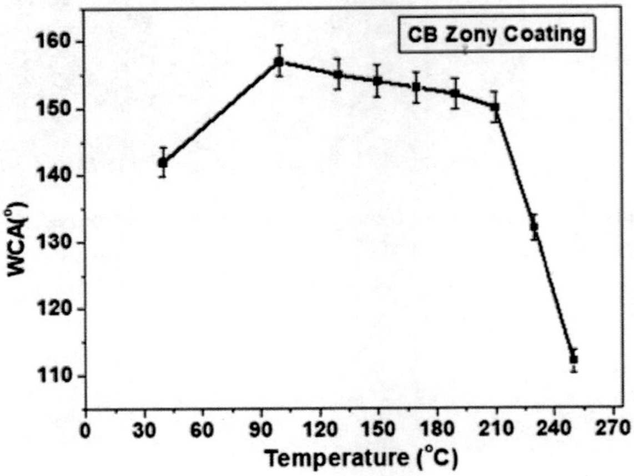

3.4 Stability of the Superhydrophobic CB Coating in Water

The effect of water immersion on WCA of modified CB coatings is shown in Fig. 8. It is observed that the modified CB coating retains superhydrophobicity even after immersion in water for 5–6 days. After 4 days of continuous immersion, the superhydrophobicity starts decreasing. The reduced WCA after 4 days of water immersion may be due to the dissolution of water-soluble zonyl surfactant in water.

The contact angle on the carbon black superhydrophobic surface with different pH (1–13) liquids was measured in order to study the stability of the coating under a corrosive environment. Figure 9 shows that the contact angle was above the superhydrophobic threshold for acidic as well as basic pH ranges.

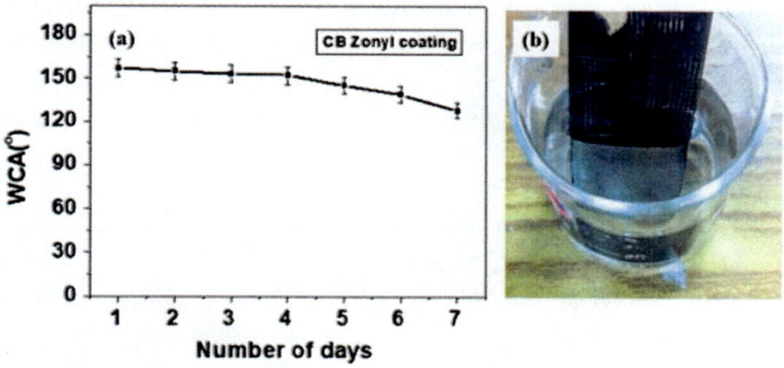

Fig. 8 **a** Effect of water immersion on WCA of the carbon black coating and **b** plastron layer formed on the surface of modified coating during water immersion

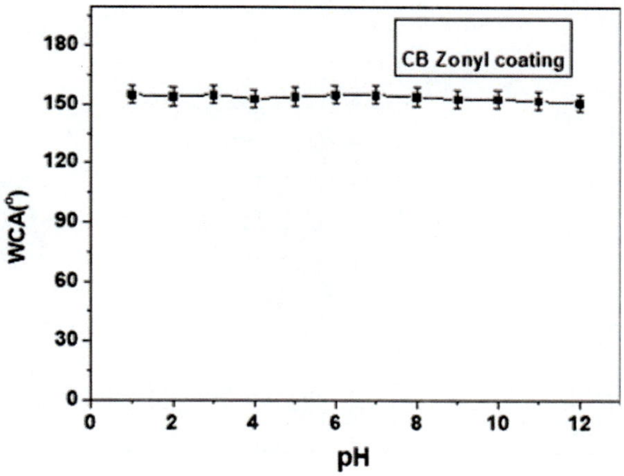

Fig. 9 Effect of pH on WCA of the modified carbon black coating

3.5 *Effect of Coating Thickness on Surface Roughness and Superhydrophobicity*

The thickness of the coating has a great influence on the roughness as well as on the wetting of the superhydrophobic coating because the thickness is a measure of the number of nanoparticles loaded on the surface. Figure 10 shows the 3D profiles of the coatings with different coating thicknesses. An average roughness of 1866 nm is observed for a surface with a coating thickness of 12 μm. It is apparent that the sample roughness slightly increased with the coating thickness which could be roughly controlled by the number of spray passes sprayed over the glass slides. This rise is mainly due to the aggregation of CB nanoparticles. With increase in coating roughness, the WCA of the coatings increases. This tendency confirms the behavior of hydrophobic surfaces, showing an increase in water contact angle with an increase in surface roughness. Without any CB nanoparticles, the zonyl coating shows weak hydrophobicity with a WCA of 102°. With the increase in the number of spray passes (CB loading), the roughness of the coating increases making the surface superhydrophobic with a WCA of 157º.

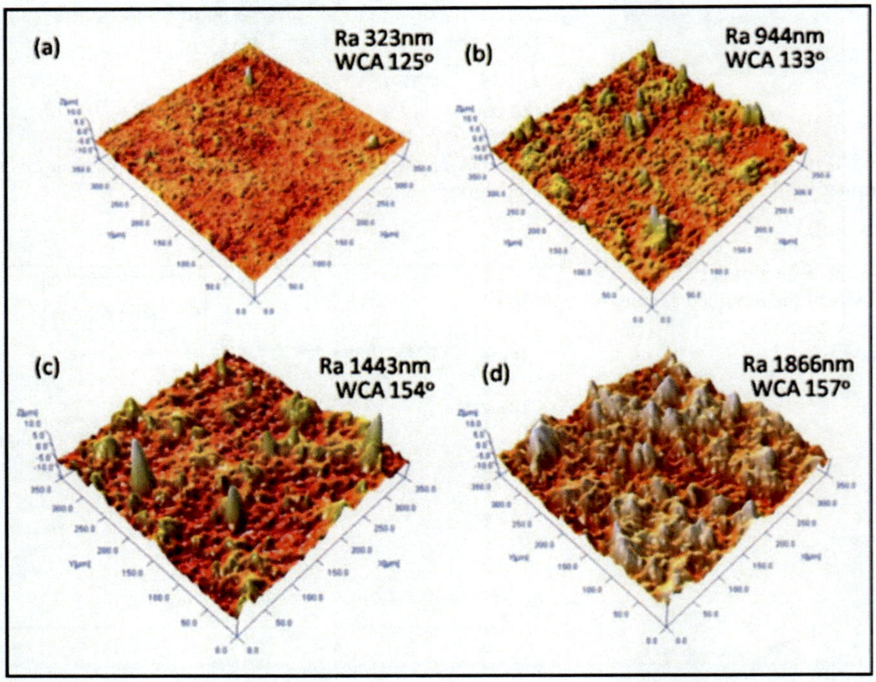

Fig. 10 3D profiles showing the effects of roughness of the CB coatings on WCA as a function of the coating thickness **a** 9 μm, **b** 10 μm, **c** 12 **μm,** and **d** 13 μm

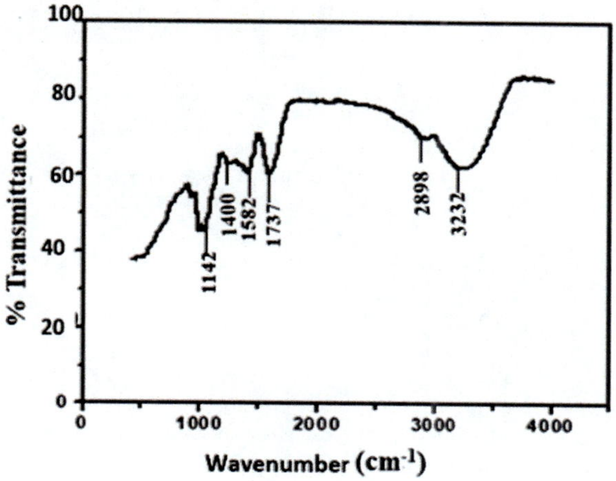

Fig. 11 FTIR spectrum of CB/zonyl composite coating

3.6 FTIR Spectrum of the CB/Zonyl Coating

The FTIR spectrum (Fig. 11) of the superhydrophobic coating shows an absorption band at 3232 cm $^{-1}$ corresponding to the presence of traces of moisture in the fabric. It can be attributed to the stretching and bending vibration of either free OH groups or free H_2O molecules. The peak at 2898 cm^{-1} is assigned to –C-H vibrations. The peak at 1582 cm^{-1} is due to the vibrations of CF_3-CF_2 groups. The peak at 1737 cm^{-1} corresponds to the C = O group of the methacrylate group of the fluorosurfactant. The peak at 1400 cm^{-1} is attributed to symmetric stretching of COO^- on the surface of carbon black stretching.

3.7 Conductivity of the CB/Zonyl Coating

The resistivity of the superhydrophobic carbon black coating with a thickness of 12 μm and a resistance of 7.53 kΩ was calculated to be 0.2 Ωcm. The corresponding conductivity was 5 Scm^{-1}. The resistivity of a carbon black coating without any modification was found to be 0.1 Ωcm, and its corresponding conductivity measured was 10 Scm^{-1}. This clearly indicates that after chemical modification of CB with fluorosurfactant, the conductivity of CB coatings was slightly lowered.

3.8 Separation of Petroleum Ether and Water

As shown in Fig. 12, when a mixture of petroleum ether (PE) and water (dyed with methylene blue for easy observation) was poured onto the superhydrophobic fabric,

Fig. 12 **a** Mixture of petroleum ether and water dyed with methylene blue for easy observation **b** and **c** water dyed with methylene blue held on the surface

Table 1 Calculated separation efficiency (η) values for different water/petroleum ether ratios

Total volume (mL)	Water/Petroleum ether (Volume ratio)	Separation efficiency (η) (%)
20	1:4	97
20	1:3	98
20	1:1	85
20	3:1	75
20	4:1	68

only the PE could pass through the fabric and water was held on the surface. The water was collected and weighed to investigate the separation efficiency. The results are listed in Table 1. The separation efficiency is >97% for water/petroleum ether volume ratios ranging from 1:4 to 1:3. The stability of the superhydrophobic fabric can provide more opportunities for many practical applications.

4 Conclusions

A cost-effective and conductive superhydrophobic coating was fabricated on a glass substrate by spraying a mixture of carbon black and a fluorosurfactant. The coating was stable for a wide range of pH, and it also withstood the water immersion test for more than 5 days. The conductivity of the coating was measured by a four-point probe method, and it was found that the conductivity decreased with the surface modification. Finally, a stable superhydrophobic cotton fabric modified with carbon black–zonyl mixture was used to separate the water/petroleum ether mixture effectively.

Acknowledgements The authors thank Director, NAL and Head, SED for their constant encouragement. The authors thank Mr. Siju and Mr. Praveen for the help received in FESEM and 3D profilometry.

References

1. Kako T, Nakajima A, Irie H, Kato Z, Uematsu K, Watanabe T, Hashimoto KJ (2004) Adhesion and sliding of wet snow on a super-hydrophobic surface with hydrophilic channels. Mater Sci 39:547–555. https://doi.org/10.1023/B:JMSC.0000011510.92644.3f
2. Bao Y, Li Q, Xue P, Huang J, Wanga J, Guo W, Wu C (2011) Tailoring the morphology of raspberry-like carbon black/polystyrene composite microspheres for fabricating superhydrophobic surface. Mater Res Bull 46:779–785. https://doi.org/10.1016/j.materresbull.2010.09.043
3. Das A, Hayvaci HT, Tiwari MK, Bayer IS, Erricolo D, Megaridis CM (2011) Superhydrophobic and conductive carbon nanofiber/PTFE composite coatings for EMI shielding. J Colloid Interface Sci 353:311–315. https://doi.org/10.1016/j.jcis.2010.09.017
4. Zhang T, Yan H, Fang Z, Yuping E, Wu T, Chen F (2014) Superhydrophobic and conductive properties of carbon nanotubes/polybenzoxazine nanocomposites coated ramie fabric prepared by solution-immersion process. App Surf Sci 309:218–224. https://doi.org/10.1016/j.apsusc.2014.05.013
5. Sansotera M, Navarrini W, Gola M, Dotelli G, Stampino PG, Bianchi CL (2012) Conductivity and superhydrophobic effect on PFPE-modified porous carbonaceous materials. Int J Hydrog Energy 37:6277–62840. https://doi.org/10.1016/j.ijhydene.2011.07.041
6. Shen L, Ding H, Cao Q, Jia W, Wang W, Guo Q (2012) Fabrication of Ketjen black-high density polyethylene superhydrophobic conductive surfaces. Carbon 50:4284–42900. https://doi.org/10.1016/j.carbon.2012.05.018
7. Mittal N, Kumar R, Mishra G, Deva D, Sharma A (2016) Mesoporous carbon nanocapsules based coatings with multifunctionalities. Adv Mater Interf 3:1500708. https://doi.org/10.1002/admi.201500708
8. Caffrey PO, Gupta MC (2014) Electrically conducting superhydrophobic microtextured carbon nanotube nanocomposite. Appl Surf Sci 314:40–45. https://doi.org/10.1016/j.apsusc.2014.06.055
9. Asthana A, Maitra T, Büchel R, Tiwari MK, Poulikakos D (2014) Multifunctional superhydrophobic polymer/carbon nanocomposites: graphene, carbon nanotubes, or carbon black? ACS Appl Mater Interf 6:8859–8867. https://doi.org/10.1021/am501649w
10. Tian H, Wang F, Ge S, Ou J, Li W, Yu S (2016) A simple and effective way to fabricate mechanical robust superhydrophobic surfaces. RSC Adv 6:28563–28569. https://doi.org/10.1039/C6RA04364J
11. Pan QM, Wang M, Wang HB (2008) Separating small amount of water and hydrophobic solvents by novel superhydrophobic copper meshes. Appl Surf Sci 25:6002–6006. https://doi.org/10.1016/j.apsusc.2008.03.034
12. Li J, Shi L, Chen Y, Zhang YB, Guo ZG, Su BL, Liu WM (2012) Stable superhydrophobic coatings from thiol-ligand nanocrystals and their application in oil/water separation. J Mater Chem 22:9774–9781. https://doi.org/10.1039/C2JM30931A
13. Lakshmi RV, Bharathidasan T, Bera P, Basu BJ (2012) Fabrication of superhydrophobic and oleophobic sol–gel nanocomposite coating. Surf Coat Technol 206:3888–3894. https://doi.org/10.1016/j.surfcoat.2012.03.044

Green Synthesis and Characterization of Zinc Ferrite and Lanthanum- Doped Zinc Ferrite

Chikkappa Udagani, H. R. Mahalakshmi, N. Kumar Govind, and J. Keerthiveni

1 Introduction

Zinc ferrite belongs to a class of spinel ferrite. The general formula of spinel ferrite is MFe_2O_4 (where M is a divalent metal ion). In spinel ferrite, M^{2+} and Fe^{3+} occupy the tetrahedral (A) and octahedral (B) interstitial sites of the fcc lattice formed by O^{2-} ions, respectively [1, 2]. The unit cell of a spinel ferrite consists of eight divalent metal ions, 16 trivalent ions, and 32 oxygen atoms [3]. The magnetic ferrites are useful in a variety of fields such as gas sensing [4, 5], diagnostic medicine [6, 7], data storage [8, 9], and transformer cores [10]. The ferrites are essential inductive components in electronic circuits such as filters, low-noise amplifiers, voltage-controlled oscillators, and impedance matching networks. In recent days, zinc ferrite ($ZnFe_2O_4$) and its composites are drawing greater attention due to their wide range of applications. The zinc ferrite is used in commercial applications such as magnetic resonance imaging (MRI), Li-ion batteries [11], and gas sensors [12]. The zinc ferrite ($ZnFe_2O4$) seems to be an ideal candidate to use as a soft magnet and low loss material at high frequency [13]. The doping of rare-earth metal ions would change the texture of magnetic ferrites [14]. Due to the high electrical insulation property of rare-earth materials, the inclusion of rare-earth ions could alter the magnetic and electrical properties [15, 16]. Due to the excellent applications of zinc ferrite and its nanocomposites, zinc ferrite and lanthanum-doped zinc ferrite have been prepared by the green synthesis route.

In recent years there is a great interest in preparing nanoparticles using biomolecules as it is cost-effective, the input energy for synthesis is less and the precursors used are nontoxic. It is also known to reduce particle aggregation. In the present study, $ZnFe_2O_4$ and lanthanum-doped $ZnFe_2O_4$ nanoferrites were prepared

C. Udagani (✉) · H. R. Mahalakshmi · N. Kumar Govind · J. Keerthiveni
Department of PG Studies and Research in Physics, University College of Science, Tumkur University, Tumkur 572103, India

U. K. Mudali et al. (eds.), *Recent Trends in Electrochemical Science and Technology*, Springer Proceedings in Materials 15,
https://doi.org/10.1007/978-981-16-7554-6_14

using biomolecules extracted from Tulsi (Ocimum sanctum) leaf extract. To understand the electrical behavior of the synthesized $ZnFe_2O_4$ and lanthanum-doped $ZnFe_2O_4$ nanoferrites, the DC conductivity measurement was also carried out.

2 Experimental

2.1 *Preparation of $ZnFe_2O_4$*

The chemicals used in the present study were analytical grade zinc nitrate hexahydrate [Zn $(NO_3)_2.6H_2O$], iron nitrate [$Fe(NO_3)_{3.}9H_2O$], and lanthanum nitrate [$La(NO_3)_2.6H_2O$]. The Tulsi leaf extract was used as a reducing agent for the synthesis of $ZnFe_2O_4$ and lanthanum-doped $ZnFe_2O_4$ nanoferrites. The Tulsi (*Ocimum sanctum*) leaves were collected locally, first washed with tap water and then with distilled water. The leaves were dried in a cool place for seven days. The dried leaves were then ground to powder using a pestle and mortar. About 4 g of Tulsi leaf powder was weighted using an electronic balance and added to 100 mL of distilled water and the mixture was stirred well using a magnetic stirrer for 1 h at 50 °C. The leaf extract was filtered using filter paper and pure leaf extract was obtained. The pH of the obtained leaf extract was about 6.5. The pH plays an important role in controlling the morphology and yield of nanoparticle synthesis. About 1.070 g of zinc nitrate and 3.232 g of iron nitrate were added to the leaf extract. The solution was stirred using a magnetic stirrer and a few drops of ammonia solution were added to adjust the pH between 7 and7.5 and the stirring was continued at 90 °C until all the water evaporated completely. Finally, the product was heated at 250 °C in a laboratory oven to facilitate a self-propagating combustion reaction of redox mixture to form crystalline $ZnFe_2O_{4.}$ The steps involved in the synthesis of $ZnFe_2O_4$ and lanthanum-doped $ZnFe_2O_4$ are shown in Fig. 1.

2.2 *Preparation of lanthanum- Doped $ZnFe_2O_4$*

About 4 g of Tulsi leaf powder was added to 200 mL of distilled water and was stirred using a magnetic stirrer for 1 h at 50 °C. The mixture was filtered using filter paper and to get pure leaf extract. About 1.070 g of zinc nitrate, 3.232 g of iron nitrate, and 0.057 g (5 wt% relative to zinc nitrate) lanthanum nitrate were added to the leaf extract and finally, ammonia solution was added to maintain the pH between 7 and 7.5. The stirring was continued at 90 °C until the water is evaporated completely. The final product was heated at 250 °C and crystallized lanthanum-doped $ZnFe_2O_4$ was obtained as the combustion product.

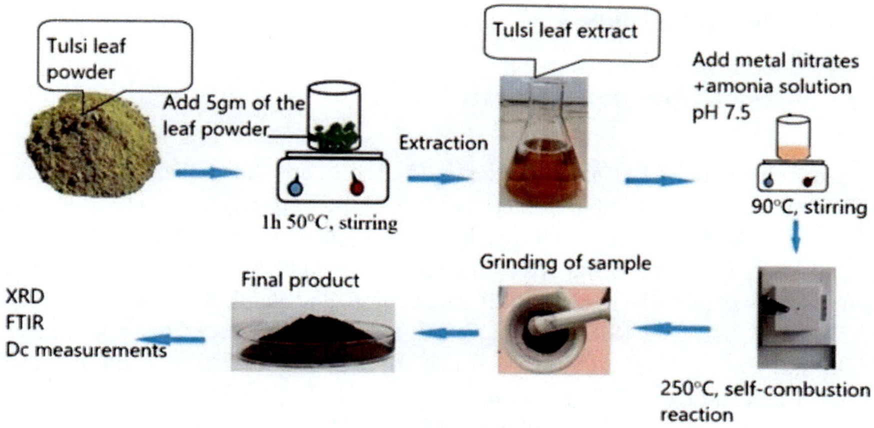

Fig. 1 Schematic showing the steps involved in the synthesis of zinc ferrites

2.3 *Characterization*

The phase purity of the powders was checked using an X-ray diffractometer (XRD, Bruker D8 Advance) and Expo-2014 crystal structure solution. The full width at half maxima (β) was estimated using Fityk software [17]. The crystallite size was found from substituting the value of β in the Scherrer's equation:

$$D = \frac{K \times \lambda}{\beta \times \cos\theta}$$

where D—Average crystallite size (nm); K—Scherrer constant. K varies from 0.68 to 2.08. $K = 0.94$ for spherical crystallites with cubic symmetry; λ-X-ray wavelength; β- FWHM (Full width half maximum) obtained from XRD peak.

The FTIR analysis of $ZnFe_2O_4$ and lanthanum-doped $ZnFe_2O_4$was carried out using a Bruker FTIR spectrometer. The DC conductivity study was carried out with the four-probe setup (Mars Edpal Pvt (Ltd), India) and constant current power supply (SES instruments, India). For the DC measurement, the nanoferrite powder was ground into a fine powder using mortar and pestle. The ground fine powder of nanoferrite was pelletized using a hydraulic press at a pressure of 2 tons. The nanoferrite pellet was cut into a square shape using a sharp razor. The square-shaped pellet was then polished smoothly for good electrical contact. The four-probe setup consists of four equally spaced spring-loaded pen-type probes with a spacing of 0.2 cm. The pellet under study was placed in between a thin mica sheet and the probes. The four-probe setup was manually adjusted with gentle pressure. The current and voltage were measured using digital multimeters. The DC measurement was carried out within the temperature range of 318–473 K while cooling.

3 Results and Discussion

3.1 X-Ray Diffraction Analysis

Figure 2 shows the XRD pattern of $ZnFe_2O_4$.The figure shows major XRD peaks at 2θ values of 30.26, 35.64, 37.28, 43.32, 53.75, 57.30, 62.93, 71.41, 74.47 and 75.48°. The XRD pattern matches well with the patterns reported in the literature [18, 19]. Figure 3 shows the XRD indexing for lanthanum-doped $ZnFe_2O_4$. The figure shows major XRD peaks at 2θ values of 30.17, 35.54, 37.18, 43.19, 47.29, 53.59, 57.13, 62.74, 71.18, 74.23 and 75.23°. The 2θ values, interplanar spacing (d), Miller indices, (h, k, l) and lattice constant (a) obtained for $ZnFe_2O_4$ on the basis of XRD data and Expo-2014 are summarized in Table 1 [20]. In Table 2, 2θ values, interplanar spacing (d), Miller indices, (h, k, l) and lattice constant (a) are listed for lanthanum-doped $ZnFe_2O_4$ on the basis of XRD data and Expo-2014. The values of (h, k, l), d-spacing, and lattice constant obtained by indexing the XRD data using the Expo-2014 reveal the formation of cubic spinel structure for both $ZnFe_2O_4$ and lanthanum-doped $ZnFe_2O_4$.

The Fityk analysis was made with the main XRD peak at $2\theta = 35.42°$. Figure 4 shows the Gaussian fit to the XRD peak at $2\theta = 35.42°$ for $ZnFe_2O_4$ and Fig. 5 shows the Gaussian fit to the XRD peak at $2\theta = 35.38°$ for lanthanum—doped $ZnFe_2O_4$. From the Fityk analysis, the full width half maximum (β) was estimated to be 0.409041 for $ZnFe_2O_4$ and 0.286329 for lanthanum-doped $ZnFe_2O_4$. The crystallite size of $ZnFe_2O_4$ was found to be 21.31 nm and for lanthanum-doped $ZnFe_2O_4$ it was found to be 30.43 nm.

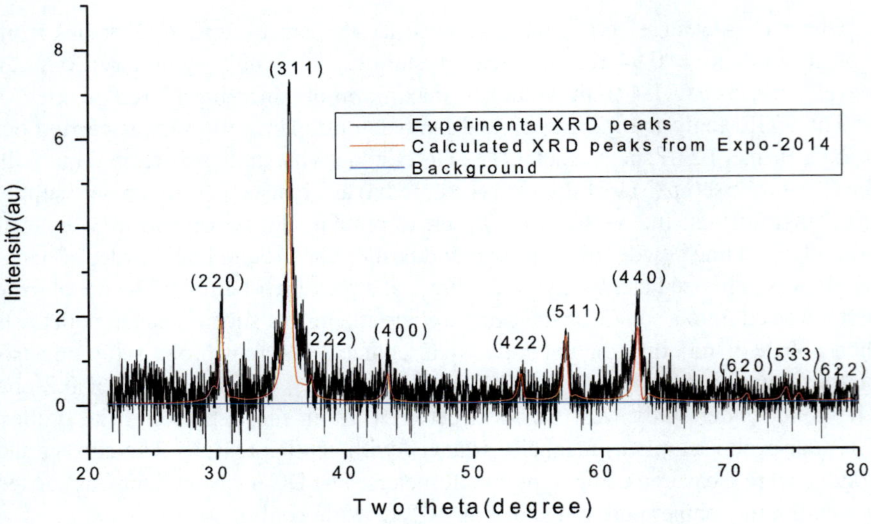

Fig. 2 XRD pattern of $ZnFe_2O_4$

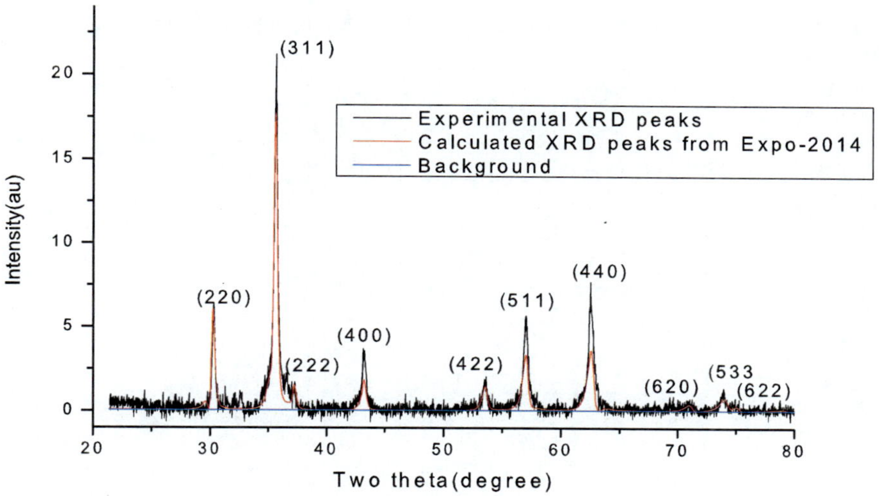

Fig. 3 XRD pattern of lanthanum-doped $ZnFe_2O_4$

Table 1 The interplanar spacing, miller indices, and lattice constant obtained for $ZnFe_2O_4$

Two theta	Interplanar spacing, d(Å)	Miller indices			Lattice constant, a(Å)
		h	*k*	*l*	
30.25908	2.95124	2	2	0	8.34737
35.64296	2.51683	3	1	1	
37.28486	2.40968	2	2	2	
43.32182	2.08684	4	0	0	
53.75302	1.70390	4	2	2	
57.30441	1.60645	5	1	1	
62.93360	1.47562	4	4	0	
71.41093	1.31983	6	2	0	
74.47334	1.27296	5	3	3	
75.48364	1.25841	6	2	2	

3.2 FTIR Analysis

The FTIR absorption spectra of the synthesized ferrite samples are shown in Figs. 6 and 7. The characteristics absorption peak at 603 cm^{-1} for $ZnFe_2O_4$is attributed to Fe–O vibration of regular spinel sample and the characteristics absorption peak at 404 cm^{-1} for $ZnFe_2O_4$ is attributed to Zn–O vibration of spinel ferrite sample [21, 22]. The characteristics absorption peak at 557 cm^{-1} for lanthanum-doped $ZnFe_2O_4$ is attributed to Fe–O vibration of regular spinel sample. The characteristics absorption peak at 423 cm^{-1} for lanthanum-doped $ZnFe_2O_4$ is attributed to Zn–O vibration of

Table 2 The interplanar spacing, miller indices, and lattice constant obtained for lanthanum-doped $ZnFe_2O_4$

Two theta	Interpalanr spacing, d(Å)	Miller indices			Lattice Constant, a(Å)
		h	*k*	*l*	
30.17132	2.95962	2	2	0	8.37108
35.53862	2.52398	3	1	1	
37.17537	2.41652	2	2	2	
43.19293	2.09277	4	0	0	
47.29292	1.92046	3	3	1	
53.58856	1.70874	4	2	2	
57.12711	1.61101	5	1	1	
62.73504	1.47981	4	4	0	
71.17779	1.32358	6	2	0	
74.22681	1.27658	5	3	3	
75.23259	1.26199	6	2	2	

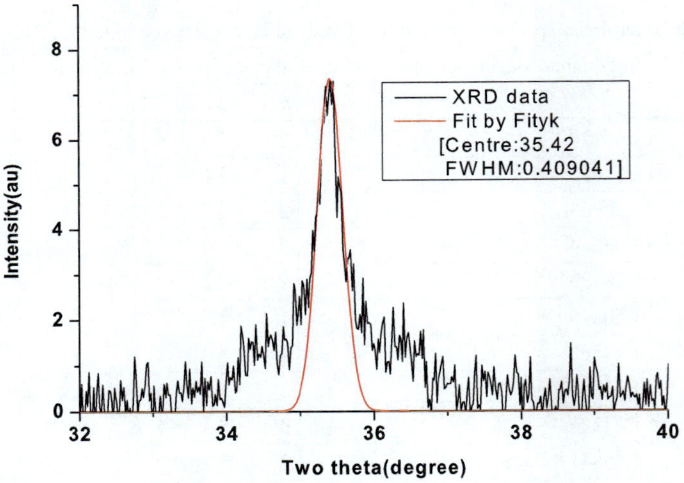

Fig. 4 FityK analysis of undoped $ZnFe_2O_4$

spinel ferrite sample. The FTIR analysis results so obtained are in accordance with the results obtained from the XRD analysis.

3.3 DC Conductivity Study

Figure 8 depicts the variation of electrical conductivity with varying temperatures and

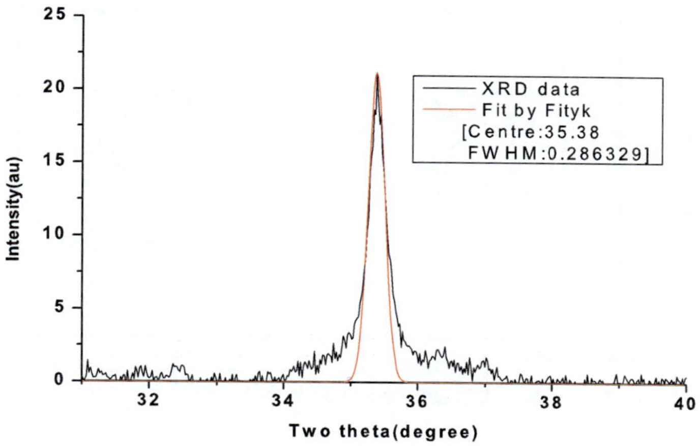

Fig. 5 FityK analysis of lanthanum-doped $ZnFe_2O_4$

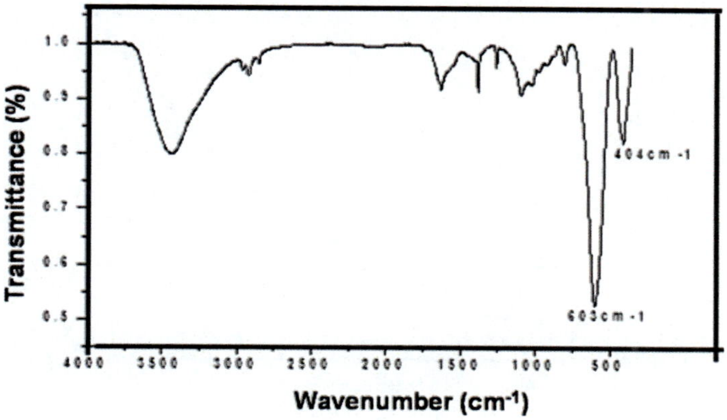

Fig. 6 FTIR spectrum of $ZnFe_2O_4$

Fig. 9 shows the plot of $\log_{10}(\rho)$ versus 1/T. It is evident from the figure that electrical conductivity increases with increasing temperature for both $ZnFe_2O_4$ and lanthanum-doped $ZnFe_2O_4$. The increase of electric conductivity of ferrites with temperature shows the semiconducting nature of the ferrites. This nature can be attributed to the thermal enhancement of carrier mobility [23, 24]. However, the conductivity of lanthanum-doped $ZnFe_2O_4$ was lower than that of $ZnFe_2O_4$. This may be due to the occupation of lanthanum ions in the octahedral sites. The occupation of La^{3+} ions in octahedral sites reduces electron hopping. The increase of conductivity with temperature indicates the semiconductor behavior of the $ZnFe_2O_4$ and lanthanum-doped $ZnFe_2O_4$. For both $ZnFe_2O_4$ and lanthanum-doped $ZnFe_2O_4$, the plot of $\log_{10}\rho$ versus $1000/T$ is linear. By finding the slope of the $\log_{10}(\rho)$ versus $1000/T$ line, the activation energy E_a can be determined using the relation:

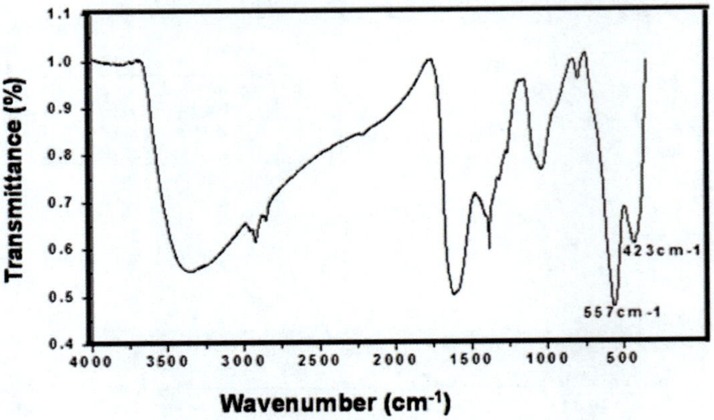

Fig. 7 FTIR spectrum of lanthanum-doped $ZnFe_2O_4$

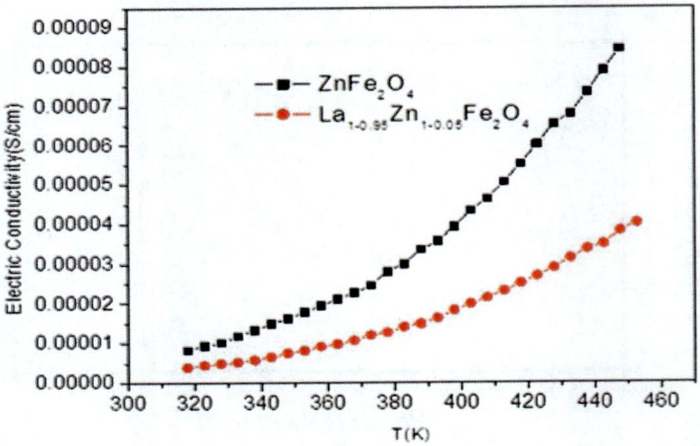

Fig. 8 Plot of electrical conductivity versus temperature for doped and undoped zinc ferrite

$$E_a = 2.303 \times 2 \times 8.617 \times 10^{-5} \times \text{Slope}$$

For $ZnFe_2O_4$, E_a was estimated as 0.352 eV. This value is in agreement with the reported value 0.331 eV [25]. For lanthanum-doped $ZnFe_2O_4$ the E_a was estimated as 0.364 eV. The activation energy of lanthanum-doped $ZnFe_2O_4$ is greater than that of $ZnFe_2O_4$. This reveals the fact that doping of rare-earth ions reduces the number of Fe^{3+} ions in octahedral site which also slow down the electron hopping between Fe^{3+} and Fe^{2+} [26].

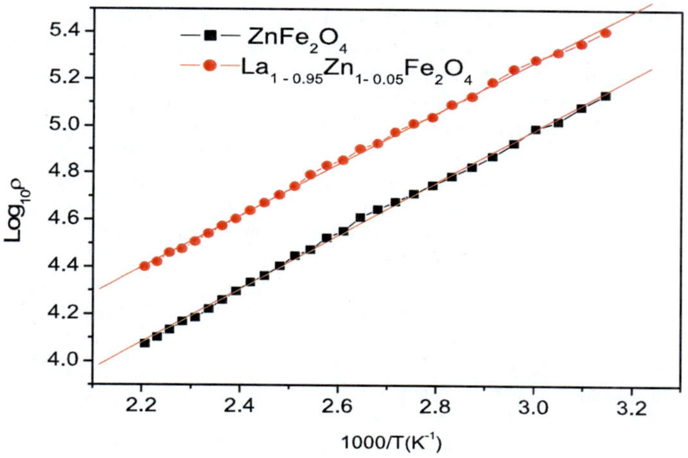

Fig. 9 Plot of $Log_{10}\ \rho$ versus $\frac{1000}{T}(K^{-1})$ for doped and undoped zinc ferrite

4 Conclusions

Zinc ferrite ($ZnFe_2O_4$) and lanthanum-doped zinc nanoferrite were synthesized via a green synthesis route using Tulsi leaf extract. The XRD and FTIR analysis revealed the normal spinel structure for both $ZnFe_2O_4$ and lanthanum-doped $ZnFe_2O_4$. The XRD structural analysis shows higher lattice constant and crystallite size for lanthanum-doped $ZnFe_2O_4$ compared to $ZnFe_2O_4$.This is attributed to the accumulation of La^{3+} in the octahedral lattice sites of the $ZnFe_2O_4$. The DC conductivity study confirmed semiconducting behavior for both $ZnFe_2O_4$ and lanthanum-doped $ZnFe_2O_4$ nanoferrites. The conductivity of lanthanum-doped $ZnFe_2O_4$ decreased in comparison to $ZnFe_2O_4$ within the measured temperature range. The reason for decreased conductivity for lanthanum-doped $ZnFe_2O_4$ is the occupation of La^{3+} in the octahedral sites. From this study, it was observed that by doping of La^{3+}ions in $ZnFe_2O_4$ the resistivity increases and thus it reveals that doping of lanthanum increases the electrical insulation of $ZnFe_2O_4$. Future studies are aimed at the cyclic voltammetric studies of the synthesized ferrites.

Acknowledgements The authors are thankful to DST-SAIF, KOCHI for XRD measurements.

References

1. Kooti M, Sedeh AN (2013) Synthesis and characterization of $NiFe_2O_4$ magnetic nanoparticles by combustion method. J Mater Sci Technol 29:34–38. https://doi.org/10.1016/j.jmst.2012.11.016

2. Sadri F, Ramazani A, Ahankar H, Taghavi Fardood S, Azimzadeh Asiabi P, Khoobi M, Joo SW, Dayyani N (2016) Aqueous- phase oxidation of alcohols with green oxidants (Oxone and hydrogen peroxide) in the presence of $MgFe_2O_4$ magnetic nanoparticles as an efficient and reusable catalyst. J Nanostruct 6:264–272
3. Prasad BBVSV, Ramesh KV, Srinivas A (2019) Structural and magnetic studies on Co-Zn nano-ferrite synthesized via sol-gel and combustion methods. Mater Sci-Poland 37:39–54. https://doi.org/10.2478/msp-2019-0013
4. You J, Chen X, Zheng B, Geng X, Zhang C (2017) Suspension plasma-sprayed $ZnFe_2O_4$ nanostructured coatings for ppm-level acetone detection. J Therm Spray Technol 26:728–734. https://doi.org/10.1007/s11666-017-0536-7
5. Jeseentharani V, George M, Jeyaraj B, Dayalan A, Nagaraja K (2013) Synthesis of metal ferrite (MFe_2O_4, M = Co, Cu, Mg, Ni, Zn) nanoparticles as humidity sensor materials. J Exp Nanosci 8:358–370. https://doi.org/10.1080/17458080.2012.690893
6. Issa B, Obaidat I, Albiss B, Haik Y (2013) Magnetic nanoparticles: surface effects and properties related to biomedicine applications. Int J Mol Sci 14:21266–21305. https://doi.org/10.3390/ijms141121266
7. Niemirowicz K, Markiewicz KH, Wilczewska AZ, Car H (2012) Magnetic nanoparticles as new diagnostic tools in medicine. Adv Med Sci 57:196–207. https://doi.org/10.2478/v10039-012-0031-9
8. Galloway JM, Talbot JE, Critchley K, Miles J, Bramble J (2015) Developing biotemplated data storage: room temperature biomineralization of L10CoPt magnetic nanoparticles. Adv Funct Mater 25:4590–4600. https://doi.org/10.1002/adfm.201501090
9. S-ichi O, Namai A, Yoshikiyo M, Imoto K, Tamazaki K, Matsuno K, Inoue O, Ide T, Masada K, Goto M, Goto T, Yoshida T, Miyazaki T (2016) Multimetal-substituted epsilon-iron oxide ϵ-Ga0.31Ti0.05Co0.05Fe1.59O3 for next-generation magnetic recording tape in the big-data era. Angew Chem Int 55:11403–11406. https://doi.org/10.1002/anie.201604647
10. Ngoua Teu Magambo JS, Bakri R, Margueron X, Le Moigne P, Mahe A, Guguen S, Bensalah T (2017) Planar magnetic components in more electric aircraft: review of technology and key parameters for DC–DC power electronic converter. IEEE Trans Transp Electrification 3:831–842. https://doi.org/10.1109/tte.2017.2686327
11. Zhong X-B, Yang Z-Z, Wang H-Y, Lu L, Jin B, Zha M, Jiang Q-C (2016) A novel approach to facilely synthesize mesoporous $ZnFe_2O_4$ nanorods for lithium ion batteries. J Power Sourc 306:718–723. https://doi.org/10.1016/j.jpowsour.2015.12.102
12. Srivastava R, Yadav BC (2014) Nanostructured $ZnFe_2O_4$thick film as room temperature liquefied petroleum gas sensor. J Exp Nanosci 10:703–717. https://doi.org/10.1080/17458080.2013.880001
13. Sathiyamurthy K, Rajeevgandhi C, Bharanidharan S, Sugumar P, Subashchandrabose S (2020) Electrochemical and magnetic properties of zinc ferrite nanoparticles through chemical co-precipitation method. Chem Data Collect 28:100477. https://doi.org/10.1016/j.cdc.2020.100477
14. Kumar L, Kar M (2012) Effect of Ho^{3+} substitution on the cation distribution, crystal structure and magnetocrystalline anisotropy of nanocrystalline cobalt ferrite. J Exp Nanosci 9:362–374. https://doi.org/10.1080/17458080.2012.661474
15. Xavier S, Thankachan S, Jacob BP, Mohammed EM (2013) Effect of samarium substitution on the structural and magnetic properties of nanocrystalline cobalt ferrite. J Nanosci 2013:1–7. https://doi.org/10.1155/2013/524380
16. Zhou B, Zhang Y-W, Liao C-S, Yan C-H, Chen L-Y, Wang S-Y (2004) Rare-earth-mediated magnetism and magneto-optical Kerr effects in nanocrystalline $CoFeMn_{0.9}RE_{0.1}O_4$ thin films. J Magn Magn Mater 280:327–333. https://doi.org/10.1016/j.jmmm.2004.03.031
17. Wojdyr M (2010) Fityk: a general-purpose peak fitting program. J Appl Crystallogr 43:1126–1128. https://doi.org/10.1107/s0021889810030499
18. Abbas SI, John HT, Fraih AJ (2017) Preparation of nano crystalline zinc—ferrite as material for micro waves absorption by sol-gel methods. Indian J Sci Technol 10:1–6. https://doi.org/10.17485/ijst/2017/v10i21/113197

19. Kanagesan S, Hashim M, AB Aziz S, Ismail I, Tamilselvan S, Alitheen N, Swamy M, Purna Chandra Rao B (2016) Evaluation of antioxidant and cytotoxicity activities of copper ferrite ($CuFe_2O_4$) and zinc ferrite ($ZnFe_2O_4$) Nanoparticles synthesized by sol-gel self-combustion method. Appl Sci 6:184. https://doi.org/10.3390/app6090184
20. Altomare A, Cuocci C, Giacovazzo C, Moliterni A, Rizzi R, Corriero N, Falcicchio A (2013) EXPO2013: a kit of tools for phasing crystal structures from powder data. J Appl Crystallogr 46:1231–1235. https://doi.org/10.1107/s0021889813013113
21. Din IU, Tasleem S, Naeem A, Shaharun MS, Al GMJ, Kaisy, (2013) Zinc ferrite nanoparticle synthesis and characterization; effects of annealing temperature on the size of nanoparticles. Aust J Basic Appl Sci 7:154–162
22. Vinosha PA, Mely LA, Jeronsia JE, Monica FH, Raja K, Das SJ (2017) Study of structural, optical, dielectric and magnetic properties of zinc ferrite synthesized by co-precipitation. Nano Hybrids Compos 17:1–9. https://doi.org/10.4028/www.scientific.net/nhc.17.1
23. Hankare P, Patil N, Patil R, Patil D, Delekar S (2013) Structural, electrical and magnetic properties of copper substituted Zn–Mn ferrites. J Mater Sci Mater Electron 24:4028–4032
24. Yadav SP, Shinde SS, Kadam AA, Rajpure KY (2013) Structural, morphological, dielectrical and magnetic properties of Mn substituted cobalt ferrite. J Semicond 34 093002. https://doi.org/10.1088/1674-4926/34/9/093002
25. Sekulic DL, Lazarevic ZZ, Sataric MV, Jovalekic CD, Romcevic NZ (2015) Temperature-dependent complex impedance, electrical conductivity and dielectric studies of MFe_2O_4 (M = Mn, Ni, Zn) ferrites prepared by sintering of mechanochemical synthesized nanopowders. J Mater Sci Mater Electron 26:1291–1303. https://doi.org/10.1007/s10854-014-2491-0
26. Pervaiz E, Gul IH (2013) Influence of rare earth (Gd^{3+}) on structural, gigahertz dielectric and magnetic studies of cobalt ferrite. J Phys Conf Ser 439:012015. https://doi.org/10.1088/1742-6596/439/1/012015

Dielectric Properties of $Ni_xCu_{1-x}Fe_2O_4$:$PbZr_{0.52}Ti_{0.48}O_3$ Multilayered Nanocomposites

Shaik Sabira Begum and S. S. Bellad

1 Introduction

In the last decade, interest has been focused on layered composite structures based on lead zirconate titanate (PZT) associated with different ferromagnetic materials such as nickel ferrites, cobalt ferrites, or lanthanum manganite. Although nickel ferrites exhibit much smaller magnetostriction than cobalt ferrites, they show a better magnetoelectric (ME) effect [1–5]. Giant magnetostrictive material such as Terfenol-D (Terbium Dysprosium iron alloy) has been used in laminated configurations associated with PZT at resonant or sub-resonant frequencies. Magnetoelectric (ME) materials that exhibit coupling between magnetic properties and dielectric properties are of great interest due to their potential use in smart electronic applications viz., sensors, waveguides, actuators, modulators, etc. The ME effect refers to a change in electric polarization when a magnetic field is applied (direct effect) or, conversely, a change in magnetization when an electric field is applied (converse effect). Intrinsic (single-phase) ME coupling was first detected in Cr_2O_3 at low temperature, but the effect was so weak that it could not find any use in electronic devices [1]. General interactions between the various phases of these composites give rise to the so-called product properties [2]. These properties are found in the behavior of the macroscopic composite but are usually absent from the constituent behavior of the individual phases. Examples of product properties are magnetoelectric, pyromagnetic, and pyroelectric properties [3, 4].

ME bulk composites have been made first by sintering together $BaTiO_3$ or PZT with spinel ferrites. However, the low resistivity of the ferrite that short-cut the piezoelectric phase acts towards a lowering of the ME effect [5]. This difficulty is overcome by using laminated structures. So, in the last decade, interest has been focused on layered composite structures based on PZT associated with different ferromagnetic

S. S. Begum · S. S. Bellad (✉)
Materials Research Center, Maharani Science College for Women, Bengaluru 560001, India

U. K. Mudali et al. (eds.), *Recent Trends in Electrochemical Science and Technology*, Springer Proceedings in Materials 15,
https://doi.org/10.1007/978-981-16-7554-6_15

materials such as nickel ferrites, cobalt ferrites, or lanthanum manganite. Although nickel ferrites exhibit much smaller magnetostriction than cobalt ferrites, they show a better ME effect. The strongest ME effects were obtained for co-sintered thin layers of PZT/nickel-zinc ferrite in the range of 10 μm using the tape-casting route [6]. Giant magnetostrictive material such as Terfenol-D (Terbium Dysprosium iron alloy), were also used in laminated configurations associated with PZT or PMN-PT at resonant or sub-resonant frequencies [7].

In the present investigation, the synthesis of $Ni_xCu_{1-x}Fe_2O_4$-PZT was carried out by solution combustion method. Multilayered structures were fabricated using the screen printing method [8, 9]. The dielectric properties of these multilayered thick films have been reported.

2 Experimental Details

2.1 *Synthesis of Ferromagnetic Phase ($Ni_xCu_{1-x}Fe_2O_4$) and Ferroelectric Phase (Pb ($Zr_{0.52}Ti_{0.48}$)O_3) by Solution Combustion Method*

The combustion method involves exothermic and self-sustaining thermally-induced anionic redox reaction of the redox mixture, which is obtained from an aqueous solution containing desired metal salts (oxidizer) and organic complexing agent (fuel) [10]. Proportions between the oxidizer and fuel are calculated according to the valences of the reacting elements in order to maintain the ratio of oxidizer/fuel as unity [11]. In this study, nitrates were used as oxidizers, and glycine was used as fuel, the ratio of oxidizer to fuel was taken according to stoichiometric proportion [12]. The magnetic phase $Ni_xCu_{1-x}Fe_2O_4$ and ferroelectric phase $Pb(Zr_{0.52}Ti_{0.48})O_3$ were prepared by solution combustion method as it gives very fine powder with particles in the nanosize range.

To synthesize $Ni_xCu_{1-x}Fe_2O_4$ ($x = 0.3, 0.4, 0.5, 0.6, 0.7, 0.9$) ferrite phase AR grade $Ni(NO_3)_2$, $Cu(NO_3)_2$ and $Fe(NO_3)_3$ were used as oxidizers. Glycine ($C_2H_5NO_2$) was chosen as a fuel in the present study as it contains nitrogen which further helps in the solubility of oxidizers and the end product will be a uniform mixture of all the metal ions, and it also gives better control over the composition. The total valence of $Ni(NO_3)_2$, $Cu(NO_3)_2$, and $Fe(NO_3)_3$ are -10, -10, and -15. The total valence of the fuel glycine is $+9$. Using this data, the ratio of oxidizer: fuel was taken as 1:2.

The ferroelectric phase ($Pb(Zr_{0.52}Ti_{0.48})O_3$) was prepared using lead nitrate, zirconyl nitrate, and titanium dioxide as oxidizers and glycine as fuel [13]. The ratio between oxidizer and fuel (1:2) was taken according to stoichiometric proportion same as that of the ferrite phase. Once the fine powder of PZT was obtained it was sintered at around 500 °C for 4 h for completion of the crystallization process.

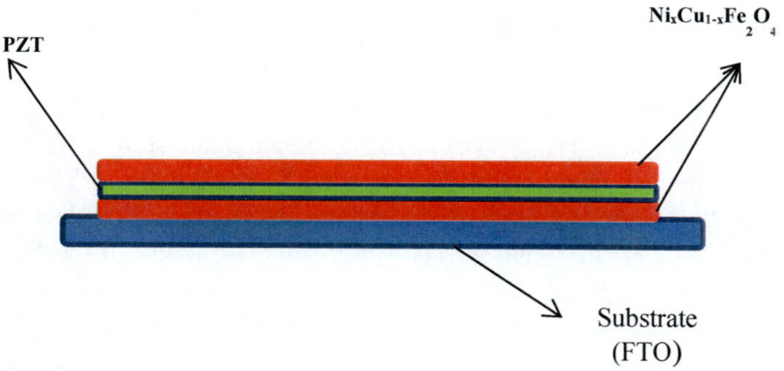

Fig. 1 Schematic representation of multilayered thick film $Ni_xCu_{1-x}Fe_2O_4$/PZT/$Ni_xCu_{1-x}Fe_2O_4$

2.2 Film Deposition Technique

Screen printing is a simple and easy method. The major advantages of this method are large areas can be coated and the thickness of the deposited film can be controlled. For the thick film deposition, a composite paste was formed using composite powder (96%), and organic vehicle (3%). The organic vehicle was the solution of ethyl cellulose (3%) in 2–2 butoxy ethyl acetate (98%) and was used to maintain the viscosity of the thick film paste. The fluorine-doped tin oxide (FTO) substrate was kept on a substrate holder and the screen was suspended about 0.5 mm above the substrate. The appropriate amount of paste was then placed on the screen at one side of the substrate and the screen was squeezed under pressure, bringing it into contact with the substrate and at the same time driving the paste through the open mashes of the screen. The first layer was dried for 10 h, then another layer of PZT powder was deposited on the substrate uniformly over the ferrite layer and it was dried for 10 h. Finally, the top layer of $Ni_xCu_{1-x}Fe_2O_4$ was deposited on the PZT layer and dried for 10 h. The so fabricated tri-layered structure was heat treated at 300 °C for 1 h. About 6 samples were prepared by depositing the layers as shown in Fig. 1. The thickness of the film was less than 2 mm approximately.

2.3 Dielectric Property Measurement

The dielectric properties were measured as a function of frequency using an impedance analyzer (Wayne Kerr 6500B) at room temperature in the frequency range of 100 Hz to 1 MHz. Silver paste (conducting paste) was applied on the top surface of the multilayered structure. Impedance analyzer probes were connected to the bottom and the top layer of the structure.

3 Results and Discussion

A typical X-ray diffraction pattern of the sample with $x = 0.5$ ($Ni_{0.5}Cu_{0.5}Fe_2O_4$) of ferrite phase and $Pb(Zr_{0.52}Ti_{0.48})O_3$ of ferroelectric phase are shown in Figs. 2 and 3. The peaks are the characteristics of the ferrite (JCPDS Card No. 742081 for nickel ferrite and JCPDS Card No. 34-0425 for copper ferrite) and ferroelectric (JCPDS Card No. 33-0784 for PZT) phases. The cubic spinel structure of the ferrite phase and the perovskite tetragonal structure of the ferroelectric phase is clearly seen. As per the JCPDS Card No. 742081 and 34-0425, the dominant (311) peak of ferrite phase is rightly placed at 2θ value of ~36°. The high intensity of all the peaks in

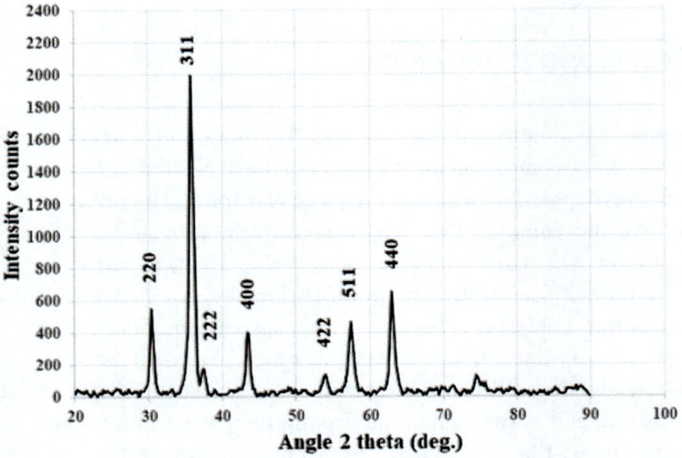

Fig. 2 XRD pattern of $Ni_{0.5}Cu_{0.5}Fe_2O_4$

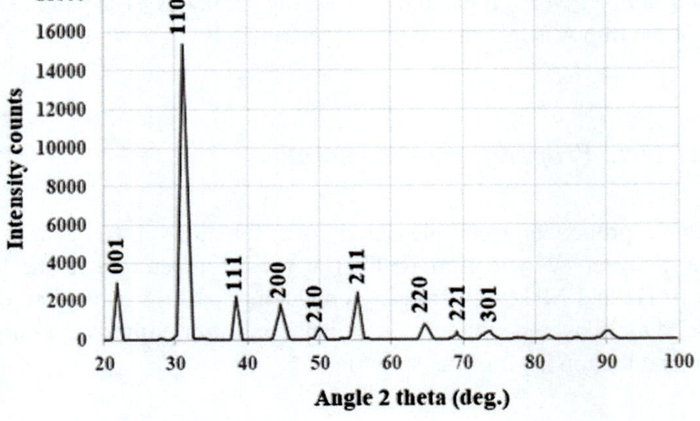

Fig. 3 XRD pattern of PZT

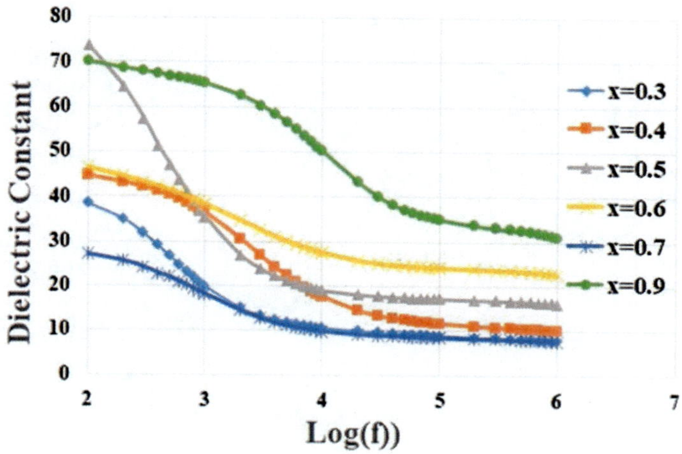

Fig. 4 Variation of dielectric constant with frequency for ($Ni_xCu_{1-x}Fe_2O_4/(Pb(Zr_{0.52}Ti_{0.48})O_3)/Ni_xCu_{1-x}Fe_2O_4$) multilayered thick films at room temperature

ferrite phase indicates that the crystallization process is completed. The broadening of the peaks at baseline also reveals that the particle size is sufficiently small. As per JCPDS Card No. 33-0784, the dominant (110) peak of ferroelectric phase is rightly placed at the 2θ value ~31°. All other peaks of both the phases are also rightly placed at their 2θ values [8].

The variation of dielectric constant with frequency of these multilayered structures is as shown in Fig. 4. It is seen from the figure that the dielectric constant initially decreases rapidly with a rise in frequency and then remains constant at a higher frequency. The dielectric dispersion observed at low frequency could be due to a difference in the composition of the ferrite phase [14, 15]. The decrease in dielectric constant with frequency can be explained based on Koop's theory [16]. At higher frequency, the hopping frequency of the charge carriers does not follow the applied frequency and hence the dielectric response remains constant.

The variation of dielectric loss tangent [tan (δ)] with frequency is as shown in Fig. 5. The tan (δ) represents the capability of dissipating the energy. The dielectric loss continuously decreases with an increase in frequency. All the samples show maxima in the frequency range of 100 Hz to 100 kHz. These kinds of maxima are generally observed when the hopping frequency of electrons between ionic sites matches the applied frequency [17].

Variation of AC resistivity with frequency for these multilayered structures is shown in Fig. 6. The AC resistivity decreases rapidly with a rise in frequency then it reaches a saturation level. The rapid decrease of resistivity with an increase of frequency is due to an increase in the hopping frequency of the electrons. Thereafter, at higher frequencies, the hopping frequencies do not follow the applied frequency [17]. Variation of conductivity with frequency is shown in Fig. 7. From the figure,

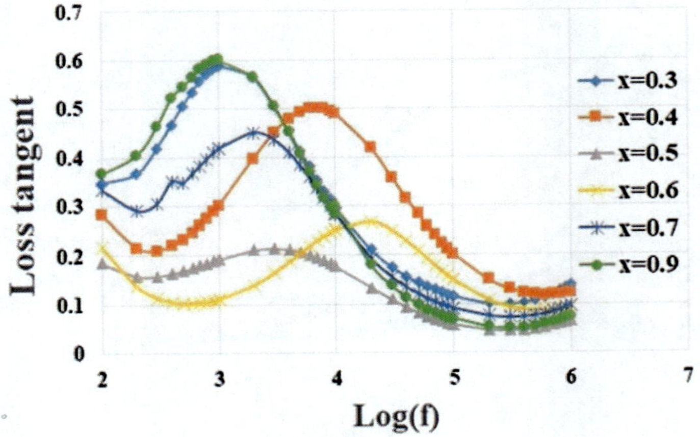

Fig. 5 Variation of loss tangent with frequency for ($Ni_xCu_{1-x}Fe_2O_4$/($Pb(Zr_{0.52}Ti_{0.48})O_3$)/$Ni_xCu_{1-x}Fe_2O_4$) multilayered thick films at room temperature

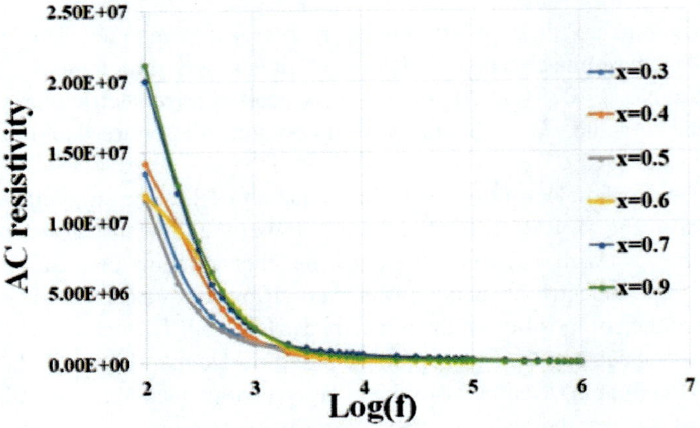

Fig. 6 Variation of AC resistivity with frequency for ($Ni_xCu_{1-x}Fe_2O_4$/($Pb(Zr_{0.52}Ti_{0.48})O_3$)/$Ni_xCu_{1-x}Fe_2O_4$) multilayered thick films at room temperature

it can be noted that plots are linear and straight lines which indicate a small polaron conduction mechanism [18].

4 Conclusions

A ferrite phase $Ni_xCu_{1-x}Fe_2O_4$ and ferroelectric phase $Pb(Zr_{0.52}Ti_{0.48})O_3$ ($x = 0.3$, 0.4, 0.5, 0.6, 0.7, 0.9) multilayered thick films were fabricated on FTO coated substrate by the screen printing technique. The X-ray diffraction pattern confirms

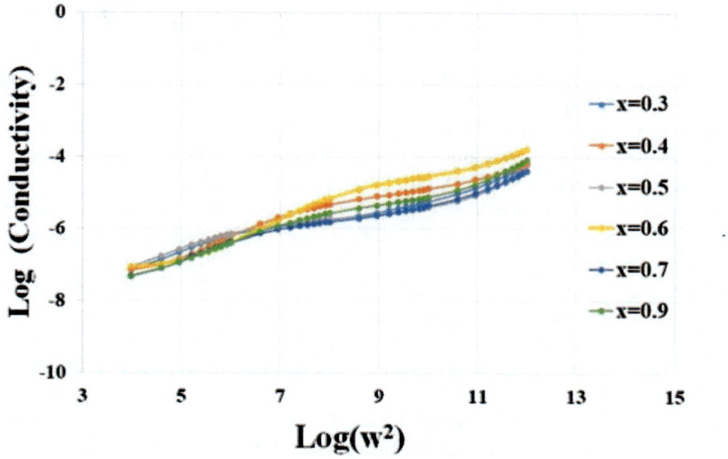

Fig. 7 Variation of log (σ_{ac}) with frequency for ($Ni_xCu_{1-x}Fe_2O_4$/(Pb($Zr_{0.52}$ $Ti_{0.48}$)O_3)/$Ni_xCu_{1-x}Fe_2O_4$) multilayered thick films at room temperature

the individual ferrite and ferroelectric phases without any impurity. The dielectric constant and loss tangent decreases with an increase in frequency, which is due to dielectric dispersion and interfacial polarization. The AC resistivity decreases rapidly with a rise in frequency. The linearity in conductivity versus frequency graph shows that the conduction process in these composite thick films is due to the small polaron hopping mechanism.

Acknowledgements Authors are thankful to Dr. G. Ram Gopal, Principal, Dr. Sriprakash G. HoD, Physics, and Dr. N. Hanumantharaju of Maharani's Science College for Women, Palace Road, Bangalore for providing dielectric measurement facilities.

References

1. Srinivasan G, Rasmussen ET, Levin BJ, Hayes R (2002) Magnetoelectric effects in bilayers and multilayers of magnetostrictive and piezoelectric perovskite oxides. Phys Rev 66:029901. https://doi.org/10.1103/PhysRevB.65.134402
2. Newnham RE, Skinner DP, Cross LE (1978) Connectivity and piezoelectric- pyroelectric composites. Mater Res Bull 13:525–536. https://doi.org/10.1016/0025-5408(78)90161-7
3. Nan CW, Bichurin MI, Dong S, Viehland D, Srinivasan G (2008) Multiferroic Magneto electric composites: historical perspective, status, and future directions. J Appl Phys 103:031101. https://doi.org/10.1063/1.2836410
4. Bichurin M, Petrov V, Priya S, Bhalla A (2012) Multiferroic magneto electric composites and their applications. Adv Condens Matter Phys 129794. https://doi.org/10.1155/2012/129794
5. Shi Z, Nan CW, Liu JM, Filippov DA, Bichurin MI (2004) Influence of mechanical boundary conditions and microstructural features on magnetoelectric behavior in a three-phase multiferroic particulate composite Phys Rev B 70:134417. https://doi.org/10.1103/PhysRevB.70.134417

6. Dong SX, Zhai J, Bai F, Li JF, Viehland D (2005) Fe-Ga/Pb($Mg_{1/3}Nb_{2/3}$)O_3—$PbTiO_3$ magneto electric laminate composites. Appl Phys 87:222504. https://doi.org/10.1063/1.2137455
7. Handley BO, Huang JK (2006) UTA Workshop on Piezoelectric Energy Harvesting, The University of Texas Arlington, TX
8. Bush AA, Ya V, Shkuratov CIA, Fetisov YK (2010) Lead zirconate titanate–nickel zink ferrite thick film composites: obtaining by the screen-printing technique and magneto electric properties. Tech Phys 55:387–394. https://doi.org/10.1134/S1063784210030096
9. Patil N, Velhal N B, Pawar R and Vijaypuri (2014) Electric magnetic and high frequency properties of screen printed ferrite-ferroelectric composite thick films on alumina substrate, Microelectron Int 32:25–31. https://doi.org/10.1108/MI-12-2013-0080
10. Waghuley SA, Yenorkar SM, Yawale SS, Yawale SP (2007) SnO_2/PPy screen-printed multilayer CO_2 gas sensor. Sens Transducers 79:1180–1185
11. Hwang C-C, Tsai J-S, Huang T-H, Peng C-H, Chen S-Y (2005) Combustion synthesis of Ni–Zn ferrite powder—influence of oxygen balance value. J Solid State Chem 178:382–389. https://doi.org/10.1016/j.jssc.2004.10.045
12. Venkateswara Rao K, Sunandana CS (2008) Co_3O_4 nanoparticles by chemical combustion: effect of fuel to oxidizer ratio on structure, microstructure and EPR. Solid State Commun 148:32–37. https://doi.org/10.1016/j.ssc.2008.07.020
13. Patil N, Velhal NB, Pawar R, Puri V (2015) Electric, magnetic and high frequency properties of screen printed ferrite-ferroelectric thick films on alumina substrate. Microelectron Int 32:25–31. https://doi.org/10.1108/MI-12-2013-0080
14. Maxwell JC (1973) Electric and magnetism. Oxford University Press I, New York, pp 828
15. Wagner KW (1913) The distribution of relaxation times in typical dielectrics. Annal D Physik 40:817–819
16. Koops CG (1951) On the dispersion of resistivity and dielectric constant of some semiconductors at audio frequencies, Phys Rev 121–124. https://doi.org/10.1103/PhysRev.83.121
17. Durgadsimi SU, Chougule S, Chougale BK, Bhosale CH, Bellad SS (2011) Dielectric and magnetoelectric Properties of Li0.5 $Ni_{0.75-x/2}Zn_{x/2}$ Fe_2O_4+$Ba_{0.5}$ $Sr_{0.5}$ TiO_3 ME composites. Mater Chem Phys 131:199–203. https://doi.org/10.1016/j.matchemphys.2011.08.069
18. Mahajan RP, Patankar KK, Kothale MB, Chaudhari SC, Mathe VL, Patil SA (2002) Magneto electric effect in cobalt ferrite—barium titanate composites and their electrical properties. Pramana-J Phys 58:1115–1124. https://doi.org/10.1007/s12043-002-0227-9